BIOCHEMISTRY LABORATORY: MODERN THEORY AND TECHNIQUES

Rodney Boyer

Hope College

San Francisco • Boston • New York
Cape Town • Hong Kong • London • Madrid • Mexico City
Montreal • Munich • Paris • Singapore • Sydney • Tokyo • Toronto

Disclaimer

The laboratory methods described in this book have been exhaustively tested for safety and all attempts have been made to select the least hazardous chemicals and procedures possible. However, the author and publisher cannot be held liable for any injury or damage which may occur during the performance of the procedures. It is assumed that before any experiment is initiated, a Material Safety Data Sheet (MSDS) for each chemical used will have been studied by the instructor and students to ensure its safe handling and disposal.

Acquisitions Editor	*Jim Smith*
Project Manager	*Cinnamon Hearst*
Managing Editor, Production	*Erin Gregg*
Production Supervisor	*Vivian McDougal*
Production Management and Art Coordination	*Progressive Publishing Alternatives*
Illustrations	*Progressive Information Technologies*
Marketing Manager	*Scott Dustan*
Compositor	*Progressive Information Technologies*
Text and Cover Design	*Progressive Publishing Alternatives*
Manufacturing Buyer	*Michael Early*
Cover Printer	*Phoenix Color*
Printer	*RR Donnelley*

ISBN 0-8053-4613-9

Library of Congress Cataloging-in-Publication Data

Boyer, Rodney F.
    Biochemistry laboratory : modern theories and techniques / Rodney Boyer.
        p. cm.
    Includes bibliographical references and index.
    ISBN 0-8053-4613-9
    1. Biochemistry—Laboratory manuals.   I. Title.
QD415.5.B69 2006
572'.078—dc22   2005005103

5 6 7 8 9 10—DOC—09

www.aw-bc.com

To my wife, Christel

TO THE STUDENT AND INSTRUCTOR

Biochemistry is an experimental science. Our understanding of the molecular nature of life processes comes from the laboratory, where data on biomolecules and their actions are collected, analyzed, and interpreted. It is imperative that students gain an appreciation for the compulsory link between laboratory activities and the growth and maturing of the scientific knowledge base of biochemistry. Thus, students in the discipline must be given opportunities to learn and practice the experimental methods accompanying biochemistry. A formal biochemistry lab course is now offered at most colleges and universities in the world and has become an essential component in the training of students for careers in biochemistry, chemistry, molecular biology, and related molecular life sciences such as cell biology, genetics, and physiology.

The purpose of this book is to serve as a resource to enhance student learning of the theories, techniques, and methodologies practiced in the biochemistry/molecular biology teaching and research laboratory. The extensive availability of laboratory experiments published in journals and the desire of instructors to design their own projects and teaching styles have lessened the need for laboratory manuals. However, because published experiments and homemade lab manuals usually consists only of procedures, there is an increased need for a companion text to explain the theories and principles that underpin laboratory activities.

ORGANIZATION OF THE BOOK

The book is divided into two parts: One part, presented in the traditional format of a textbook, which introduces the theory and experimental techniques used in the modern biochemistry laboratory, and the second part, a treatise, on a dedicated Web site, describing how to design and teach a biochemistry/molecular biology laboratory.

The first part begins with an introduction to skills and concepts that students must master including safety, communicating lab results, preparation of solutions, pipetting, statistics, buffers and pH, measurement of protein and nucleic acid solutions, radioisotopes, use of the computer and the Internet, and other general laboratory procedures/principles. The historical development of general techniques is explored and followed by discussion of current applications (Chapters 1–3).

Chapters 4–11 provide an introduction to the core techniques and instrumentation that may be applied to the study of all biomolecules and biological processes: centrifugation, chromatography, electrophoresis, spectroscopy, ligand-protein binding, methods in molecular biology, protein purification, and Internet databases. An important premise in this section is that the expansion of our knowledge in biochemistry and molecular biology is dependent upon the continued development of powerful analytical techniques, especially instrumentation.

The second part, a dedicated Web site (www.aw-bc.com/boyer), provides an introduction to teaching the biochemistry lab. Topics covered include a discussion of different teaching methods and the concepts and skills that should be part of a biochemistry laboratory course. This section is highlighted by a listing of proven experiments/projects that have been published in biochemistry education journals. The list gives instructors the opportunity to select laboratory exercises that are compatible with their backgrounds/expertise and with available instrumentation and facilities. Special efforts were made to include projects integrating traditional topics in biochemistry with the modern topics of genomics and proteomics. The Web site list begins with about 250 experimental projects and will be updated on a periodic basis.

The concepts and techniques incorporated in this book have been selected by reviewing undergraduate laboratory curricula recommended by the American Society for Biochemistry and Molecular Biology (ASBMB), the Biochemical Society (United Kingdom), and the American Chemical Society (ACS). In addition, the author's opinion, seasoned by 30 years of teaching and research, was an important, but perhaps biased, resource.

PEDAGOGICAL FEATURES OF THE BOOK

The book has been written with a special focus on student learning in the teaching and research laboratory. Several features are present that will assist students in mastering laboratory concepts.

Use of Computers and the Internet

The computer is now being applied to all aspects of the collection, analysis, and management of biochemical data; hence, computer use is integrated thoroughly into all sections of the book. Chapter 2 introduces students to the computer and to Internet Web sites that maintain directories and databases for biochemistry and molecular biology. All chapters have a special section on computer applications and often tables listing Web sites pertinent to topics in the chapter. In addition, Appendix I contains a complete listing of Web sites and software associated with topics in each chapter.

End-of-Chapter Study Problems

Several study problems are provided for student practice at the end of each chapter. Questions deal with both theoretical and procedural aspects of the chapter. To enhance student awareness, an icon ▶ is used to indicate questions that are answered in Appendix IX.

Study Problems within Chapters

Several study problems have been incorporated into the text of each chapter. These exercises give students the opportunity to review a topic and check their knowledge before they move on to the next section.

Further Reading and Study

Each chapter ends with an abundant list of literature references that provide either a more detailed theoretical background or an expanded explanation of procedures and techniques. The listing also includes Web sites related to chapter topics.

ACKNOWLEDGMENTS

Writing and publishing this textbook required the assistance of many talented and dedicated individuals. My thanks go to Jim Smith, Acquisitions Editor at Benjamin Cummings, for his unwavering support of the project. From just a brief prospectus and rough sample chapter, he was able to envision the birth of a textbook. Special thanks go to Project Editor, Cinnamon Hearst, for her hard work and continuous vigilance over the project. Other individuals who assisted at various stages were Erin Gregg (Managing Editor of Production), Vivian McDougal (Production Supervisor), Mike Early (Manufacturing Buyer), Scott Dustan (Marketing Manager), and Crystal Clifton (Project Manager).

The preparation of "Biochemistry Laboratory: Modern Theory and Techniques" was dependent on the many reviewers who, with busy research and teaching schedules, still found time to critique drafts of the manuscript. The knowledgeable scientists and dedicated educators who served as reviewers of the manuscript include:

Donald Becker, University of Nebraska

Jeannie Collins, University of Southern Indiana

Tim M. Dwyer, Towson University

David P. Goldenberg, University of Utah

Frank R. Gorga, Bridgewater State College

Mark E. Hemric, Oklahoma Baptist University

Scott Lefler, Arizona State University East

Mary E. Peek, Georgia Institute of Technology

Margaret Rice, California Polytechnic University

Veronique Vouille, Smith College

I also wish to thank my wife, Christel, who patiently tolerated the lifestyle changes associated with writing a book. In addition, she completed many of the necessary chores including searching the literature, proofreading, photocopying, and preparing art manuscript. After 16 years on this Earth, my writing companion, Mausi, our blue-point Himalayan, is now taking a well-deserved nap under a lamp in kitty heaven. My desk lamp has a new occupant, Mohrchen, a beautiful black, golden-eyed, domestic short-hair adopted from the local humane society.

I encourage all users of this book to send comments that will assist in the preparation of future editions.

Rodney Boyer
boyer@hope.edu

Rod Boyer served on the faculty at Hope College, Holland, MI, where he taught, researched, and wrote biochemistry for nearly 30 years. He earned his B.A. in chemistry and mathematics at Westmar College (Iowa) and his Ph.D. in physical organic chemistry at Colorado State University. After three years as an NIH Postdoctoral Fellow with M. J. Coon (cytochromes P-450) in the Department of Biological Chemistry at the University of Michigan Medical School, he joined the chemistry faculty at Hope. There he directed the work of over 75 undergraduate students in research supported by the NIH, NSF, Dreyfus Foundation, Howard Hughes Medical Institute, and the Petroleum Research Fund (ACS). With his students, he published numerous papers in the areas of ferritin iron storage and biochemical education. He spent a sabbatical year as an American Cancer Society Scholar in the lab of Nobel laureate, Tom Cech at the University of Colorado, Boulder. Rod is also the author of Modern Experimental Biochemistry (third edition, 2000, Benjamin-Cummings) and Concepts in Biochemistry (third edition, 2006, John Wiley & Sons) and serves as an Associate Editor for the journal Biochemistry and Molecular Biology Education (BAMBED). He is a member of the American Society for Biochemistry and Molecular Biology (ASBMB) and its Education and Professional Development Committee that recently designed the undergraduate biochemistry degree recommended by the ASBMB. Rod has recently retired from teaching and research and resides in Bozeman, Montana, where he continues to write and consult in biochemical education.

TABLE OF CONTENTS

APPENDICES

INTRODUCTION TO THE BIOCHEMISTRY LABORATORY

Welcome to the biochemistry laboratory! You are reading this book for one of the following reasons: (1) you are enrolled in a formal biochemistry lab course at a college or university and you will use the book as a guide to procedures; or (2) you have started a research project in biochemistry and desire an understanding of the theories and techniques you will use in the lab; or (3) you have started a job in a biochemistry lab and wish to review theory and techniques. Whether you are a novice or experienced in biochemistry, I believe you will find the subject matter and lab work to be exciting and dynamic. Most of the experimental techniques and skills that you have acquired and mastered in other laboratory courses will be of great value in your work. However, you will be introduced to many new concepts, procedures, and instruments that you have not used in chemistry or biology labs. Your success in biochemistry lab activities will depend on your mastery of these specialized techniques, use of equipment, and understanding of chemical/biochemical principles.

As you proceed through this text, you will no doubt compare your activities with previous laboratory experiences. In organic lab you ran reactions, isolated and purified several hundred milligrams or a few grams of solid or liquid products, and characterized the material. In biochemistry lab you will work with milligram or even microgram quantities, and in most cases the biomolecules will be extracted from biological sources and dissolved in solution, so you never really "see" the materials under study. But you will observe the dynamic chemical and biological changes brought about by biomolecules. The procedures and instruments introduced in the lab will be your "eyes" and will monitor the occurrence of biochemical events.

This chapter is an introduction to procedures that are of utmost importance for the safe and successful completion of a biochemical project. It is

1

recommended that you become familiar with the following sections before you begin laboratory work.

A. SAFETY IN THE LABORATORY

Safety First

The concern for laboratory safety can never be overemphasized. Most students who are involved in biochemistry laboratory activities have progressed through several years of college lab work without even a minor accident. This record is, indeed, something to be proud of; however, it should not lead to overconfidence. You must always be aware that chemicals used in the laboratory are potentially toxic, irritating, or flammable. However, such chemicals are a hazard only when they are mishandled or improperly disposed of. It is my experience as a lab instructor for 30 years that accidents happen least often to those who come to each lab session mentally prepared and with a complete understanding of the experimental procedures to be followed. Because dangerous situations can develop unexpectedly, though, you must be familiar with general safety practices, facilities, and emergency action. When we work in the lab, we must also have a special concern for the safety of lab mates. Carelessness on the part of one person can often cause injury to others.

Material Safety Data Sheets

The procedures in this book are designed and described with an emphasis on safety. However, no amount of planning or pretesting of procedures substitutes for awareness and common sense on the part of the student. All chemicals used in the procedures outlined here must be handled with care and respect. The use of chemicals in all U.S. workplaces, including academic research and teaching labs, is regulated by the Federal Hazard Communication Standard, a document written by the Occupational Safety and Health Administration (OSHA).[1] Specifically, the OSHA standard requires all workplaces where chemicals are used to do the following: (1) develop a written hazard communication program, (2) maintain files of **Material Safety Data Sheets (MSDS)** on all chemicals used in that workplace (an MSDS is a detailed description of the properties of a chemical substance, the potential health hazards, and the safety precautions that must be taken when handling it), (3) label all chemicals with information regarding hazardous properties and procedures for handling, and (4) train employees in the proper use of these chemicals. Several states have passed "right-to-know" legislation that amends and expands the federal OSHA standard. If you have an interest in or concern about any chemical used in the laboratory, the MSDS may be obtained from your instructor or laboratory director or from the Internet (http://research.nwfsc.noaa.gov/msds.html). The actual form of an MSDS for a chemical may vary, but certain specific information must be present. Figure 1.1 is a partial copy of the MSDS for glacial acetic

---

[1]*Federal Register,* Vol. 48, Nov. 25, 1983, p. 53280; Vol. 50, Nov. 27, 1985, p. 48758.

Figure 1.1

Partial MSDS for glacial acetic acid. *Courtesy of Sigma-Aldrich Corp., St. Louis, MO; www.sigma-aldrich.com/*

Section 2—Composition/Information on Ingredient

Substance Name	CAS #	SARA 313
ACETIC ACID	64-19-7	No

Formula   C2H4O2

Synonyms   Acetic acid (ACGIH:OSHA), Acetic acid, glacial, Acide acetique (French), Acido acetico (Italian), Azijnzuur (Dutch), Essigsaeure (German), Ethanoic acid, Ethylic acid, Glacial acetic acid, Kyselina octova (Czech), Methanecarboxylic acid, Octowy kwas (Polish), Vinegar acid

Section 4—First Aid Measures

Oral Exposure
If swallowed, wash out mouth with water provided person is conscious. Call a physician immediately.

Inhalation Exposure
If inhaled, remove to fresh air. If not breathing give artificial respiration. If breathing is difficult, give oxygen.

Dermal Exposure
In case of skin contact, flush with copious amounts of water for at least 15 minutes. Remove contaminated clothing and shoes. Call a physician.

Eye Exposure
In case of contact with eyes, flush with copious amounts of water for at least 15 minutes. Assure adequate flushing by separating the eyelids with fingers. Call a physician.

Section 7—Handling and Storage

Handling
  User Exposure
  Do not breathe vapor. Do not get in eyes, on skin, on clothing. Avoid prolonged or repeated exposure.

Storage
  Suitable
  Keep tightly closed. Store in a cool dry place.

Section 9—Physical/Chemical Properties

Appearance	Color	Form
	Colorless	Clear liquid

Molecular Weight: 60.05 AMU

Property	Value	At Temperature or Pressure
pH	N/A	
BP/BP Range	117-118°C	760 mmHg
MP/MP Range	4°C	
Freezing Point	N/A	
Vapor Pressure	11.4 mmHg	20°C
Vapor Density	2.07 g/f	
Saturated Vapor Conc.	N/A	
SG/Density	1.06 g/cm3	

Section 11—Toxicological Information

Route of Exposure
  Skin Contact
  Causes burns.
  Skin Absorption
  Harmful if absorbed through skin.
  Eye Contact
  Causes burns.
  Inhalation
  May be harmful if inhaled.
  Ingestion
  May be harmful if swallowed.

Target Organ(s) or System(s)
Teeth. Kidneys.

Signs and Symptoms of Exposure
Material is extremely destructive to tissue of the mucous membranes and upper respiratory tract, eyes, and skin. Inhalation may result in spasm, inflammation and edema of the larynx and bronchi, chemical pneumonitis, and pulmonary edema. Symptoms of exposure may include burning sensation, coughing, wheezing, laryngitis, shortness of breath, headache, nausea, and vomiting. Ingestion or inhalation of concentrated acetic acid causes damage to tissues of the respiratory and digestive tracts. Symptoms include: hematemesis, bloody diarrhea, edema and/or perforation of the esophagus and pylorus, hematuria, anuria, uremia, albuminuria, hemolysis, convulsions, bronchitis, pulmonary edema, pneumonia, cardiovascular collapse, shock, and death. Direct contact or exposure to high concentrations of vapor with skin or eyes can cause: erythema, blisters, tissue destruction with slow healing, skin blackening, hyperkeratosis, fissures, corneal erosion, opacification, iritis, conjunctivitis, and possible blindness. To the best of our knowledge, the chemical, physical, and toxicological properties have not been thoroughly investigated.

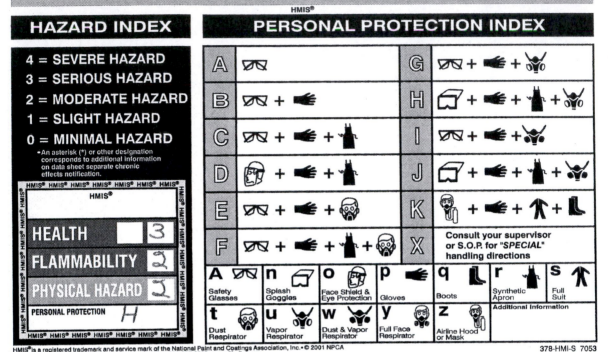

Figure 1.2

HMIS® III label for glacial acetic acid. The bottom left corner displays the hazard index for the chemical. HMIS® III is a registered trademark of the National Paint & Coatings Association, Inc. (NPCA). *It is used with permission and may not be further reproduced. NPCA has granted an exclusive license to produce and distribute HMIS® III materials to J. J. Keller & Associates, Inc. Those wishing to utilize the HMIS® III system should contact J. J. Keller at 1-800-327-6868 or www.jjkeller.com.*

acid, a reagent often used in biochemical research. All chemical reagent bottles in a workplace, lab, or stockroom must be labeled to identify potential hazardous materials and to specify personal protection necessary for handling. One standard hazard communication system used for this purpose is the Hazardous Materials Identification System (HMIS® III; shown in Figure 1.2). The health, flammability, physical hazard, and personal protection codes for the chemical reagent are summarized on the bottle label for quick identification.

Safe Practices in the Biochemistry Laboratory

It is easy to overlook some of the potential hazards of working in a biochemistry laboratory. Students often have the impression that they are working less with chemicals and more with natural biomolecules; therefore, there is less need for caution. However, this is not true; many reagents used are flammable and toxic. In addition, materials such as fragile glass (disposable pipets), sharp objects (needles), and potentially infectious biological materials (blood, bacteria, viruses) must be used and disposed of with caution. The extensive use of electrical equipment, including hot plates, stirring motors, and high-voltage power supplies for electrophoresis, presents special hazards.

Proper disposal of all waste chemicals, sharp objects, and infectious agents is essential not only to maintain safe laboratory working conditions, but also to protect the general public and your local environment. Some of the liquid chemical reagents and reaction mixtures from experiments are relatively safe and may be disposed of in the laboratory drainage system without causing environmental damage. However, special procedures must be followed for the use and disposal of most reagents and materials. Often this means that your instructor will provide detailed information on proper use procedures. In some cases, proper disposal will require the collection of waste materials from each laboratory worker and the institution will be responsible for removal. For each procedure described in this book, appropriate handling of all reagents, materials, and equipment will be recommended.

It is essential that all students be aware of the potential hazards of working in a biochemistry laboratory. A set of rules is an appropriate way to communicate the importance of practicing safe science. The general rules outlined in Figure 1.3 serve as guidelines. Your institution and instructor may have their own list of rules or may want to add guidelines for specific activities. Rules of laboratory safety and chemical handling are not designed to impede productivity, nor should they instill a fear of chemicals or of laboratory procedures. Rather, their purpose is to create a healthy awareness of potential laboratory hazards, to improve the efficiency of each student worker, and to protect the general public and the environment from waste contamination. The list of references at the end of this chapter includes books, journal articles, manuals, and Web sites describing proper and detailed safety procedures.

B.   KEEPING RECORDS AND COMMUNICATING EXPERIMENTAL RESULTS

The Laboratory Notebook

The biochemistry laboratory experience is not finished when you complete the experimental procedure and leave the laboratory. All scientists, including students, have the obligation to prepare and present written and oral reports on the results of their experimental work. Because these reports may be read and heard by many other professional scientists, they must be completed in a clear, concise, orderly, and accurate manner. Reports are most easily prepared outside of the lab using notes taken in a laboratory notebook *while* the experiment is in progress. These notes usually include procedural details, preparation of all reagents and solutions, setup of equipment, collection of

Figure 1.3

Guidelines for Safety in
the Biochemistry Laboratory.

1. Some form of eye protection is required at all times. Safety glasses with wide side shields or goggles are recommended, but normal eyeglasses with safety lenses may be permitted under some circumstances. Your instructor will inform you of the type of eye protection required. A statement regarding the wearing of contact lenses in the laboratory has been made by the American Chemical Society.[1] In general, contact lenses may be acceptable for wear in the laboratory, but the student, of course, must also wear safety glasses with side shields or goggles like all other students. Students who wear contacts must report to their lab instructor to determine the local rules for the lab.
2. Wear appropriate clothes — comfortable, well-fitting, older clothes that cover most of your skin. Sandals or bare feet are never allowed.
3. Never work alone in the laboratory.
4. Be familiar with the properties of all chemicals used in the laboratory. This includes their flammability, reactivity, toxicity, and proper disposal. This information may be obtained from your instructor, from the HMIS® III label, and from the MSDS. Always wear disposable gloves when using potentially dangerous chemicals or infectious agents.
5. Be familiar with your local rules for the safe handling and disposal of all non-chemical hazards. These include broken glass, "sharps" (needles, syringes, etc.), and biohazards (blood, bacteria, etc.).
6. Be especially careful with electrical equipment like stirrers, hot plates, and power supplies (electrophoresis, etc.). Always unplug before handling and avoid contact with water.
7. If open flames like Bunsen burners are necessary, make sure there are no flammable solvents in the area.
8. Eating, drinking, and smoking in the laboratory are strictly prohibited at all times.
9. Unauthorized experiments are not allowed.
10. Mouth suction should never be used to fill pipets or to start siphons.
11. Become familiar with the location and use of standard safety features in your laboratory. All laboratories should be equipped with fire extinguishers, eyewashes, safety showers, fume hoods, chemical spill kits, fire blankets, first-aid supplies, and containers for chemical disposal. Receptacles should also be available for disposal of dangerous materials like glass, biohazards, and sharps. Any questions regarding the use of these features should be addressed to your instructor, teaching assistant, or lab director.
12. Report all chemical spills, presence of biohazards, accidents, and injuries (even minor) to your instructor.

[1]American Chemical Society, Washington, DC; Committee on Chemical Safety, 1998.

data, and your thought processes and observations during the experiment. Experiments are often complex and move rather quickly, and it would be impossible to write down your data and observations after you have completed the experiments and left the lab. It is also not a good practice to record results on scraps of paper or on paper towels that may easily become lost or destroyed. The lab notebook will also come in handy if you need to troubleshoot or repeat an experiment because of inconsistent results.

Your instructor may have his or her own rules for preparation of the lab notebook, but here are some useful guidelines:

• The notebook should be hardbound with quadrille-ruled (gridded) pages; writing should be done with pen. This provides a permanent, durable record and the potential for construction of tables, graphs, charts, etc. Number each page of the book.

- Save the first few pages of the book for construction of a table of contents. Keep this up-to-date by entering the name of each experiment and page number.
- Use the right-hand pages only for writing your experimental notes. The left-hand pages may be used as scratch paper for your own personal notes, reminders, or calculations not appropriate for the main entry.
- Each entry for an experiment or project must begin with a title and date. The general outline required by many instructors for the written material is shown in Figure 1.4 and described below. Note that Parts I–IIc are labeled Prelab and should be completed before you begin the actual procedures in the lab.

Details of the Experimental Write-Up (see Figure 1.4)

Below is an outline that may be used as a guideline to write a complete report on an experiment.

Introduction

This section begins with a three- or four-sentence statement of the objective or purpose of the experiment. For preparing this statement, ask yourself, "What are the goals of this experiment?" This statement is followed by a brief discussion of the theory behind the experiment. If a new technique or instrumental method is introduced, give a brief description of the method. Include chemical and biochemical reactions and structures of reagents when appropriate.

Figure 1.4

General outline for experimental write-up.

Prelab
I. Introduction
    (a) Objective or purpose
    (b) Theory
II. Experimental
    (a) Table of materials and reagents
    (b) List of equipment
    (c) Flowchart
    (d) Record of procedure
III. Data and Calculations
    (a) Record of all raw data including printouts
    (b) Method of calculation with statistical analysis
    (c) Present final data in tables, graphs, or figures when appropriate
IV. Results and Discussion
    (a) Conclusions
    (b) Compare results with known values
    (c) Discuss the significance of the data
    (d) Was the original objective achieved?
    (e) Literature references

Experimental

Begin this section with a list of all reagents and materials used in the experiment. The sources of all chemicals and the concentrations of solutions should be listed. Instrumentation is listed with reference to company name and model number. A flowchart to describe the stepwise procedure for the experiment should be included after the list of equipment.

The write-up to this point is to be completed as a Prelab assignment. The experimental procedure followed is then recorded in your notebook as you proceed through the activities. The details should be sufficient so that a fellow student could use your notebook to repeat the experiment. You should include observations, such as color change or gas evolution, made during the experiment. If you obtain a computer printout of numbers, a spectrum from a spectrophotometer, or a photograph, these records must be saved with the notebook.

Data and Calculations

All raw data from the experiment are to be recorded directly in your notebook, not on separate sheets of paper that can easily become lost. Calculations involving the data must be included for at least one series of measurements. All data numbers should be analyzed by appropriate statistical methods described in Section E, Chapter 1.

For many experiments, the clearest presentation of data is in a tabular or graphical form. A graph may be prepared directly on the gridded pages of your notebook, or one prepared by computer software.

Results and Discussion

This important section of your write-up answers the questions, "Did you achieve your proposed goals and objectives?" and "What is the significance of the data?" Any conclusion that you make must be supported by experimental results. It is often possible to compare your data with known values and results from the literature. If problems were encountered in the experiments, these should be outlined with possible remedies for future experiments.

All library references (books, journal articles, and Web sites) that were used to complete the experiment should be listed at the end of the write-up. It is especially important to report references used for laboratory procedures. The standard format to follow for a reference listing is shown at the end of this chapter in the Further Reading section.

Everyone has his or her own writing style. Because there is always room for improvement, it is imperative that you continually try to enhance your writing skills. When your instructor reviews your notebook, he or she should include helpful writing tips. References at the end of this chapter provide further instructions in scientific writing. Your instructor may accept, as a final report, the experimental write-up as described above and in Figure 1.4. However, she or he may request that you present your experimental results in one of the more formal written or oral modes described next.

Communicating Results from Biochemistry Research

A scientific project is not complete until its discoveries have been communicated to colleagues around the world. The three most important methods or tools for communication are: the **scientific paper,** the **oral presentation,** and the **poster.** Although there are many differences in how to prepare for these three common methods of introducing new biochemical information, they all have one thing in common–the sharing of experimental results and conclusions. The distinct rules and traditions of each of the methods will be described and compared here.

The Scientific Paper

A paper published in a biochemical journal is a formal way to report research results to colleagues in the international biochemical community. Before writing such a document, one must first determine the journal to which the article will be submitted. There are hundreds of journals that accept manuscripts in the field of biochemistry (Figure 1.5). Some have very high rank, prestige, and status, based on the significance of research results published, reputation of authors, numbers of citations, whether or not manuscripts are peer reviewed, and numbers of readers. Most journals are peer reviewed, indicating that before a manuscript is published, it is studied by members of the journal's editorial board to assure that the manuscript is scientifically significant, that it appears to be accurate, and that it is useful and of value to readers of the journal. Some journals accept manuscripts in all areas of biochemistry, but the screening done in rigorous peer review by specialists is often very competitive. Other journals, which are also reputable, are more specialized and accept peer-reviewed articles only in certain areas of biochemistry. Perhaps the best advice is to submit the manuscript to the most prestigious journal that has a large audience interested in your specialized topic. Publishing a paper in a reputable, peer-reviewed journal offers historic permanence for your work, status, and exposure as a scientist; however, because the lag time between acceptance and publication of a manuscript can sometimes stretch up to one year, the data reported can become insignificant.

Your instructor may require that you write up the results from an experiment in the form of a journal article, so it is important to understand the conventions used in preparing a manuscript for publication. Most biochemical journal articles have the same basic organization: Title, Abstract, Introduction, Experimental Methods, Results, Discussion, and References. The specific requirements for each of these sections vary among the many journals, so it is important to review several articles in different journals to get a flavor of what is expected in writing and submitting an article. All scientific journals publish an "Instructions to Authors" section at least once a year. Although your instructor will most likely assign the organization of a specific journal, it is instructive to study articles in the following high-ranking journals that publish biochemistry topics: *The Journal of Biological*

Figure 1.5

Some journals that publish research articles in biochemistry.

Accounts of Chemical Research
Analytical Biochemistry
Annual Review of Biochemistry
Archives of Biochemistry and Biophysics
Biochemical and Biophysical Research Communications
Biochemical and Molecular Medicine
Biochemical Journal
Biochemistry
Biochemistry and Molecular Biology Education
Biochimica et Biophysica Acta: General Subjects; Molecular and Cell Biology;
    Protein Structure and Molecular Enzymology
BioEssays
Bioorganic Chemistry
Biophysical Journal
Canadian Journal of Biochemistry
Cell
ChemBioChem
Chemistry and Biology
Current Opinion in Structural Biology
DNA Research Online
Electrophoresis
European Journal of Biochemistry
FASEB Journal
Glycobiology
Glycoconjugate Journal
Journal of Biochemistry
Journal of Biological Chemistry
Journal of Chemical Education
Journal of Lipid Research
Journal of Molecular Biology
Journal of Neurochemistry
Journal of Plant Physiology
Macromolecules
Methods: A Companion to Methods in Enzymology
Molecular and Cellular Biochemistry
Molecular and Cellular Proteomics
Nature
Nature Reviews Molecular Cell Biology
Nature Structural Biology
Nucleic Acid Research
Phytochemistry
Proceedings of the National Academy of Sciences USA
Prostaglandins, Leukotrienes and Essential Fatty Acids
Protein Science
Proteomics
RNA
Science
Scientific American
Trends in Biochemical Sciences

Chemistry (published by the American Society for Biochemistry and Molecular Biology), *Biochemistry* (published by the American Chemical Society), *Science*, and *Nature*.

The Oral Presentation

The purpose and mechanics of an oral presentation are quite different from preparing and publishing a paper. You may write a paper over a period of days, weeks, and even months, and the published work is available as a permanent record for readers to study anytime in the future. In an oral presentation, you have a fleeting moment to present data and attempt to convince your audience of the importance of your work. One advantage of the oral presentation, however, is the more direct contact with your audience than with a paper; thus the opportunity exists for immediate questions and feedback.

Presentations usually range from 15 to 60 minutes. Shorter presentations cover a much smaller unit of a research project, whereas 60-minute talks (often called seminars) can give a broader exposure to the research area.

Scientific presentations involve mixed media—oral and visual. The important verbal points are reinforced with the use of a visual aid such as a figure, graph, or other element. Scientific presenters today often use Power-Point, computer software that projects electronic slides onto a screen, although overhead transparencies are also acceptable and efficient. Whatever the type of visual aid, the slides must be carefully constructed with special concern for the total number of slides and the amount of information on each. Some presenters use the approximate ratio of one–two slides per minute of presentation.

The organization of a talk is similar to that of a paper—Introduction, Experimental Methods, Results, Discussion, Conclusions, Questions/Comments. If your instructor expects you to present a talk, he or she will provide specific information regarding length of time, range of topic, type of visual aids, multimedia, etc.

The Scientific Poster

The scientific poster is a communication method that may be considered a hybrid, as it combines elements of the oral presentation (verbal expression and visual aids) with elements of a paper (printed text and figures). The poster has become the primary medium by which new scientific information is exchanged at all professional conferences including local, regional, national, and international meetings. At meetings, posters that consist of text and figures arranged in panels on a thin piece of cardboard (average 3′ × 5′) are set up in designated areas, during specified times (usually for a day or two), and there is often an official time when the presenter is to be in attendance. The poster, however, may be available to readers for long periods of time in the absence of the creator. Some of the specific characteristics that describe a poster include (Figure 1.6):

Introduction

This is a Microsoft Powerpoint template that has column widths and font sizes optimized for printing a 36 x 56" poster—just replace the "tips" and "blah, blah, blah" repeat motifs with actual content. Try to keep your total word count under 1100. More tips can be found at the companion site, "Advice on designing scientific posters," located at,

http://www.swarthmore.edu/NatSci/cpurrin1/posteradvice.htm

Fig. 1. Use a photograph or drawing here to quickly introduce a viewer to your question, organism, or allele du jour. Use a non-serif font for figure legends to provide subtle cue to reader that he/she is not reading normal text section. Color can also be used as a cue.

Materials and methods

This paragraph has "justified" margins, but be aware that simple left-justification (all other paragraphs) is infinitely better if your font doesn't "space" nicely when fully justified. Sometimes spacing difficulties can be fixed by manually inserting hyphens into longer words (Powerpoint doesn't do this automatically).

Your main text is easier to read if you use a "serif" font such as Palatino or Times. Use a non-serif font for title and section headings (and for figure legends, graph text, etc.).

Be brief, and opt for photographs or drawings whenever possible to illustrate organism, protocol, or experimental design.

Fig. 2. Photograph or drawing of organism, chemical structure, or whatever focus of study is. Don't use graphics from the web (they look terrible when printed).

Fig. 3. Illustration of important piece of equipment, or perhaps a flow chart summarizing experimental design. Scanned, hand-drawn illustrations are often preferable to computer generated ones.

Results

The overall layout for this section can, and probably *should*, be modified from this template, depending on the size and number of charts and photographs your specific experiment generated. You might want a single, large column to accommodate a large map, or perhaps you could arrange 6 figures in a circle in the center of the poster: do whatever it takes to make your results *graphically* clear. To see examples of how others have abused this template to fit their presentation needs, perform a Google search for "powerpoint template for scientific posters."

Paragraph format is fine, but sometimes a simple list of "bullet" points can communicate results more effectively:

- 9 out of 12 brainectomized rats survived.
- Control rats completed maze faster, on average, than rats without brains (**Fig. 3**) ($t = 9.84$, df $= 21$, $p = 0.032$).

Fig. 4(a-c). Make sure legends have enough detail to fully explain to the viewer what the results are. Note that for posters it is good to put some "Materials and methods" information within the figure legends or onto the figures themselves—it allows the M&m section to be shorter, and gives viewer a sense of experiment(s) even if they have skipped directly to figures. Don't be tempted to reduce font size in figure legends, axes labels, etc.— your viewers are probably most interested in reading your figures and their legends! Font size in graphs should be same size as text in body of section (e.g., easily legible from 6' away).

Often you will have some more text-based results between your figures. This text should *explicitly* guide the reader through the figures.

Blah, blah, blah (**Figs. 4a,b**). Blah, blah, blah. Blah, blah, blah. Blah, blah, blah. Blah, blah, blah. Blah, blah, blah. Blah, blah, blah. Blah, blah, blah.

Blah, blah, blah. Blah, blah, blah. Blah, blah, blah. Blah, blah, blah. Blah, blah, blah. Blah, blah, blah. Blah, blah, blah. Blah, blah, blah (**Fig. 4c**). Blah, blah, blah. Blah, blah, blah. Blah, blah, blah. Blah, blah, blah. Blah, blah, blah. Blah, blah, blah (data not shown).

Blah, blah, blah. Blah, blah, blah. Blah, blah, blah. Blah, blah, blah. Blah, blah, blah. Blah, blah, blah. Blah, blah, blah. Blah, blah, blah. Blah, blah, blah. Blah, blah, blah (God, personal communication).

Figure 1.6

Template for a poster in biochemistry. *Courtesy of Professor Colin Purrington, Department of Biology, Swarthmore College, www.swarthmore.edu/NatSci/cpurrin1/posteradvice.htm*

- usually composed of small units of a research project and most often are based on preliminary results and conclusions.

- contain many of the same organizational elements as a paper or talk– Title/Authors, Abstract, Introduction, Methods, Results, Discussion, Conclusions, References–but in a much briefer form.

- posters are often enhanced with a brief, oral summary given by the presenter. Only the main points such as the purpose, results, conclusions, and future experiments for the project should be included in this concise summary. As a presenter, during your official time at the poster

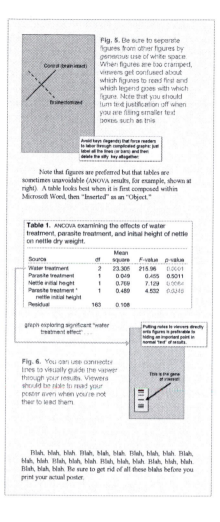

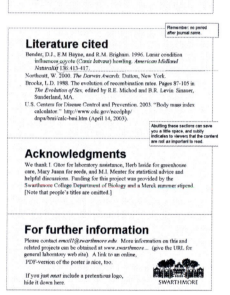

you will be visited by individuals or small groups who will spend an average of about 10–12 minutes at your poster.

- posters must be completely self-explanatory, as you will not always be present to answer a reader's questions.
- the environment is usually interactive and informal, allowing for one-on-one contact with other researchers.
- poster sessions at scientific meetings are very democratic and inclusive, as the presenters and audience may consist of all levels of scientists—tenured research professors and undergraduate research students.

Your instructor may request that you prepare a poster for local display at your institution or for presentation at a regional or national meeting. Specific details will not be given here, as all organizations sponsoring poster sessions at meetings publish their own rules and regulations (poster size, font sizes, etc.) for preparing posters. Many colleges and universities now schedule local meetings where students may obtain experience preparing and presenting posters about their research results. Attend one of these local meetings or walk around the halls of your chemistry and biology departments looking for posters made by research students at your institution. These serve as very good models for your own creation. You may also find useful information about the specific details of poster construction by searching the Internet. Some helpful Web sites with poster templates are listed in the Further Reading section at the end of this chapter.

C. USING BIOCHEMICAL REAGENTS AND SOLUTIONS

Water Quality

The most common solvent for solutions used in the biochemistry laboratory is water. Ordinary tap water contains a variety of impurities including particulate matter (sand, silt, etc.); dissolved organics, inorganics, and gases; and microorganisms (bacteria, viruses, protozoa, algae). In addition, the natural degradation of microorganisms leads to the presence of by-products called pyrogens. Tap water should never be used for the preparation of any reagent solutions. For most laboratory procedures, it is recommended that some type of purified water be used.

There are five basic water purification technologies—distillation, ion exchange, activated carbon adsorption, reverse osmosis, and membrane filtration. Most academic and industrial research laboratories are equipped with "in-house" purified water, which typically is produced by a combination of the above purifying technologies. For most procedures carried out in the biochemistry lab, water purified by deionization, reverse osmosis, or distillation usually is acceptable. For special procedures such as buffer standardization, liquid chromatography, and tissue culture, ultrapure water, usually bottled and available commercially, should be used.

The water quality necessary will depend on the solutions to be prepared and on the biochemical procedures to be investigated. Water that is purified only by ion exchange will be low in metal-ion concentration, but may contain certain organics that are washed from the ion-exchange resin. These contaminants will increase the ultraviolet absorbance properties of water. If sensitive ultraviolet absorbance measurements are to be made, distilled water (especially glass distilled) is better than deionized.

Cleaning Laboratory Glassware

The results of your experimental work will depend, to a great extent, on the cleanliness of your equipment, especially glassware used for preparing and transferring solutions. There are at least two important reasons for this. (1) Many of the chemicals and biochemicals will be used in milligram, microgram, or even nanogram amounts. Any contamination, whether on the inner walls of a beaker, in a pipet, or in a glass cuvette, could be a significant percentage of the total experimental sample. (2) Many biochemicals and biochemical processes are sensitive to one or more of the following common contaminants: metal ions, detergents, and organic residues. In fact, the objective of many experiments is to investigate the effect of a metal ion, organic molecule, or other chemical agent on a biochemical process. Contaminated glassware will virtually ensure failure in these activities.

Many contaminants, including organics and metal ions, adhere to the inner walls of glass containers. Washing glassware, including pipets, with dilute detergent (0.5% in water) followed by 5–10 water rinses is probably sufficient for most purposes. The final rinse should be with distilled or deionized water. Metal-ion contamination of glassware can be greatly reduced by rinsing with concentrated nitric acid followed by extensive rinsing with purified water.

Dry equipment is required for most processes carried out in the biochemistry laboratory. When you needed dry glassware in the organic laboratory, you probably rinsed the piece of equipment with acetone, which rapidly evaporated, leaving a dry surface. Unfortunately, that surface is coated with an organic residue consisting of nonvolatile contaminants in the acetone. Because this residue could interfere with your experiments, it is best to refrain from acetone washing. Glassware and plasticware should be rinsed well with purified water and dried in an oven designated for glassware, not one used for drying chemicals.

Never clean cuvettes or any optically polished glassware with ethanolic KOH or other strong base, as this will cause etching. All glass cuvettes should be cleaned carefully with 0.5% detergent solution, in a sonicator bath, or in a cuvette washer, followed by thorough rinsing with purified water.

Solutions: Concentrations and Calculations

The concentrations for solutions used in biochemistry laboratory may be expressed in several different units. The most common units are:

* *Molarity (M):* concentration based on the number of moles of solute per liter of solution. A 1 M solution of the amino acid alanine (MW = 89.1), contains 1 mole or 89.1 g of alanine in a solution volume of 1 liter. In biochemistry, it is more common to use concentration ranges that are millimolar (mM, $1 \times 10^{-3} M$), micromolar (μM, $1 \times 10^{-6} M$), or nanomolar (nM, $1 \times 10^{-9} M$). A 1 mM solution of alanine contains 0.089 g or 89 mg (89.1 $\times$ 0.001) of alanine in a solution volume of 1 liter.

How many grams of alanine are present in 100 mL of the 1 mM alanine solution? (Ans: 0.0089 g). How many milligrams of alanine are present in 100 mL? (Ans: 8.9 mg).

- *Percent by weight (% wt/wt):* concentration based on the number of grams of solute per 100 g of solution. A 5% wt/wt solution of alanine contains 5 g of alanine in 100 g of solution. How many grams of alanine are present in 10 g of this solution? (Ans: 0.5 g).

- *Percent by volume (% wt/vol):* concentration based on the number of grams of solute per 100 mL of solution. A 10% wt/vol solution of alanine contains 10 g of alanine in 100 mL of solution. How many grams of alanine are present in 50 mL of this solution? (Ans: 5 g).

- *Weight per volume (wt/vol):* concentration based on the number of grams, milligrams, or micrograms of solute per unit volume; for example, mg/mL, g/L, mg/100 mL, etc. A solution of alanine, concentration wt/vol = 5 g/L, contains 5 g of alanine in a liter of solution. How many grams of alanine would be present in 2 liters of this solution? In 10 mL? (Ans: 10 g; 0.005 g or 5 mg).

➤ **Study Exercise 1.1**

(a) Many solutions you use will be based on molarity. For practice, assume you require 1 liter of solution that is 0.1 M (100 mM) glucose:

MW of glucose = 180.2

1 mole of glucose = 180.2 g

0.1 mole of glucose = 18.02 g

To prepare a 0.1 M glucose solution, weigh 18.02 g of glucose and transfer to a 1-liter volumetric flask. Add about 700–800 mL of purified water and swirl to dissolve. Then add water so that the bottom of the meniscus is at the etched line on the flask. Stopper and mix well. The flask must be labeled with solution contents, 0.1 M glucose, date prepared, and name of preparer.

(b) Assume that you need only 250 mL of 0.10 M glucose. Explain how you would prepare the solution. Emphasize any changes from Part (a).

➤ **Study Exercise 1.2 Concentration Unit Conversions**   It is often necessary in your biochemistry lab work to convert one concentration unit to another. For example, you may need to know the concentration of the 0.1 M glucose solution (Study Exercise 1.1) in concentration terms of mg/mL. Here are some basic calculations for practice.

(a) Convert the concentration units of 0.1 M glucose to the units of mg of glucose in 1 mL (mg/mL).

According to the procedure described in Study Exercise 1.1(a), the 1000-mL solution of glucose contains 18.02 g of glucose or it is 18.02 g/1000 mL. This is equivalent to 1.80 g/100 mL or 0.018 g/mL. Therefore, the 0.1 M glucose solution concentration is equivalent to about 18 mg/mL.

(b)   Convert the concentration units of 1 M alanine to the units of g/100 mL.

(c)   Convert the concentration units of 1 M alanine to the units of % wt/vol.

(d)   Calculate the concentration of a 0.1 M glucose solution to the units of % wt/vol.

Preparing and Storing Solutions

In general, solid solutes should be weighed on weighing paper or plastic weighing boats, with the use of an electronic analytical or top-loading balance. Liquids are more conveniently dispensed by volumetric techniques; however, this assumes that the density is known. If a small amount of a liquid is to be weighed, it should be added to a tared flask by means of a disposable Pasteur pipet with a latex bulb. The hazardous properties of all materials should be known before use (read MSDS) and the proper safety precautions obeyed.

The storage conditions of reagents and solutions in the biochemistry lab are especially critical. Although some will remain stable indefinitely at room temperature, it is good practice to store all solutions in a closed container. Often it is necessary to store some solutions in a refrigerator at 4°C. This inhibits bacterial growth and slows decomposition of the reagents. Some solutions may require storage below 0°C. If these are aqueous solutions or others that will freeze, be sure there is room for expansion inside the container. Stored solutions must always have a label containing the name and concentration of the solution, the date prepared, and the name of the preparer.

All stored containers, whether at room temperature, 4°C, or below freezing, must be properly sealed. This reduces contamination by bacteria and vapors in the laboratory air (carbon dioxide, ammonia, HCl, etc.). Volumetric flasks, of course, have glass stoppers, but test tubes, Erlenmeyer flasks, bottles, and other containers should be sealed with screw caps, corks, or hydrocarbon foil (Parafilm). Remember that hydrocarbon foil, a wax, is dissolved by solutions containing nonpolar organic solvents like chloroform, diethyl ether, and acetone.

Bottles of pure chemicals and reagents should also be properly stored. Many manufacturers now include the best storage conditions for a reagent on the label. The common conditions are: store at room temperature; store at 0–4°C; store below 0°C; or store in a desiccator at room temperature, 0–4°C, or below 0°C. Many biochemical reagents form hydrates by taking up moisture from the air. If the water content of a reagent increases, the molecular weight and purity of the reagent change. For example, when the coenzyme nicotinamide adenine dinucleotide (NAD^+) is purchased, the label usually reads "Anhydrous molecular weight = 663.5; when assayed, contained 3 H_2O per mole." The actual molecular weight that should be used for solution preparation is $663.5 + (18)(3) = 717.5$. However, if this reagent is stored in a moist refrigerator or freezer outside a desiccator, the moisture content may increase to an unknown value.

D. QUANTITATIVE TRANSFER OF LIQUIDS

Practical biochemistry is highly reliant on analytical methods. Many analytical techniques must be mastered, but few are as important as the quantitative transfer of solutions. Some type of pipet will almost always be used in liquid transfer. Because students may not be familiar with the many types of pipets and the proper techniques in pipetting, this instruction is included here.

Pipets and Pipetting

Pipet Fillers

Figure 1.7 illustrates the various types of pipets and fillers. The use of any pipet requires some means of drawing reagent into the pipet. *Liquids should never be drawn into a pipet by mouth suction on the end of the pipet!* Small latex bulbs are available for use with disposable pipets (Figure 1.7A). For volumetric and graduated pipets, two types of bulbs are available. One type (Figure 1.7B) features a special conical fitting that accommodates common sizes of pipets. To use these, first place the pipet tip below the surface of the liquid. Squeeze the bulb with the left hand (if you are a right-handed pipettor) and then hold it tightly to the end of the pipet. Slowly release the pressure on the bulb to allow liquid to rise to 2 or 3 cm above the top graduated mark. Then, remove the bulb and quickly grasp the pipet with your index finger over the top end of the pipet. The level of solution in the pipet will fall slightly, but should not fall below the top graduated mark. If it does fall too low, use the bulb to refill.

Safety Pipet Fillers

Mechanical pipet fillers (made of silicone and sometimes called safety pipet fillers, propipets, or pi-fillers) are more convenient than latex bulbs. As shown in Figure 1.7C,D, these fillers are equipped with a system of hand-operated valves and can be used for the complete transfer of a liquid. The use of a safety pipet filler is outlined in Figure 1.8. *Never allow any solvent or solution to enter the pipet bulb!* To avoid this, two things must be kept in mind:

1. Always maintain careful control while using valve S to fill the pipet.
2. Never use valve S unless the pipet tip is below the surface of the liquid. If the tip moves above the surface of the liquid, air will be sucked into the pipet and solution will be flushed into the bulb.

Figure 1.7

Examples of pipets and pipet fillers. *Courtesy of Sargent-Welch, VWR International; www.vwr.com/*
A Latex bulb, **B** Pipet filler, **C** Mechanical pipet filler, **D** Pipettor pump, **E** Pasteur pipet, **F** Volumetric pipet, **G** Mohr pipet, **H** Serological pipet.

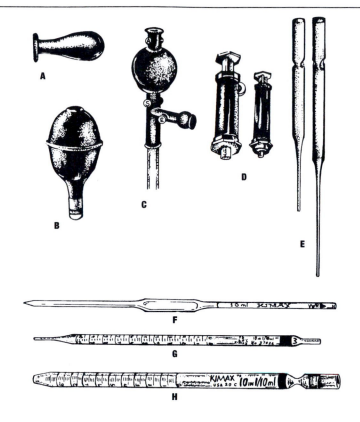

Disposable Pasteur Pipets

Often it is necessary to perform a semiquantitative transfer of a small volume (1–10 mL) of liquid from one vessel to another. Because pouring is not efficient, a **Pasteur pipet** with a small latex bulb may be used (Figure 1.7A,E). Pasteur pipets are available in two lengths (15 and 23 cm) and hold about 2 mL of solution. These are especially convenient for the transfer of nongraduated amounts to and from test tubes. Typical recovery while using a Pasteur pipet is 90 to 95%. If dilution is not a problem, rinsing the original vessel with a solvent will increase the transfer yield. Used disposable pipets should be discarded in special containers for broken glass.

Calibrated Pipets

Although most quantitative transfers are now done with automatic pipetting devices, which are described later in the chapter, instructions will be given for the use of all types of pipets. If a quantitative transfer of a specific and accurate volume of liquid is required, some form of calibrated pipet must be used.

Figure 1.8

How to use a safety pipet filler. *Courtesy of Sargent-Welch, VWR International.*

1. Using thumb and forefinger, press on valve A and squeeze bulb with other fingers to produce a vacuum for aspiration. Release valve A leaving bulb compressed.

2. Insert pipet into liquid. Press on valve S. Suction draws liquid to desired level.

3. Press on valve E to expel liquid.

4. To deliver the last drop, maintain pressure on valve E, cover E inlet with middle finger, and squeeze the small bulb.

- *Volumetric pipets* (Figure 1.7F) are used for the delivery of liquids required in whole-milliliter amounts (1, 2, 5, 10, 15, 20, 25, 50, and 100 mL). To use these pipets, draw liquid with a latex bulb or mechanical pipet filler to a level 2–3 cm above the fill line. Release liquid from the pipet until the bottom of the meniscus is directly on the fill line. Touch the tip of the pipet to the inside of the glass wall of the container from which it was filled. Transfer the pipet to the inside of the second container and release the liquid. Hold the pipet vertically, allow the solution to drain until the flow stops, and then wait an additional 5–10 seconds. Touch the tip of the pipet to the inside of the container to release the last drop from the outside of the tip. Remove the pipet from the container. Some liquid may still remain in the tip. Most volumetric pipets are calibrated as "TD" (to deliver), which means the intended volume is transferred *without final blow-out*, that is, the pipet delivers the correct volume.

Fractional volumes of liquid are transferred with **graduated pipets,** which are available in two types:

- *Mohr pipets* (Figure 1.7G) are available in long- or short-tip styles. Long-tip pipets are especially attractive for transfer to and from vessels with small openings. Virtually all Mohr pipets are TD, and they are available in many sizes (0.1 to 10 mL). The marked subdivisions are usually 0.01 or 0.1 mL, and the markings end a few centimeters from the tip. Selection of the proper size is especially important. For instance, do not try to transfer 0.2 mL with a 5- or 10-mL pipet. Use the smallest pipet that is practical. The use of a Mohr pipet is similar to that of a volumetric pipet. Draw the liquid into the pipet with a pipet filler to a level about 2 cm above the 0 mark. Lower the liquid level to the 0 mark. Remove the last drop from the tip by touching it to the inside of the glass container. Transfer the pipet to the receiving container and release the desired amount of solution. The solution should not be allowed to move below the last graduated mark on the pipet. Touch off the last drop.

- *Serological pipets* (Figure 1.7H) are similar to Mohr pipets, except that they are graduated downward to the very tip and are designed for blow-out. Their use is identical to that of a Mohr pipet except that the last bit of solution remaining in the tip must be forced out into the receiving container with a rubber bulb. This final blow-out should be done after 15–20 seconds of draining.

Cleaning and Drying Pipets

Special procedures are required for cleaning glass pipets. Immediately after use, every pipet should be placed, tip up, in a vertical cylinder containing a dilute detergent solution (less than 0.5%). The pipet must be completely covered with solution. This ensures that any reagent remaining in the pipet is forced out through the tip. If reagent solutions are allowed to dry inside a pipet, the tips can easily become clogged and are very difficult to open. After several pipets have accumulated in the detergent solution, the pipets should be transferred to a pipet rinser. Pipet rinsers continually cycle fresh water through the pipets. Immediately after detergent wash, tap water may be used to rinse the pipets, but distilled water should be used for the final rinse. Pipets may then be dried in an oven.

Automatic Pipetting Devices

For most quantitative transfers, including many repeated small-volume transfers, a mechanical microliter pipettor (i.e., Eppendorf type, Pipetman) is ideal. This allows accurate, precise, and rapid dispensing of fixed volumes from 1 to 10,000 μL (0.001 to 10 mL). The pipet's push-button system can be operated with one hand, and it is fitted with detachable polypropylene tips (Figure 1.9). The advantage of polypropylene tips is that the reagent film remaining in the pipet after delivery is much less than for glass tips. Mechanical pipettors are available in many different sizes. Newer models offer continuous volume adjustment, so a single model can be used for delivery of specific volumes within a certain range.

To use the pipettor, choose the proper size and place a polypropylene pipet tip firmly onto the cone, as shown in Figure 1.9. Tips for pipets are available in several sizes, for 1–20, 20–250, 200–1000, 1000–5000, and 10,000-μL capacities. Details of the operation of an adjustable pipet are shown in Figure 1.9.

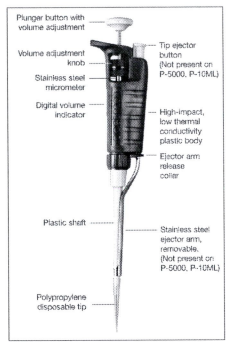

Pipetman P–200

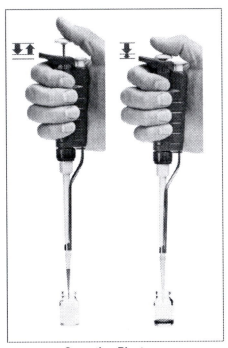

Operating Pipetman

Figure 1.9

How to use an adjustable pipetting device. Set the desired volume with the digital micrometer or plunger button. Attach a new disposable tip to the shaft of the pipet. Press on firmly with a slight twisting motion. Depress the plunger to the first positive stop, immerse the disposable tip into the sample liquid to a depth of 2–4 mm, and allow the pushbutton to return slowly to the up position and wait 1–2 seconds. To dispense sample, place the tip end against the side wall of the receiving vessel and depress the plunger slowly to the first stop. Wait 2–3 seconds, and then depress the plunger to the second stop to achieve final blowout. Withdraw the device from the vessel carefully with the tip sliding along the inside wall of the vessel. Allow the plunger to return to the up position. Discard the tip by depressing the tip ejector button. *Photos courtesy of Rainin Instrument Company, Inc., Woburn, MA. Pipetman is a registered trademark of Gilson Medical Electronics. Exclusive license to Rainin Instrument Company, Inc. www.rainin.com/*

For rapid and accurate transfer of volumes greater than 5 mL, automatic repetitive dispensers are commercially available. These are particularly useful for the transfer of corrosive materials. The dispensers, which are available in several sizes, are simple to use. The volume of liquid to be dispensed is mechanically set; the syringe plunger is lifted for filling and pressed downward for dispensing. Hold the receiving container under the spout while depressing the plunger. Touch off the last drop on the inside wall of the receiving container.

E.   STATISTICAL ANALYSIS OF EXPERIMENTAL DATA

The purpose of most laboratory exercises is to observe and measure characteristics of a biomolecule or a biological process. The characteristic is often quantitative–a single number or a group of numbers. These measured quantities may be the molecular weight of a protein, the pH of a buffer solution, the absorbance of a colored solution, the rate of an enzyme-catalyzed reaction, the concentration of a protein in solution, or the radioactivity associated with a molecule. If you measure a quantitative characteristic many times under identical conditions, a slightly different result will most likely be obtained each time.

For example, if a radioactive sample is counted twice under identical experimental and instrumental conditions, the second measurement immediately following the first, the probability is very low that the numbers of counts will be identical. If the absorbance of a solution is determined several times at a specific wavelength, the value of each measurement will surely vary from the others. If an assay for cholesterol is performed several times on a blood serum sample from the same individual, the values will probably be close, but not all will be the same (see Study Problem 1.12). Which measurements, if any, are correct? Before this question can be answered, you must understand the source and treatment of numerical variations in experimental measurements.

Defining Statistical Analysis

An **error** in an experimental measurement is defined as a deviation of an observed value from the true value. There are two types of errors, **determinate** and **indeterminate.** Determinate errors are those that can be controlled by the experimenter and are associated with malfunctioning equipment, improperly designed experiments, and variations in experimental conditions. These are sometimes called human errors because they can be corrected or at least partially alleviated by careful design and performance of the experiment. Indeterminate errors are those that are random and cannot be controlled by the experimenter. Specific examples of indeterminate errors are variations in radioactive counting and small differences in the successive measurements of glucose in a serum sample.

Two statistical terms involving error analysis that are often used and misused are **accuracy** and **precision.** Precision refers to the extent of agreement among repeated measurements of an experimental value. Accuracy is defined as the difference between the experimental value and the true value for the quantity. Because the true value is seldom known, accuracy is better defined as the difference between the experimental value and the accepted true value. Several experimental measurements may be precise (that is, in close agreement with each other) without being accurate.

If an infinite number of identical, quantitative measurements could be made on a biosystem, this series of numerical values would constitute a **statistical population.** The average of all of these numbers would be the **true value** of the measurement. It is obviously not possible to achieve this in practice. The alternative is to obtain a relatively small **sample of data,** which is a subset of the infinite population data. The significance and precision of these data are then determined by statistical analysis.

Most quantitative biological measurements can be made in duplicate, triplicate, or even quadruplicate, but it would be impractical and probably a waste of time and materials to make numerous determinations of the same measurement. Rather, when you perform an experimental measurement in the laboratory, you will collect a small sample of data from the population of infinite values for that measurement. To illustrate, imagine that an infinite number of experimental measurements of the pH of a buffer solution are made, and the results are written on slips of paper and placed in a container. It is not feasible to calculate an average value of the pH from all of these numbers, but it is possible to draw five slips of paper, record these numbers, and calculate an average pH. By doing this, you have collected a sample of data. By proper statistical manipulation of this small sample, it is possible to determine whether it is representative of the total population and the amount of confidence you should have in these numbers.

The Mean, Sample Deviation, and Standard Deviation

Radioactive decay with emission of particles is a random process. It is impossible to predict with certainty when a radioactive event will occur. Therefore, a series of measurements made on a radioactive sample will result in a series of different count rates, but they will be centered around an **average** or **mean** value of counts per minute. Table 1.1 contains such a series of count rates obtained with a scintillation counter on a single radioactive sample. A similar table could be prepared for other biochemical measurements, including the rate of an enzyme-catalyzed reaction or the protein concentration of a solution as determined by the Bradford method. The arithmetic average or mean of the numbers is calculated by totaling all the experimental values observed for a sample (the counting rates, the velocity of the reaction, or protein concentration) and dividing the total by

Table 1.1	
The Observed Counts and Sample Deviation from a Typical Radioactive Sample	
Counts per Minute	Sample Deviation $x_i - \bar{x}$
1243	+21
1250	+28
1201	−21
1226	+4
1220	−2
1195	−27
1206	−16
1239	+17
1220	−2
1219	−3
Mean = 1222	

the number of times the measurement was made. The mean is defined by Equation 1.1.

>>
$$\bar{x} = \frac{\sum\limits^{n} x_i}{n}$$

Equation 1.1

where

$\bar{x}$ = arithmetic average or mean

x_i = the value for an individual measurement

n = the total number of experimental determinations

Σ = sum of all the values

The mean counting rate for the data in Table 1.1 is 1222. If the same radioactive sample were again counted for a series of ten observations, that series of counts would most likely be different from those listed in the table, and a different mean would be obtained. If we were able to make an infinite number of counts on the radioactive sample, then a **true mean** could be calculated. The true mean would be the actual amount of radioactivity in the sample. Although it would be desirable, it is not possible experimentally to measure the true mean. Therefore, it is necessary to use the average of the counts as an approximation of the true mean and to use statistical analysis to evaluate the precision of the measurements (that is, to assess the agreement among the repeated measurements).

Because it is not usually practical to observe and record a measurement many times as in Table 1.1, what is needed is a way to determine the reliability of an observed measurement. This may be stated in the form of a question. How close is the result to the true value? One approach to this analysis is to calculate the **sample deviation,** which is defined as the difference between the value for an observation and the mean value, $\bar{x}$ (Equation 1.2). The sample deviations are also listed for each count in Table 1.1.

>> Sample deviation $= x_i - \bar{x}$ **Equation 1.2**

A more useful statistical term for error analysis is **standard deviation,** a measure of the spread of the observed values. Standard deviation, s, for a sample of data consisting of n observations may be estimated by Equation 1.3.

>> $$s = \sqrt{\frac{\sum(x_i - \bar{x})^2}{n - 1}}$$ **Equation 1.3**

It is a useful indicator of the probable error of a measurement. Standard deviation is often transformed to **standard deviation of the mean** or **standard error.** This is defined by Equation 1.4, where n is the number of measurements.

>> $$s_m = \frac{s}{\sqrt{n}}$$ **Equation 1.4**

It should be clear from this equation that as the number of experimental observations becomes larger, s_m becomes smaller, or the precision of a measurement is improved.

Standard deviation may also be illustrated in graphical form (Figure 1.10). The shape of the curve in Figure 1.10 is closely approximated by the **Gaussian distribution** or **normal distribution curve.** This mathematical treatment is based on the fact that a plot of relative frequency of a given event yields a dispersion of values centered about the mean, $\bar{x}$. The value of $\bar{x}$ is measured at the maximum height of the curve. The normal distribution curve shown in Figure 1.10 defines the spread or dispersion of the data. The probability that an observation will fall under the curve is unity or 100%. By using an equation derived by Gauss, it can be calculated that for a single set of sample data, 68.3% of the observed values will occur within the interval $\bar{x} \pm s$, 95.5% of the observed values within $\bar{x} \pm 2s$, and 99.7% of the observed values within

Figure 1.10

The normal distribution curve.

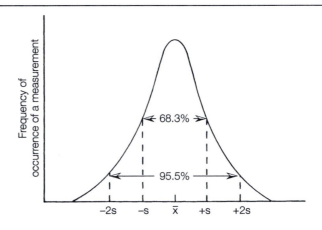

$\bar{x} \pm 3s$. Stated in other terms, there is a 68.3% chance that a single observation will be in the interval $\bar{x} \pm s$.

For many experiments, a single measurement is made so a mean value, $\bar{x}$, is not known. In these cases, error is expressed in terms of s but is defined as the **percentage proportional error,** $\%E_x$, in Equation 1.5.

>> $$\%E_x = \frac{100k}{\sqrt{n}}$$ *Equation 1.5*

The parameter k is a proportional constant between E_x and the standard deviation. The percent proportional error may be defined within several probability ranges. **Standard error** refers to a confidence level of 68.3%; that is, there is a 68.3% chance that a single measurement will not exceed the $\%E_x$. For standard error, $k = 0.6745$. **Ninety-five hundredths error** means there is a 95% chance that a single measurement will not exceed the $\%E_x$. The constant k then becomes 1.45.

The previous discussion of standard deviation and related statistical analysis placed emphasis on estimating the reliability or precision of experimentally observed values. However, standard deviation does not give specific information about how close an experimental mean is to the true mean. Statistical analysis may be used to estimate, within a given probability, a range within which the true value might fall. The range or confidence interval is defined by the experimental mean and the standard deviation. This simple statistical operation provides the means to determine quantitatively how close the experimentally determined mean is to the true mean. **Confidence limits** (L_1 and L_2) are created for the sample mean as shown in Equations 1.6 and 1.7.

>> $$L_1 = \bar{x} + (t)(s_m)$$ *Equation 1.6*

>> $$L_2 = \bar{x} - (t)(s_m)$$ *Equation 1.7*

where

> $t = $ a statistical parameter that defines a distribution between a sample mean and a true mean

The parameter t is calculated by integrating the distribution between percent confidence limits. Values of t are tabulated for various confidence limits (Table 1. 2). Each column in the table refers to a desired confidence level (0.05 for 95%, 0.02 for 98%, and 0.01 for 99% confidence). The table also includes the term **degrees of freedom,** which is represented by $n - 1$, the number of experimental observations minus 1. The values of $\bar{x}$ and s_m are calculated as previously described in Equations 1.1 and 1.4.

Table 1.2

Values of *t* for Analysis of Statistical Confidence Limits

Probability of Larger Value of t, Sign Ignored

d.f.	0.05	0.02	0.01	d.f.	0.05	0.02	0.01
1	12.706	31.821	63.657	14	2.145	2.624	2.977
2	4.303	6.096	9.925	15	2.131	2.602	2.947
3	3.182	4.541	5.841	16	2.120	2.583	2.921
4	2.776	3.747	4.604	17	2.110	2.567	2.898
5	2.571	3.365	4.032	18	2.101	2.552	2.878
6	2.447	3.143	3.707	19	2.093	2.539	2.861
7	2.365	2.998	3.499	20	2.086	2.528	2.845
8	2.306	2.896	3.355	21	2.080	2.518	2.831
9	2.262	2.821	3.250	22	2.074	2.508	2.819
10	2.228	2.764	3.169	23	2.069	2.500	2.807
11	2.201	2.718	3.106	24	2.064	2.492	2.797
12	2.179	2.681	3.055	25	2.060	2.485	2.787
13	2.160	2.650	3.012				

Spreadsheet Statistics

It is common practice today to use computer spreadsheet programs for statistical analysis of biochemical data. A spreadsheet provides a means to collect and enter data in the form of numbers and text. Perhaps the most versatile and easy-to-use spreadsheet software is Microsoft Excel, although more specialized statistical software programs including SPSS and SyStat are also very useful (see Chapter 2 and Appendix I). Using Excel to estimate statistical terms for experimental data is relatively straightforward. Launching the Microsoft Excel program on your computer brings up the Excel spreadsheet, which consists of rows (number headings) and columns (letter headings). The square at the intersection of a row and column is called a **cell.** Each set of experimental data is entered in a separate column or row. Excel has built-in statistical functions that are accessed through the "Function Wizard" tool found on the toolbar. A statistical summary of the experimental data entered is obtained by clicking on "Tools," then "Data analysis," then "Descriptive statistics." Then select the range of cells and click on "Summary statistics." This process may be used to estimate mean, standard deviation, variance, standard error of the mean, and other important terms. More detailed instructions for the statistical applications of Excel are found at www.microsoft.com/ and in the book *Investigations, A Handbook for Biology and Chemistry Courses at Grinnell College* (www.grinnell.edu/academic/biology/BC_manual.pdf).

Statistical Analysis in Practice

The equations for statistical analysis that have been introduced in this section are of little value if you have no understanding of their practical use, meaning, and limitations. A set of experimental data will first be presented, and then several statistical parameters will be calculated using the equations. This example will serve as a summary of the statistical formulas and will also illustrate their application.

➤ | **Study Exercise 1.3**   Ten identical protein samples were analyzed by the Bradford method for protein analysis. The following values for protein concentration were obtained.

Observation Number	Protein Concentration (mg/mL), x
1	1.02
2	0.98
3	0.99
4	1.01
5	1.03
6	0.97
7	1.00
8	0.98
9	1.03
10	1.01

Sample mean

$$\bar{x} = \frac{\sum x}{n} = \frac{10.02}{10} = 1.00 \text{ mg/mL}$$

Sample deviation

Sample deviation $= x_i - \bar{x}$

Observation	$x_i - \bar{x}$
1	+0.02
2	−0.02
3	−0.01
4	+0.01
5	+0.03
6	−0.03
7	0.00
8	−0.02
9	+0.03
10	+0.01

Calculation of the sample deviation for each measurement gives an indication of the precision of the determinations.

Standard deviation

$$s = \sqrt{\frac{\sum (x_i - \bar{x})^2}{n - 1}}$$

$$s = 0.02$$

The mean can now be expressed as $\bar{x} \pm s$ (for this specific example, 1.00 ± 0.02 mg/mL). The probability of a single measurement falling within these limits is 68.3%. For 95.5% confidence ($2s$), the limits would be 1.00 ± 0.04 mg/mL.

Standard error of the mean

$$s_m = \frac{s}{\sqrt{n}}$$

$$s_m = \frac{0.02}{\sqrt{10}}$$

$$s_m = 0.006$$

This statistical parameter can be used to gauge the precision of the experimental data.

Confidence limits

The desired confidence limits will be set at the 95% confidence level. Therefore we will choose a value for t from Table 1.2 in the column labeled $t_{0.05}$ and $n - 1 = 9$.

$$L_1 = \bar{x} + (t_{0.05})(s_m)$$

$$L_2 = \bar{x} - (t_{0.05})(s_m)$$

$$s_m = 0.006$$

$$t_{0.05} = 2.262$$

$$\bar{x} = 1.00$$

$$L_1 = 1.00 + (2.262)(0.006)$$

$$L_1 = 1.01$$

$$L_2 = 0.99$$

We can be 95% confident that the true mean falls between 0.99 and 1.01 mg/mL.

Study Problems

1. Define each of the following terms.
   (a) OSHA
   (b) MSDS
   (c) Flowchart
   (d) Pasteur pipet
   (e) Purified water
   (f) Error
   (g) Standard deviation
   (h) Molarity

➤ 2. What personal protection items must be worn when handling glacial acetic acid?

3. Draw a schematic picture of your biochemistry lab and mark locations of the following safety features: eyewashes, first-aid kit, shower, fire extinguisher, chemical spill kits, and direction to nearest exit.

➤ 4. Describe how you would prepare a 1-liter aqueous solution of each of the following reagents:
    (a) 1 M glycine
    (b) 0.5 M glucose
    (c) 10 mM ethanol
    (d) 100 nM hemoglobin

➤ 5. Describe how you would prepare just 10 mL of each of the solutions in Problem 4.

➤ 6. If you mix 1 mL of the 1 M glycine solution in Problem 4 with 9 mL of water, what is the final concentration of this diluted solution in mM?

➤ 7. Convert each of the concentrations below to mM and μM.
    (a) 10 mg of glucose per 100 mL
    (b) 100 mL of a solution 2% in alanine

➤ 8. You have just prepared a solution by weighing 20 g of sucrose, transferring it to a 1-liter volumetric flask, and adding water to the line. Calculate the concentration of the sucrose solution in terms of mM, mg/mL, and % (wt/vol).

➤ 9. The concentrations of cholesterol, glucose, and urea in blood from a fasting individual are listed below in units of mg/100 mL (sometimes called mg%). These are standard concentration units used in the clinical chemistry lab. Convert the concentrations to mM.
    (a) cholesterol–200 mg%
    (b) glucose–75 mg%
    (c) urea–20 mg%

➤ 10. The following optical rotation readings were taken by a polarimeter on a solution of an unknown carbohydrate. Use Excel to estimate statistical terms.
    (a) Calculate the sample mean.
    (b) Calculate the standard deviation.
    (c) Calculate the 95% confidence levels for the measurement.

α_{obs} (degrees)			
+3.24	+3.20	+3.17	+3.25
+3.15	+3.21	+3.23	
+3.30	+3.19	+3.20	

11. Describe how you would prepare 100 mL of a single solution containing all of the following reagents at the designated concentrations.
   (a) 0.1 *M* NaCl
   (b) 0.05 *M* glucose
   (c) 5% wt/vol alanine
   (d) 1 mg/mL urea

12. A technician at a clinical laboratory received a blood serum sample for cholesterol analysis. In order to check the reliability of the procedure, the technician repeated the assay five times and obtained the results below. Use a spreadsheet statistical analysis to estimate the mean, standard deviation, variance, and standard error of the mean for the data.

   Total cholesterol in blood serum sample in mg/100 mL (mg%): 157, 154, 155, 152, 155.

Further Reading

Pipetting

http://www.biology.lsu.edu/introbio/tutorial/Pipets/1208.pipet.html
   Pipets and pipetting

http://www.gilson.com
   Information on automatic pipets, procedures for use, and hints.

http://www.rainin.com
   Instruction manuals for operation of the Pipetman.

Preparation of Solutions

M. Caspers and E. Roberts-Kirchhoff, *Biochem. Mol. Biol. Educ.* **31**, 303–307 (2003). "An Undergraduate Biochemistry Laboratory Course with an Emphasis on Research Experience."

S. Kegley and J. Andrews, *The Chemistry of Water* (1997), University Science Books (Sausalito, CA). A discussion of water purity and analysis.

J. Risley, *J. Chem. Educ.* **68**, 1054–1055 (1991). "Preparing Solutions in the Biochemistry Lab."

Safety

American Chemical Society, *Safety in Academic Chemical Laboratories, Volume I: Accident Prevention for College and University Students; Volume II: Accident Prevention for Faculty and Administrators*, 7th ed. (2003), ACS (Washington, DC).

M. Armour, *Hazardous Laboratory Chemicals Disposal Guide,* 3rd ed. (2003), CRC Press (Boca Raton, FL).

K. Barker, Editor, *At the Bench: A Laboratory Navigator* (1998), Cold Spring Harbor Laboratory Press (Cold Spring Harbor, NY).

National Research Council, *Prudent Practices in the Laboratory: Handling and Disposal of Chemicals* (1995), National Academy Press (Washington, DC).

http://www.hendrix.edu/chemistry/chemsafe.htm
Information on chemical hygiene and safety with links to MSDS searches.

http://www.osha.gov/
Review of functions and regulatory procedures of OSHA.

http://www.practicingsafescience.org
Advice from the Howard Hughes Medical Institute.

Statistical Analysis of Data

R. Boyer, *Modern Experimental Biochemistry,* 3rd ed. (2000), Benjamin-Cummings (San Francisco), pp. 18–25.

P. Meier and R. Zund, *Statistical Methods in Analytical Chemistry,* 2nd ed. (2000), John Wiley & Sons (New York).

J. Miller and J. Miller, *Statistics and Chemometrics for Analytical Chemistry,* 4th ed. (2000), Prentice-Hall (Englewood Cliffs, NJ).

C. S. Tsai, *An Introduction to Computational Biochemistry* (2002), John Wiley & Sons (New York), pp. 11–40.

http://www.graphpad.com/prism/Prism.htm
Software for statistics and curve fitting.

http://www.statistics.com/
Click on Free Web-based software for data analysis.

Writing Laboratory Reports and Communicating Science

K. Barker, Editor, *At the Bench: A Laboratory Navigator* (1998), Cold Spring Harbor Laboratory Press (Cold Spring Harbor, NY). Covers lab orientation, keeping a notebook, and lab procedures.

H. Beall and J. Trimbur, *A Short Guide to Writing about Chemistry,* 2nd ed. (2001), Benjamin-Cummings (San Francisco).

R. Boyer, *Biochem. Mol. Biol. Educ.* **31,** 102–105 (2003). "Concepts and Skills in the Biochemistry/Molecular Biology Lab."

J. Dodd, Editor, *The ACS Style Guide: A Manual for Authors and Editors,* 2nd ed. (1997), Oxford University Press (New York).

H. Ebel, C. Bliefert, and W. Russey, *The Art of Scientific Writing: From Student Reports to Professional Publications in Chemistry and Related Fields,* 3rd ed. (2004), John Wiley & Sons (New York).

P. Frey, *Biochem. Mol. Biol. Educ.* **31,** 237–241 (2003). "Guidelines for Writing Research Papers."

J. Kovac and D. Sherwood, *Writing Across the Chemistry Curriculum: An Instructor's Handbook* (2001), Prentice-Hall (Upper Saddle River, NJ).

C. Lobban and M. Schefter, *Successful Lab Reports: A Manual for Science Students* (1992), Cambridge University Press (Cambridge, UK).

A. Penrose and S. Katz, *Writing in the Sciences: Exploring Conventions of Scientific Discourse,* 2nd ed. (2004), Benjamin-Cummings (San Francisco)

I. Valiela, *Doing Science: Design, Analysis, and Communication of Scientific Research* (2001), Oxford University Press (New York).

J. Walker, *Biochem. Educ.* **19,** 31–32 (1991). "A Student's Guide to Practical Write-ups."

http://www.ce.umn.edu/~smith/supplements/poster/guide.htm
Preparing professional scientific posters.

http://www.swarthmore.edu/NatSci/cpurrin1/posteradvice.htm
Advice for designing scientific posters.

http://home.okstate.edu/homepages.nsf/toc/PFFmakeposter
Guide to making poster presentation

http://www.dsc.dixie.edu/owl/
Procedures for making and using a lab notebook.

http://www.grinnell.edu/academic/biology/BC_manual.pdf
A useful handbook for biology and chemistry labs.

The Computer as a Tool in Biochemistry and Molecular Biology

The computer has become an important tool in biochemistry and molecular biology for use in the routine jobs of word processing, graphing, and statistical analysis of research data. In addition, if a computer is connected to the Internet, then it may be used for literature searching, analyzing and graphing complex data, accessing databases containing information about nucleic acid and protein sequences, predicting protein structure, seeking research methodology, designing computer models for biological processes, and much more. This new application of computer technology to the analysis, management, and manipulation of biochemical data is called **computational biochemistry.** An important part of computational biochemistry that we will use extensively in this text is **bioinformatics,** defined as the application of computer technology to the storage and use of biological data, especially protein and nucleic acid sequence and structural information. The primary purpose of this chapter is to learn how to solve problems in biochemistry and molecular biology using computer and software technology. Here we will outline general Internet resources that may be used in later chapters. Each chapter of the text will also describe more specific Internet resources that can be applied to the topics covered in that chapter.

A.   COMPUTERS IN BIOCHEMISTRY AND MOLECULAR BIOLOGY

The modern computer has revolutionized the way we live. Not surprisingly, the computer has also changed the way we do scientific research. The computer has become an important tool for the analysis of all aspects of biomolecules—structure, function, reactions, and information. The need for computers in biochemistry and molecular biology is growing because of two factors: (1) the fields are becoming increasingly quantitative, so there is a requirement for complex and accurate calculations–the computer with its access to Internet software and application programs is ideal for this work; and (2) because of the ease of determining protein and nucleic acid sequences, there has been a proliferation of biological information that needs to be stored and made readily available to researchers. To do research in biochemistry and molecular biology, one must become familiar with the use of the computer databases and database management systems.

Your first encounter with a computer in the laboratory was probably while you were using an instrument that had a computer to control its operation, to collect data, and to analyze data. All major pieces of scientific equipment, including UV-VIS spectrophotometers, high-performance liquid chromatographs, gas chromatographs, nuclear magnetic resonance spectrometers, and DNA sequencers are now controlled by computers. But your use of the computer will not end in the lab. You will use a computer to prepare each laboratory report, including graphical and statistical analysis of experimental data. If the computer is connected to the Internet, you will greatly broaden its use to some of the following: (1) searching the biochemical literature for pertinent books and journal articles, (2) using free software for analyzing experimental data, and (3) accessing biological databases that provide nucleic acid and protein sequences, structures, and more.

Personal Computing in Biochemistry

It is now possible for most students to purchase a basic computer system at low cost. If a personal computer is not in the budget, most colleges and universities provide students access to campus-wide computer systems as part of tuition and fees. By this point in your studies, you are familiar with the use of a computer, but a few introductory comments are made just to help you get started with computing in the biochemistry laboratory. In terms of equipment, you will need a computer, monitor, printer, and some basic software. Laptop computers are becoming increasingly popular. Some recommendations for specific hardware and software will be given here, but one must be aware that new products and important upgrades are continually being developed.

For word processing (writing lab reports), basic software programs including Microsoft Word and Word Perfect are most widely used. Software specialized for scientific writing is available, but probably not necessary at this level. For many experiments you will need to prepare data in a spreadsheet or graphical form. Current software programs for graphing and spreadsheet with graphing capability include Lotus 123, Excel, Sigmaplot, Quattropro, Kaleidagraph, and CricketGraph. Other software programs (such as DynaFit or Leonora) that are appropriate for specific types of graphing will be listed as needed in later chapters. Some graphs that you prepare (e.g., Michaelis-Menten) are nonlinear, so special computer programs like Excel Solver are recommended for their analysis. To reduce the cost of computing, most software suggested for use will be available as freeware on the Internet.

It is important for your education and career that you become knowledgeable and skilled in the use of the computer and the Internet. These are tools that you will use every day. Many terms that may be new are introduced in this chapter. All words in bold print in this chapter are defined in a glossary at the end of the chapter.

The Computer and the Internet

If you are using the computer as described above, you are saving time and also preparing good-looking lab reports. However, if your computer is not linked to the **Internet,** then you are not tapping into a vast wealth of biochemical tools and information available. The Internet can be defined, in simple terms, as a worldwide matrix that allows all computers and networks to communicate with each other. If the computer you are using is college owned, then it is probably already linked to the Internet, and the college pays the costs for that service. For your own home computer, you may need to subscribe to an Internet service provider (ISP) and obtain a **modem** to transmit computer signals through a telephone line. Many colleges and universities are now installing wireless systems so there is no need for a cable (phone line). Once you have access to the Internet, many programs are available as **freeware,** software provided without charge by its creator.

After you are connected to the Internet, what are the basic facilities available for use? First, you will be able to communicate by **e-mail (electronic mail).** Messages containing text, files, and graphics may be sent to anyone who has a computer with an Internet link and an e-mail address. Addresses have three basic components: the user name, an @ sign, and the user's location or **domain.** Common domains that you will encounter are listed in Table 2.1. You will need an e-mail program to receive, send, and organize

Table 2.1

Common Domain Names

Inside the United States

Commercial site	.com
Educational site	.edu (.ac in the U.K. and other countries)
Government site	.gov
Gateway or network	.net
Organization site	.org

Outside the United States

Australia	.au
Britian	.uk
Canada	.ca
France	.fr
Germany	.de
Japan	.jp
Korea	.kr
Sweden	.se
Switzerland	.ch

messages. The most popular ones are Outlook, Eudora, and Pegasus. (Practice your e-mail skills by sending a message, perhaps a question, to your laboratory instructor or to the author of this book: boyer@hope.edu.) Communication among scientists is now done primarily by e-mail. Connected to the Internet, you will also be able to join in list server discussions such as USENET. One of the most widely used facilities on the Internet is the ability to place and retrieve network data by **file transfer protocol (ftp).** More detail on ftp's will be given in later sections.

B.   THE WORLD WIDE WEB

The newest and most rapidly growing component of the Internet is the **World Wide Web** (WWW, also called "the Web"). This facility, which was launched in 1992, permits the transfer of data as pages in **multimedia form** consisting of text, graphs, audio, and video. The pages are linked together by **hypertext** pointers so that data stored on computers in different locations may be retrieved via the network by your computer. Web documents are written in a special coded language called **HyperText Markup Language (HTML).** To access all of the resources on the Web, you will need a **browser,** an interface program that reads hypertext and displays Web pages on your computer. The most commonly used Web browsers are Internet Explorer and Netscape Navigator. The use of these programs will not be described in detail here, as they are constantly changing and students at this level are already familiar with their use. However, a brief summary will be presented.

To access the Web, the Web browser is activated. Displayed on the screen will be the **home page** or starting point for entry into the Web. On this page will be a dialogue box into which you can type text. The dialogue box may ask for "Address," "Netsite," "Location," or **URL (Uniform Resource Locator).** To request a specific Web page from another computer site, type in the Web page address, which is usually in the form http://www.--. (For many computer systems, it is no longer necessary to type in "http://www." Try it on yours.) The home page, with instructions on the use of the Web site, will then be displayed on the screen (see Figure 2.1 for the home page of NCBI Entrez). One important feature you will note is that some words on the page are highlighted. If you click the mouse on one of these words (called **hyperlinks**) your computer will connect to another, related, Web page that provides information on the hyperlink. This feature greatly enhances the use of the Web because related Web sites are connected or linked together and may be quickly accessed by a click of the mouse.

Web Sites for Biochemistry and Molecular Biology

Web addresses that are useful for biochemistry and molecular biology are presented in Appendix I and in Tables 2.2–2.4. Appendix I lists software programs and Web sites that are useful for each chapter in this book. Table 2.2 consists of Web addresses that contain database directories and catalogs that list the

Figure 2.1

The Home page for NCBI
Entrez (www.ncbi.nlm.nih.gov/
Entrez). Contains a complete
list of the available databases.
A click on the database name
or icon takes one directly to the
search page for that database.

services available on the Internet. This is a good place to start if you want to
become familiar with the many resources available. Table 2.3 lists Web sites
that have databases and tools that are of special interest in biochemistry and
molecular biology. A short descriptive note in the table defines the specific
databases available.

Table 2.4 lists databases that allow searches using the name of a biochemical or class of biochemicals. These Web sites are especially useful when seeking
information and properties of biomolecules, including nomenclature, physical

Table 2.2

Web Database Directories and Catalogs

Name	URL
BioResearch	http://bioresearch.ac.uk/
INFOBIOGEN Catalog of DB	http://www.infobiogen.fr/
Biology Workbench	http://biology.ncsa.uiuc.edu
Pedro's Biomolecular Research Tools	http://www.public.iastate.edu/~pedro/ research_tools.html
CMS Molecular Biology Resources	http://www.sdsc.edu/ResTools/cmshp.html
BioTech	http://biotech.icmb.utexas.edu
Protocol Online	http://www.protocol-online.net/
Chem Connection	http://chemconnection.com/news/ journals.html
Survey of Molecular Biology Databases and Servers	http://www.ai.sri.com/people/mimdb/

Table 2.3

Biochemical Databases and Tools

Name	Description	URL
Protein Data Bank (PDB)	Protein structures determined by X-ray and NMR	http://www.rcsb.org/pdb/
European Bioinformatics Institute (EBI)	DNA sequences	http://www.ebi.ac.uk/
National Center for Biotechnology Information (NCBI)	Variety of databases and resources	http://www.nlm.nih.gov/
Swiss-Protein	Protein sequences and analysis	http://www.expasy.ch/tools/
Biocatalysis/Biodegradation Database of the University of Minnesota	Microbial metabolism of many chemicals	http://www.labmed.umn.edu/umbbd/index.html
REBASE-The Restriction Enzyme Database	Restriction enzyme directory and action	http://rebase.neb.com/
Georgia Institute of Technology	Tutorials on PDB and RasMol	http://www.chemistry.gatech.edu/faculty/williams/bCourse_information/4582/labs/rasmol_pdb.html
The Institute for Genomic Research	Collection of genomic databases	http://www.tigr.org/
RasMol (RasMac)	Molecular graphics for proteins	http://www.umass.edu/microbio/rasmol/
Predict Protein	Protein sequence and structure prediction	http://www.embl-heidelberg.de/predictprotein/
Gene Quiz	Protein function analysis based on sequence	http://www.sander.ebi.dc.uk/gqsrv/submit
Entrez browser of NCBI	Database searching including PubMed literature	http://www.ncbi.nlm.nih.gov/Entrez/
National Bio-Technology Information Facility	Databases in Biotech	http://www.nbif.org/data/data.html
Protein Information Resource (PIR)	Database searching for proteins	http://pir.georgetown.edu
Munich Information Center for Protein Sequences (MIPS)	Protein sequences	http://www.mips.biochem.mpg.de/
Human Genome Project Information	Project database	http://www.ornl.gov/TechResources/Human_Genome
Journal of Chemical Education Resource Shelf	List of all current biochemistry texts	http://www.umsl.edu/divisions/artscience/chemistry/books/welcome.html
Protein Explorer (NSF)	Protein structures	http://molvis.sdsc.edu/protexpl/frntdoor.htm or www.proteinexplorer.org
Chimera Molecular Modeling System (UCSF)	Advanced Molecular Modeling	http://www.cgl.ucsf.edu/chimera/
BiomoleculesAlive (ASBMB)	Instructional material	http://www.biomoleculesalive.org
BioResearch	Searchable catalog of Internet resources in biochemistry	http://bioresearch.ac.uk
ChemDraw	Chemical structure drawing program	http://www.cambridgesoft.com
MEDLINE (PubMed)	U.S. National Library of Medicine	http://www.nlm.nih.gov/

properties, structural data, reaction characteristics, spectroscopic information, and even chromatographic data. The table also contains Web sites that have catalogs of biomolecules organized by category, including amino acids, proteins, enzymes, lipids, RNA, carbohydrates, and pharmaceuticals.

Many of the current Web sites you will need are listed here; however, what about new Web sites that have been established since publication of this book? Millions of new Web sites are created every year. To access these new sites, you need the help of a **search engine,** a searchable directory that organizes Web pages by subject classification. Major search engines include

Table 2.4

Databases of Biochemical Compounds

Name	Description	URL
International Union of Biochemistry and Molecular Biology (IUBMB)	Nomenclature	http://www.qmw.ac.uk/iubmb/
IUBMB Enzyme List	Catalog of enzymes	http://www.qmw.ac.uk/iubmb/enzyme/
Klotho	Alphabetical list of biochemicals	http://ibc.wustl.edu/klotho/
ChemFinder	Structures and properties	http://www.chemfinder.com
AAindex	Properties of amino acids	http://www.genome.ad.jp/dbget/aaindex.html
Lipid Bank	Information on lipids	http://lipid.bio.m.u-tokyo.ac.jp/
Monosaccharide Database	Information on sugars	http://www.cermav.cnrs.fr/databank/mono/
RNA Database	RNA structures	http://rnabase.org
Merck Index	Biochemical/pharmaceutical directory	http://merck.com.pubs.mmanual/
Worthington Enzyme Manual	Properties of enzymes	http://www.worthington-biochem.com/index/manual.html
Enzyme database of ExPASy	Enzyme names and numbers	http://www.expasy.ch/enzyme/
BRENDA	A comprehensive enzyme information system	http://www.brenda.uni-koeln.de
EMBL European Genetic Database	DNA sequences	http://www.ebi.ae.uk
NRL-3D	Protein structure database	http://www.gdb.org/Dan/proteins/nrl3d.html

AltaVista, Google, Excite, HotBot, Lycos, Netscape Search, and Yahoo!. Most of the search engines require the input of a keyword or terms for searching. As you surf the Web, you may find sites you wish to save and review at a later date. You may use the **bookmark** (Netscape) or **favorite** (Explorer) function to save it for the future.

➤ | **Study Exercise 2.1**   What are the categories of WWW links to information and services available on Pedro's Biomolecular Research Tools Web site?

Solution: Study the home page for Pedro's site. Here you will find the following:

Part 1: Molecular Biology Search and Analysis

Part 2: Bibliographic, Text, and WWW searches

Part 3: Guides, Tutorials, and Help Tools

What are the last two categories listed?

➤ | **Study Exercise 2.2**   Study the home page for Protocol-online. What are the subtopics available under the main heading of "Biochemistry?"

Solution: On the home page, click on "Biochemistry." Here you will find hyperlinks to three subtopics: amino acids and proteins, enzyme analysis, and lipids.

➤ | **Study Exercise 2.3**   Use the IUBMB enzyme site to find the E.C. number for the enzyme, cellulase.

Solution: Cellulase catalyzes the hydrolysis of the glycosidic bonds between glucose residues in the substrate cellulose. The E.C. number is 3.2.1.4. What information about the enzyme do you find in the Worthington Enzyme Manual?

C.   APPLICATIONS OF THE WEB

It is not necessary to have a complete, detailed understanding of the Internet in order to tap into its vast resources. The fundamental concepts provided next will allow you to take advantage of two essential activities: (1) biochemical literature searching and (2) using Web directories and biological databases.

The Biochemical Literature

Biochemists involved in research do not spend all of their working time in the laboratory. An important component of a biochemistry research project is reading the scientific literature. Much of this library work may be done on the computer. The library (and computer) should be considered a tool for experimental biochemistry in the same way as any scientific instrument. The use of the biochemical literature by the student in biochemistry laboratory is not as extensive as that of a full-time researcher, but you must be aware of what is available in the library and how to use it.

The library is used in all stages of research. Before an investigator can begin experimentation, a research idea must be generated. This idea develops only after extensive reading and study of the literature. A research project usually begins in the form of a question to be answered or problem to be solved. For ease of solution, a major project is subdivided into questions that may be answered by experimentation. Before laboratory work can begin, the researcher must have a knowledge of the past and current literature dealing with the research area. This can be reduced to two questions: What is the current state of knowledge in the area? and What are the significant unknowns? These questions can be answered only by developing a familiarity with the biochemical literature. The researcher will find that this knowledge of the literature is also invaluable for the design of experiments. The development of experiments requires knowledge of techniques and laboratory procedures. Excellent methods books and journals are available that provide experimental details. Finally, while performing experiments, the researcher often needs physical and chemical constants and miscellaneous information. Various handbooks and encyclopedias are excellent for this purpose. The beginning student in biochemistry laboratory will not be expected to proceed through all of these stages in the design of an experiment. However, a familiarity with the literature will increase your understanding of lab activities and may aid in the development of more effective methods. When you do begin a research program, you will be able to use the library to the fullest advantage.

The biochemical literature is massive and expanding rapidly. It is almost a full-time job just to maintain a current awareness of a specialized research area. There are few disciplinary boundaries in the study of biochemistry. The biochemical literature overlaps into the biological sciences, the physical sciences, the basic medical sciences, and information technology. The intent of the following discussion is to bring some order to the many textbooks, reference books, research journals, computer information-retrieval services, and handbooks that are available.

Textbooks

The student's first exposure to biochemistry is probably a lecture course accompanied by the reading of a general textbook of biochemistry. By providing an in-depth survey of biochemistry, textbooks allow students to build a strong foundation of important principles and concepts. By the time most books are in print, the information is 1 to 2 years old, but textbooks still should be considered the starting point for mastery of the fundamentals of biochemistry.

Reference Books and Review Publications

For more specialized and detailed biochemical information that is not offered by textbooks, reference books must be used. Reference works range from general surveys to specialized series. The best works are multivolume sets that continue publication of volumes on a periodic basis. Each volume usually covers a specialized area with articles written by recognized authorities in the field. It should be noted that reference articles of interest to biochemists are often found in publications that are not strictly biochemical. The best known and most widely used review publication is *Annual Review of Biochemistry*. Each volume in this series, which was introduced in 1932, contains several detailed and extensive articles written by experts in the field. For shorter reviews emphasizing current topics, *Trends in the Biochemical Sciences* (TIBS) is widely read.

Research Journals

The core of the biochemical literature consists of research journals. It is essential for a practicing biochemist to maintain a knowledge of biochemical advances in his or her field of research and related areas. Scores of research journals are published with the intent of keeping scientists up to date. With the expansion of scientific information has come the need for efficient storage and use of research journals. Many publishers are now providing journals in forms such as microcards, microfilm, microfiche, and more recently CD-ROM disks and on line. Some research journals have achieved especially excellent reputation, and articles therein are considered to be of the highest quality. A recent ranking of the biochemical journals, based on the number of citations received, produced the following order for the top six: *Journal of Biological Chemistry, Biochimica et Biophysica Acta, Biochemistry, Proceedings of the National Academy of Sciences of the United States of America, Biochemical Journal,* and *Biochemical and Biophysical Research Communications.* The core journals used by an individual depend on the area of specialty and are best determined from experience. (See Figure 1.5.)

Methodology References

The active researcher has a continuing need for new methods and techniques. Several publications specialize in providing details of research methods, and many research methods are now available on the Web. Some of the useful biochemical methodology publications are:

> *Analytical Biochemistry,* a monthly journal.
>
> *Analytical Chemistry,* a monthly journal.
>
> *Biochemical Preparations,* an annual volume.
>
> *Current Protocols in Molecular Biology,* P. Ausabel et al., Editors. A manual of techniques in two volumes that are updated quarterly.
>
> *Laboratory Techniques in Biochemistry and Molecular Biology,* T. S. Work and R. G. Burdon, Editors (formerly T. S. Work and E. Work). Each volume in the series is concentrated in an area of biochemistry and written by recognized authorities.
>
> *Methods of Enzymatic Analysis,* H. Bergmeyer, Editor. Contains methods for enzyme purification and assay, in several volumes.
>
> *Methods in Enzymology,* various editors. The most valuable methods series available. Each volume contains numerous articles describing biochemical techniques. The series is well indexed and easy to use. Over 200 volumes.
>
> *A Practical Guide to Molecular Cloning,* 2nd ed., B. Perbal. Useful for setting up research projects in molecular cloning.

Handbooks of Chemical and Biochemical Data

Students in introductory biochemistry laboratory may use methodology books more than any other type, although much of the data is now on the Web. While doing biochemical experiments, you may need physical, chemical, and biochemical information such as definition of terms, R_f values, molecular weights, and physical constants. This information is easily found in the many handbooks and collections of biochemical data. Some useful handbooks with a brief description of contents include:

Dictionary of Biochemistry and Molecular Biology, 2nd ed., J. Stenesh (1989), Wiley Interscience (New York). Contains definitions of 16,000 terms.

Glossary of Biochemistry and Molecular Biology, D. Glick (1990), Raven Press (New York). Emphasis on words and phrases that are unique to biochemistry and molecular biology.

Merck Index, S. Budavari, Editor, Merck & Co. (Rahway, NJ). An encyclopedia of chemicals, drugs, and biological substances (Web site in Table 2.4).

Practical Handbook of Biochemistry and Molecular Biology, G. Fasman, Editor (1989), CRC Press (Boca Raton, FL). Excellent source of chemical and physical data for nucleic acids, proteins, lipids, and carbohydrates.

Worthington Enzyme Manual, 2nd ed., C. Worthington, Editor (1988), Worthington Biochemical Corp. (Freehold, NJ). Contains detailed assay information and references for over 75 enzymes (Web site in Table 2.4).

Computer-Based Searches and Other Aids to the Literature

As you study and work in biochemistry, you will often need to complete a thorough literature search on some specialized area or topic. It is not practical to survey the hundreds of books, journals, and reports that may contain information related to the topic. Two publications that provide brief summaries of published articles, reviews, and patents are *Chemical Abstracts* and *Biological Abstracts.* If you are not familiar with the use of these abstracts, ask your instructor or reference librarian for assistance.

Research articles of interest to biochemists may appear in many types of research journals. Research libraries do not have the funds necessary to subscribe to every journal, nor do scientists have the time to survey every current journal copy for articles of interest. Two publications that help scientists to keep up with published articles are *Chemical Titles* (published every 2 weeks by the American Chemical Society) and the weekly *Current Contents* available in hard copy and computer disks (published by the Institute of Science Information). The Life Science edition of *Current Contents* is the most useful for biochemists. The computer revolution has reached into the chemical and biochemical literature, and most college and university libraries now subscribe to computer bibliographic search services. One such service is STN International, the scientific and technical information network. This on-line system allows direct access to some of the world's largest scientific databases. The STN databases of most value to life scientists include BIOSIS Previews/RN (produced by Bio Sciences Information Service; covers original research reports, reviews, and U.S. patents in biology and biomedicine), CA (produced by Chemical Abstracts service; covers research reports in all areas of chemistry), MEDLINE, and MEDLARS (produced by the U.S. National Library of Medicine and *Index Medicus,* respectively; cover all areas of biomedicine). Another useful service is SciFinder Scholar. These networks provide on-line service, and their databases can be accessed from personal computers in the office, laboratory, or library. Some of the computer bibliography services are freeware on the Internet, but others have user fees. For example, MEDLINE (PubMed) produced by the National Library of Medicine, available at http://www.nlm.nih.gov/, may be used free of charge.

Web Directories, Tools, and Databases

Biochemical research generates huge amounts of data of interest to all scientists. For example, thousands of genes and proteins have been sequenced during the past several years, and thousands more will be sequenced in the future. This number was greatly expanded by the Human Genome Project, which completed its goal of sequencing of the entire human genome in 2001. In addition, determining the structures of proteins by X-ray diffraction and by NMR has become routine. Sequence and structural data are now being stored in computer networks for retrieval by biochemists throughout the world. Here, we will discuss the many biological databases and provide examples of their use. Our approach will be to focus on the use of databases that are readily available, free of charge, on the Web. However, it is important to recognize that many commercial hardware and software systems for analyzing biological databases are available, but they are often very expensive and complicated to use.

A wide variety of databases are currently available including bibliographic, nucleic acid sequence, protein sequence and structure, metabolic pathways, transcription factors, enzymes, and many others (see Appendix I). One of the best ways to find the resources suited to your needs is to use a directory that collects lists of information, tools, and other services. Several very good ones are available (Table 2.2). Some of these sites are hyperlinked to the database sites. Specifically, they will include protein primary, secondary, and tertiary structure, sequence homology, sequence alignment, and structure prediction. The Web addresses for these resources are listed in Table 2.3. Because of the huge amount of data available, it is often necessary to use programs to help you analyze the data. Table 2.5 lists several software

Table 2.5

Useful Programs for Exploring Structures/Sequences

Program	Function
BLAST	Searches for similar protein and nucleic acid sequences
Chime	Protein structures on moving 3D coordinates
Entrez (NCBI)	Sequence retrieval system for cross-referencing databases
FASTA	Searches for similar protein sequences
GenBank (NCBI)	Database of gene sequences
Molecules RUs	Provides coordinates for protein 3D structure and manipulation
RasMol (Ras Mac)	Provides coordinates for protein 3D structure and manipulation
SRS (EMBL)	Sequence retrieval system for cross-referencing databases

programs that are available and usually hyperlinked to the database sites. Those that we will introduce here are FASTA (protein amino acid sequences), BLAST (comparing protein sequence data), RasMol or RasMac (coordinates for protein structure manipulation), Chime (protein structure coordinates), SWISS-MODEL (protein modeling), VAST (protein structure similarities), and Molecules R Us (protein structure coordinates).

➤ | **Study Exercise 2.4**

Searching the Biochemical Literature on MEDLINE

To illustrate the use of this search service, point your Web browser to the appropriate URL (http://www.nlm.nih.gov/). This will connect you to the National Center for Biotechnology Information. Click the mouse on the hyperlink "PubMed." Many features on display are available, but the most basic is the search capability. For bibliographic searching you may enter in the dialogue box a search term, author name, or journal name. For example, type in "bovine alpha-lactalbumin," a protein found in milk. Clicking on "Search" will then provide over 500 citations (or articles). The lists are composed of author(s), title, and reference in reverse chronological order. By clicking on the author's name (in hypertext), you can retrieve the abstract of the article. Another useful and time-saving feature is the hypertext "(see Related Articles)." Clicking on this will provide a list of papers related to the specific citation. The 500 papers or so that you obtained in your original search are too many to screen; you may change the search parameters to reduce the number. You may also practice the search method using terms, concepts, or scientists related to your biochemistry lecture class. Note that the NCBI home page offers other hypertexts, including entry to Entrez, BLAST, etc. We will use these in the next section; however, access will be through a different database Web site.

➤ | **Study Exercise 2.5**

Using Web Tools and Biological Databases

The application of the primary databases and structural analytical tools will be introduced using a protein, α-lactalbumin from bovine milk. Here you will learn about the structure of a related protein, α-lactalbumin from humans. We will search databases to find and view its primary and secondary structure and also determine if there are other proteins with a similar amino acid sequence and structure. After completion of these exercises, you will be able to apply these computer tools to proteins of your own choice.

Point your Web browser to the Protein Data Bank (PDB) and the Research Collaboratory for Structural Bioinformatics (http://www.rcsb.org/pdb/). Become acquainted with the PDB by viewing the home page and perhaps clicking on some hyperlinks. Scroll until you find the term Search on the right side of the screen. Clicking on Search will display a dialogue box for keywords. Type in "human alpha-lactalbumin" and click on Search. Your query will find at least seven structures that are listed. Click on the white square to the left and "EXPLORE" to the right of Structure 1A4V. This will display "Structure Explorer" with "Summary Information" about the structure of the protein. Clicking on the "?" will provide help if necessary. Review the functions possible on the left side of the screen. Click on "View Structure" to observe "Interactive 3D Display" and "Still Images." First, study the still images of human α-lactalbumin in ribbon or cylinder form. You may click on 250×250 or 500×500 to enlarge. Note the presence of α-helices and β-sheets in the structure. After studying the still images, click on "Chime" under Interactive 3D Display. Now you will observe the ribbon structure rotating on an axis. Use "Chime Help" at the bottom of the screen to learn Mouse Controls of the rotating structure. Click on "Sequence Details" to observe the amino acid sequence and definition of secondary structures. You may do an ftp download of this file by clicking on "Download in FASTA format." FASTA format is a listing of amino acid sequences using the standard single-letter abbreviation for each amino acid. Clicking on "Geometry" will display tables of bond angles and lengths. Similar sequence studies may be done by clicking on the function "Structural Neighbors." Several tools are available to search for similar structures. Try the VAST tool. Clicking on "Sequence Neighbors: single chain" will display a list of many proteins with sequences similar to that of human α-lactalbumin. Note that most are α-lactalbumins from other species, but if you scroll far enough, you will see the enzyme lysozyme listed. Returning to the former screen and clicking on Structure Neighbors will display about eight structures similar to human α-lactalbumin. Note again the presence of lysozyme in the list. Clicking on "Other Sources" will display other data files with references to α-lactalbumin. It is interesting to note that the proteins α-lactalbumin and lysozyme have similar primary, secondary, and tertiary structures but they have quite different biochemical activities. The two proteins, which have about 40% sequence identity, may have been derived from a common ancestral gene.

Another useful structure tool is RasMol (or RasMac). This will allow you to view the detailed structure of a protein and rotate it on coordinates so you can see it from all perspectives. A hyperlink to RasMol is present under the "View Structure" function just above "Chime." You may need to study RasMol instructions provided under Help, or you may use a RasMol tutorial listed in Table 2.3. Another useful protein viewer is the Swiss-Protein Pdv Viewer (Table 2.3). BLAST is an advanced sequence similarity tool available at NCBI. To access this, go to the NCBI home page (www.ncbi.nlm.nih.gov) and click on "BLAST" to obtain a dialogue box into which you may type the amino acid sequence of human α-lactalbumin. This process may be streamlined by downloading the amino acid sequence in FASTA format into a file and transferring the file into the BLAST dialogue box. BLAST will provide a list of proteins with sequences similar to the one entered.

Another approach to a study of protein (or nucleic acid) structure and sequence is through Entrez. This can be entered via the NCBI home page. Then click on "Proteins" to obtain a dialogue box where you can type "human alpha-lactalbumin" and then click on Search. You can retrieve about 25 documents for review. Note that you may also enter BLAST through Entrez.

Virtual Biochemistry Labs

For some colleges and universities around the world, offering a "real" biochemistry laboratory course for students is not a possibility. Some of the reasons for this include a lack of expensive scientific instruments, equipment, and reagents; a scarcity of appropriate lab space; a lack of staff expertise; and an inability to fit scheduled lab time into a tight curriculum. Alternative activities for students are now available in the form of "virtual biochemistry labs" on the Web (Table 2.6). Although simulated labs should never be considered as a replacement or substitute for the real thing, especially for student majors in biochemistry, molecular biology and other biological sciences, and chemistry, they may be suitable for some students who will never work in a lab or manipulate lab data. In addition, viewing experiments and procedures in virtual labs may actually offer benefits to students who will perform similar experiments in the lab. The virtual lab may be considered as a "prelab," where students will become acquainted with equipment, techniques, and procedures, and can more efficiently conduct real experiments in the lab. This experience is especially important when students are dealing with expensive equipment and reagents. The virtual labs may also be considered as a safety feature—students can receive training regarding the use of dangerous reagents and procedures.

Table 2.6

Virtual Biochemistry Labs

Name	Description	URL
Lab3D, University of Virginia	Experiments in amino acid titration, electrophoresis, enzyme kinetics	http://lab3d.chem.virginia.edu
The Virtual Biochemistry Lab; The Nobel Foundation	Listen to lectures and perform experiments in chromatography, electrophoresis, NMR, X-ray, protein sequencing and folding	http://www.nobel.se/
BioResearch	Searchable catalog of lab methods and teaching materials in biochemistry	http://bioresearch.ac.uk/
Biointeractive Virtual Lab, Howard Hughes Medical Institute	Labs in: Immunology, Bacterial identification, Cardiology, Neurophysiology, Transgenic Fly	http://www.biointeractive.org

Two virtual biochemistry labs available on the Web include **Lab3D** from the University of Virginia and **The Virtual Biochemistry Laboratory** from The Nobel Foundation.

Lab3D (http://lab3d.chem.virginia.edu) is a three-dimensional, virtual biochemistry laboratory that offers interactive laboratory experiences for undergraduate students in biochemistry, chemistry, and biology. The current experiments are amino acid titration, electrophoresis, and enzyme kinetics. In addition to learning basic principles, performing many important procedures, and becoming familiar with equipment, students have the opportunity to design and conduct their own experiments. The Web site also offers a complete list of references and other resources.

The **Virtual Biochemistry Lab from the Nobel Foundation** (http://www.nobel.se/) is prepared for students at the high school and undergraduate level. The fully interactive Web site allows one to move around the lab, perform experiments on biomolecules, and listen to lectures. Topics covered include chromatography, electrophoresis, NMR, X-ray diffraction, protein sequencing, and protein folding.

BioResearch (http://bioresearch.ac.uk/) is a new Web-based catalog of Internet resources covering the biological and biomedical sciences, including biochemistry. Topics that may be searched in biochemistry include analytical techniques, tutorials, carbohydrates, lipids, journals, nucleic acids, proteins, and teaching materials. (See Figure 2.2 for a copy of the home page for Bio-Research, Biochemistry Topics.)

Another Web site that is of interest to biochemistry students is the **Biointeractive Virtual Lab provided by the Howard Hughes Medical Institute:** http://www.biointeractive.org. There are five virtual laboratories related to biochemistry available at this site:

1. The Immunology Lab
2. The Bacterial Identification Lab
3. The Cardiology Lab
4. The Neurophysiology Lab
5. The Transgenic Fly Lab

Study Problems

1. Use PubMed bibliographic searches to find two recent research articles authored by Thomas R. Cech, who won the Nobel Prize in Chemistry for the discovery of catalytic RNA (ribozymes). Write brief summaries of the articles.

2. Use PubMed or an enzyme Web site to answer the following questions about the enzyme tyrosinase.
   (a) What are sources of the enzyme besides mushroom?
   (b) What metal ion is present in the native enzyme?

Figure 2.2

Home page for
BioResearch, Biochemistry
(http://bioresearch.ac.uk/, then
click on subject, "Biochemistry").
Each subtopic contains many
resources that are constantly
updated.

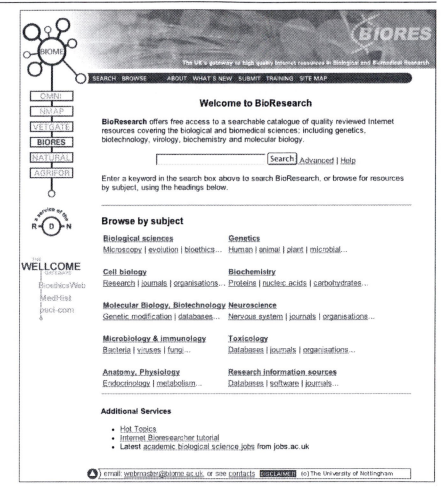

(c) Find two references that study inhibition of tyrosinase. What inhibitor molecules have been investigated?

(d) Find another substrate for the enzyme besides L-DOPA.

(e) Search for the three-dimensional structure of the enzyme.

3. The technique of immobilized metal-ion affinity chromatography (IMAC) is widely used to purify proteins. Find two proteins that have recently been purified by this procedure.

4. The Western blot procedure is now used to test human serum for the presence of antibodies to the AIDS virus. Find two publications that describe procedures for this assay.

5. Outline the pathway for microbial degradation of the detergent used in denaturing electrophoresis, sodium dodecyl sulfate (SDS). Hint: See the Web site on Biocatalysis/Biodegradation.

6. Use the REBASE site to determine the specificity of the restriction enzyme *Hin*d II.

7. Use the techniques in Study Exercise 2.4 to find the amino acid sequence of the protein human α-lactalbumin.

8. Study the nucleotide sequence for the human α-lactalbumin gene. Hint: Begin at the NCBI home page and enter Entrez. Click on "Nucleotides" and do a search on the protein. Review the GenBank report for a position of introns and exons. Obtain a FASTA report, download the files, and complete a BLAST search for related sequences.

9. Use the BLAST tool to compare the amino acid sequences for human α-lactalbumin and lysozyme. Repeat the process with the use of BLAST to compare the nucleotide sequences for the genes coding for α-lactalbumin and lysozyme.

Further Reading

M. Barnes and I. Gray, Editors, *Bioinformatics for Geneticists* (2003), John Wiley & Sons (New York).

P. Biggs, *Computers in Chemistry* (2000), Oxford University Press (Oxford).

J. Boyle, *Biochem. Mol. Biol. Educ.* **32,** 236–238 (2004). "Bioinformatics in Undergraduate Education."

J. Claverie and C. Notredame, *Bioinformatics for Dummies* (2003), John Wiley & Sons (New York).

P. Clote and R. Backofen, *Computational Molecular Biology: An Introduction* (2001), John Wiley & Sons (New York).

S. Cooper, *Biochem. Mol. Biol. Educ.* **29,** 167–168 (2001). "Integrating Bioinformatics into Undergraduate Courses."

P. Craig, *Biochem. Mol. Biol. Educ.* **31,** 151–152 (2003). "BioMoleculesAlive.org: The Biochemistry and Molecular Biology Digital Library Update."

R. Doolittle, Editor, *Methods in Enzymology,* Vol. 266 (1996), Academic Press (San Diego). "Computer Methods for Macromolecule Sequence Analysis."

D. Higgins and W. Taylor, Editors, *Bioinformatics: Sequence, Structure and Databank* (2000), Oxford University Press (Oxford).

S. Hoersch, C. Leroy, N. Brown, M. Andrade, and C. Sander, *Trends Biochem. Sci.* **25,** 33–35 (2000). "The Gene Quiz Web Server."

M. Johnson and L. Brand, Editors, *Methods in Enzymology,* Vol. 321, (2000), Academic Press (San Diego). "Numerical Computer Methods."

R. Kaspar, *Biochem. Mol. Biol. Educ.* **30,** 36–39 (2002). "Integrating Internet Assignments into a Biochemistry/Molecular Biology Laboratory Course."

L. Kirkup, *Data Analysis with Excel* (2002), Cambridge University Press (Cambridge).

D. Leon, S. Uridil, and J. Miranda, *J. Chem. Educ.* **75,** 731–734 (1998). "Structural Analysis and Modeling of Proteins on the Web."

S. Misener and S. Krawetz, *Bioinformatics: Methods and Protocols* (2000), Humana Press (Totowa, NJ).

H. Salter, *Biochem. Educ.* **26,** 3–10 (1998). "Teaching Bioinformatics."

R. Sayle and E. Milner-White, *Trends Biochem. Sci.* **20,** 374–376 (1995). "RASMOL: Biomolecular Graphics for All."

I. Schomburg et al., *Trends Biochem. Sci.* **27,** 54–56 (2002). "BRENDA: A Resource for Enzyme Data and Metabolic Information."

Trends Guide to the Internet (1997), Elsevier Science (Cambridge).

C. Tsai, *Computational Biochemistry* (2002), Wiley-Liss (New York).

D. Westhead, J. Parish, and R. Twyman, *Instant Notes in Bioinformatics* (2000), BIOS Scientific (Oxfordshire).

T. Zielinski and M. Swift, Editors, *Using Computers in Chemistry and Chemical Education* (1997), American Chemical Society (Washington, DC).

http://lab3d.chem.virginia.edu
>   Lab3D, a virtual biochemistry lab from the University of Virginia. The following experiments are available: amino acid titration, electrophoresis, and enzyme kinetics.

http://www.nobel.se
>   From the Nobel Foundation. Under the heading "Educational," visit "The Virtual Biochemistry Lab" to perform experiments on DNA and proteins and listen to lectures. Procedures include chromatography, electrophoresis, NMR, X-ray diffraction, protein sequencing, and protein folding.

http://www.umsl.edu/divisions/artscience/chemistry/books/welcome.html
>   The Journal of Chemical Education Chemical Education Resource Shelf. For a complete list of current general biochemistry texts and lab manuals, click on "Index to Chemistry Textbooks and Software" and scroll to "Biochemistry."

http://www.apple.com/scitech/stories/bioinformatics/index.html
>   "A Portable Bioinformatics Teaching Laboratory," T. Littlejohn and B. Gaeta.

http://www.biointeractive.org
>   Virtual biology labs sponsored by Howard Hughes Medical Institute.

Glossary for the Internet

biological databases–computer sites that organize, store, and disseminate files that contain information consisting of literature references, nucleic acid sequences, protein sequences, and protein structures.

bookmark–a function to save a Web site address for later use (Netscape Navigator).

browser–an interface program that reads hypertext and displays Web pages on your computer.

domain–the computer user's location or local network.

e-mail–electronic mail; a means of exchanging messages via computer.

favorites–the Internet Explorer form of a bookmark.

freeware–software that is provided free of charge by its developer.

ftp–file transfer protocol; a mechanism for transferring data across a network.

home page–the beginning page for access to a Web site.

HTML–HyperText Markup Language; a special, coded language that is used to write Web pages.

hyperlink–link or connection between related Web pages.

hypertext–a language that connects similar documents on the Web.

Internet–the worldwide matrix that allows all computers and networks to communicate with each other.

Java–a programming language that allows the incorporation of multimedia into Web pages.

modem–electronic hardware that transmits computer signals through a telephone line.

multimedia–several forms of media including text, graphics, video, and audio.

search engine–a searchable directory that organizes Web pages by subject classification.

server–a computer that acts as the storage site for retrievable data.

URL–uniform resource locator; a standard address form that identifies the location of a document on the Internet.

Web site–a collection of documents (Web pages) on a server.

WWW–World Wide Web ("the Web"); a component of the Internet that uses a hypertext-based language to provide resources.

GENERAL LABORATORY PROCEDURES

All biochemistry experimental activities, whether in the teaching or research laboratory or in an industrial biotech lab, are replete with techniques that must be carried out on an almost daily basis. This chapter outlines the theoretical and practical aspects of some of these general and routine procedures, including preparation of buffers, use of buffers, pH measurements, dialysis, membrane filtration, lyophilization, quantitative methods for protein and nucleic acid measurement, centrifugal vacuum concentration, and measurement of radioactive samples.

A. PH, BUFFERS, ELECTRODES, AND BIOSENSORS

Most biological processes in the cell take place in a water-based environment. Water is an **amphoteric** substance; that is, it may serve as a proton donor (acid) or a proton acceptor (base). Equation 3.1 shows the ionic equilibrium of water.

$$H_2O \rightleftharpoons H^+ + OH^-$$

Equation 3.1

In pure water, $[H^+] = [OH^-] = 10^{-7}M$; in other words, the pH or $-\log[H^+]$ is 7. Acidic and basic molecules, when dissolved in water in a biological cell or test tube, react with either H^+ or OH^- to shift the equilibrium of Equation 3.1 and result in a pH change of the solution.

Biochemical processes occurring in cells and tissues depend on strict regulation of the hydrogen ion concentration. Biological pH is maintained at a constant value by natural buffers. When biological processes are studied *in vitro*, artificial media must be prepared that mimic the cell's natural environment. Because of the dependence of biochemical reactions on pH, the accurate determination of hydrogen ion concentration has always been of major interest. Today, we consider the measurement and control of pH to be a simple and rather mundane activity. However, an inaccurate pH measurement or a poor choice of buffer can lead to failure in the biochemistry laboratory. You should become familiar with several aspects of pH measurement, electrodes, and buffers. A table of commonly used acids and bases is shown in Appendix II.

Measurement of pH

A pH measurement is usually taken by immersing a glass or plastic combination electrode into a solution and reading the pH directly from a meter. At one time, pH measurements required two electrodes, a pH-dependent glass electrode sensitive to H^+ ions and a pH-independent calomel reference electrode. The potential difference that develops between the two electrodes is measured as a voltage as defined by Equation 3.2.

$$V = E_{constant} + \frac{2.303RT}{F}\ pH$$

Equation 3.2

where

$$V = \text{voltage of the completed circuit}$$
$$E_{constant} = \text{potential of reference electrode}$$
$$R = \text{the gas constant}$$
$$T = \text{the absolute temperature}$$
$$F = \text{the Faraday constant}$$

A pH meter is standardized with buffer solutions of known pH before a measurement of an unknown solution is taken. It should be noted from Equation 3.2 that the voltage depends on temperature. Hence, pH meters must have some means for temperature correction. Older instruments usually have a knob labeled "temperature control," which is adjusted by the user to the temperature of the measured solution. Newer pH meters automatically display a temperature-corrected pH value.

Using the pH Electrode

Most pH measurements today are obtained using a single **combination electrode** (Figure 3.1). Both the reference and the pH-dependent electrode are contained in a single glass or plastic tube. Although these are more expensive than dual electrodes, they are much more convenient to use, especially for smaller volumes of solution. Using a pH meter with a combination electrode is relatively easy, but certain guidelines must be followed. A pH meter not in use is left in a "standby" position. Before use, check the

Figure 3.1

The combination pH
electrode. *Courtesy of Hanna
Instruments; www.hannainst.com*

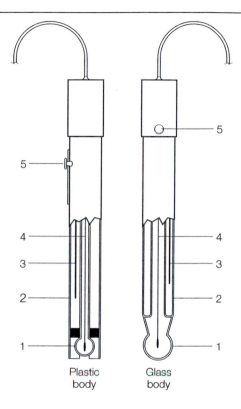

Plastic
body

Glass
body

Electrodes are housed in either
plastic or an all-glass body configuration. They can be either single
cells or as shown in the diagram,
combined into one body for ease of
use. Regardless of the configuration, there are several features common to all electrodes.

1. Sensing membrane glass:
   Performs actual measurement.

2. Reference junction: Acts as a liquid path electrical conductor.

3. Internal reference: Supplies a
   constant equilibrium voltage.

4. pH internal element: Supplies a
   voltage based on the pH value of
   the sample.

5. Reference fill hole: Used to
   replace the reference electrolyte
   solution.

level of saturated KCl in the electrode. If it is low, check with your instructor for the filling procedure. Turn the temperature control, if available, to
the temperature of the standard calibration buffers and the test solutions.
Be sure the function dial is set to pH. Lift the electrode out of the storage
solution, rinse it with distilled water from a wash bottle, and *gently* clean
and dry the electrode with a tissue. Immerse the electrode in a standard
buffer. Common standard buffers are pH 4, 7, and 10 with accuracy of
±0.02 pH unit. The standard buffer should have a pH within two pH
units of the expected pH of the test solution. The bulb of the electrode
must be completely covered with solution. Turn the pH meter to "on" or
"read" and adjust the meter with the "calibration dial" (sometimes called
"intercept") until the proper pH of the standard buffer is indicated on the
dial. Turn the pH meter to standby position. Remove the electrode and
again rinse with distilled water and carefully blot dry with tissue. Immerse
the electrode in a standard buffer of different pH and turn the pH meter to
"read." The dial should read within ±0.05 pH unit of the known value. If it
does not, adjust to the proper pH and again check the first standard pH
buffer. Clean the electrode and immerse it in the test solution. Record the
pH of the test solution.

As with all delicate equipment, the pH meter and electrode must receive proper care and maintenance. All electrodes should be kept clean and stored in solutions suggested by manufacturers. Glass electrodes are fragile and expensive, so they must be handled with care. If pH measurements of protein solutions are often taken, a protein film may develop on the electrode; it can be removed by soaking in 5% pepsin in 0.1 M HCl for 2 hours and rinsing well with water. The pH instrument should always be set on "standby" when the electrode is not in a solution.

Measurements of pH are always susceptible to experimental errors. Some common problems are:

1. **The Sodium Error**   Many glass combination electrodes are sensitive to Na^+ as well as H^+. The sodium error can become quite significant at high pH values, where 0.1 M Na^+ may decrease the measured pH by 0.4 to 0.5 unit. Several things may be done to reduce the sodium error. Some commercial suppliers of electrodes provide a standard curve for sodium error correction. Newer electrodes that are virtually Na^+ impermeable are now commercially available. If neither a standard curve nor a sodium-insensitive electrode is available, potassium salts may be substituted for sodium salts.

2. **Concentration Effects**   The pH of a solution varies with the concentration of buffer ions or other salts in the solution. This is because the pH of a solution depends on the *activity* of an ionic species, not on the concentration. **Activity,** you may recall, is a thermodynamic term used to define species in a nonideal solution. At infinite dilution, the activity of a species is equivalent to its concentration. At finite dilutions, however, the activity of a solute and its concentration are not equal.

   It is common practice in biochemical laboratories to prepare concentrated "stock" solutions and buffers. These are then diluted to the proper concentration when needed. Because of the concentration effects described above, it is important to adjust the pH of these solutions *after* dilution.

3. **Temperature Effects**   The pH of a buffer solution is influenced by temperature. This effect is due to a temperature-dependent change of the dissociation constant (pK_a) of ions in solution. The pH of the commonly used buffer Tris is greatly affected by temperature changes, with a $\Delta pK_a/C°$ of -0.031. This means that a pH 7.0 Tris buffer made up at 4°C would have a pH of 5.95 at 37°C. The best way to avoid this problem is to prepare the buffer solution at the temperature at which it will be used and to standardize the electrode with buffers at the same temperature as the solution you wish to measure.

Biochemical Buffers

Buffer ions are used to maintain solutions at constant pH values. The selection of a buffer for use in the investigation of a biochemical process is of critical importance. Before the characteristics of a buffer system are discussed, we will review some concepts in acid-base chemistry.

Weak acids and bases do not completely dissociate in solution but exist as equilibrium mixtures (Equation 3.3).

$$\text{HA} \underset{k_2}{\overset{k_1}{\rightleftharpoons}} \text{H}^+ + \text{A}^-$$

Equation 3.3

HA represents a weak acid and A^- represents its conjugate base; k_1 represents the rate constant for dissociation of the acid and k_2 the rate constant for association of the conjugate base and hydrogen ion. The equilibrium constant, K_a, for the weak acid HA is defined by Equation 3.4.

$$K_a = \frac{k_1}{k_2} = \frac{[\text{H}^+][\text{A}^-]}{[\text{HA}]}$$

Equation 3.4

which can be rearranged to define $[\text{H}^+]$ (Equation 3.5).

$$[\text{H}^+] = \frac{K_a[\text{HA}]}{[\text{A}^-]}$$

Equation 3.5

The $[\text{H}^+]$ is often reported as pH, which is $-\log [\text{H}^+]$. In a similar fashion, $-\log K_a$ is represented by pK_a. Equation 3.5 can be converted to the $-\log$ form by substituting pH and pK_a:

$$\text{pH} = pK_a + \log \frac{[\text{A}^-]}{[\text{HA}]}$$

Equation 3.6

Equation 3.6 is the familiar Henderson-Hasselbalch equation, which defines the relationship between pH and the ratio of acid and conjugate base concentrations. The salt of the acid (A^-) is also referred to as the **proton acceptor (A)** and the acid (HA) as the **proton donor (D).** The Henderson-Hasselbalch equation is of great value in buffer chemistry because it can be used to calculate the pH of a solution if the molar ratio of buffer ions ($[\text{A}^-]/[\text{HA}]$) and the pK_a of HA are known. Also, the molar ratio of HA to A^- that is necessary to prepare a buffer solution at a specific pH can be calculated if the pK_a is known.

A solution containing both HA and A^- has the capacity to resist changes in pH; i.e., it acts as a **buffer.** If acid (H^+) were added to the buffer solution, it would be neutralized by A^- in solution:

$$\text{H}^+ + \text{A}^- \rightarrow \text{HA}$$

Equation 3.7

Base (OH^-) added to the buffer solution would be neutralized by reaction with HA:

$$\text{OH}^- + \text{HA} \rightarrow \text{A}^- + \text{H}_2\text{O}$$

Equation 3.8

The most effective buffering system contains equal concentrations of the acid, HA, and its conjugate base, A⁻. According to the Henderson-Hasselbalch equation (3.6), when [A⁻] is equal to [HA], pH equals pK_a. Therefore, the pK_a of a weak acid-base system represents the center of the buffering region. The effective range of a buffer system is generally two pH units, centered at the pK_a value (Equation 3.9).

>> Effective pH range for a buffer = $pK_a \pm 1$ **Equation 3.9**

Selection of a Biochemical Buffer

Virtually all biochemical investigations must be carried out in buffered aqueous solutions. The natural environment of biomolecules and cellular organelles is under strict pH control. When these components are extracted from cells, they are most stable if maintained in their normal pH range, usually 6 to 8. An artificial buffer system is found to be the best substitute for the natural cell milieu. It should also be recognized that many biochemical processes (especially some enzyme processes) produce or consume hydrogen ions. The buffer system neutralizes these solutions and maintains a constant chemical environment.

Although most biochemical solutions require buffer systems effective in the pH range 6 to 8, there is occasionally a need for buffering over the pH range 2 to 12. Obviously, no single acid–conjugate base pair will be effective over this entire range, but several buffer systems are available that may be used in discrete pH ranges. Figure 3.2 compares the effective buffering ranges of common biological buffers. It should be noted that some buffers (phosphate, succinate, and citrate) have more than one pK_a value, so they may be used in different pH regions. Many buffer systems are effective in the usual biological pH range (6 to 8); however, there may be major problems in their use. Several characteristics of a buffer must be considered before a final selection is made. The molecular weights and pK values of several common buffer compounds are listed in Appendix III. Following is a discussion of the advantages and disadvantages of the commonly used buffers.

Phosphate Buffers

The phosphates are among the most widely used buffers. These solutions have high buffering capacity and are very useful in the pH range 6.5 to 7.5. Because phosphate is a natural constituent of cells and biological fluids, its presence affords a more "natural" environment than many buffers. Sodium or potassium phosphate solutions of all concentrations are commonly prepared with the use of the Henderson-Hasselbalch equation. The major disadvantages of phosphate solutions are (1) precipitation or binding of common biological cations (Ca^{2+} and Mg^{2+}), (2) inhibition of some biological processes, including some enzymes, and (3) limited useful pH range.

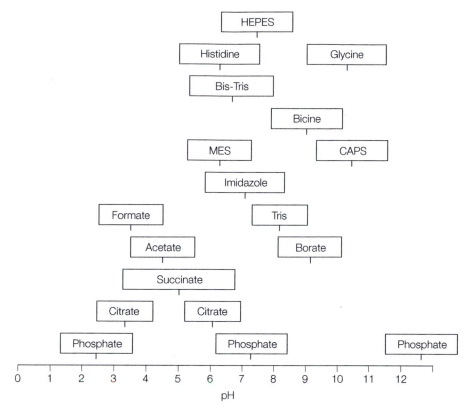

Figure 3.2

Effective buffering ranges of several common buffers. See Table 3.1
for abbreviations.

➤ | **Study Exercise 3.1**   Here we will describe the calculation method for
preparation of a sodium phosphate buffer of 0.05 M total phosphate concentration, pH of 7.0, and a total, final volume of 1 liter. The Henderson-
Hasselbalch equation (Equation 3.6) will be used to calculate the conjugate
acid and conjugate base concentrations. At pH 7.0, the two forms of phosphate present are:

$$HO-\overset{\overset{\displaystyle O}{\|}}{\underset{\underset{\displaystyle OH}{|}}{P}}-O^-Na^+ \;\rightleftharpoons\; HO-\overset{\overset{\displaystyle O}{\|}}{\underset{\underset{\displaystyle O^-Na^+}{|}}{P}}-O^-Na^+ + H^+$$

Proton donor (D)   Proton acceptor (A)

We now use the Henderson-Hasselbalch equation with pH = 7.0 and pK = 7.21 (from Appendix III) to calculate the molar ratio:

[A]/[D] = 0.62

This gives only one equation, but two unknowns, the concentrations of A and D; therefore, we need another quantitative relationship between [A] and [D]. We know that the total phosphate concentration ([A] + [D]) is 0.05 M, so:

[A] + [D] = 0.05

Using these two equations for A and D, we determine that we need 0.031 mole of D and 0.019 mole of A. The appropriate phosphate reagent forms available commercially are:

D = donor: monobasic sodium phosphate monohydrate, $NaH_2PO_4 \cdot H_2O$ (MW = 138); 0.031 mole.

A = acceptor: dibasic sodium phosphate heptahydrate, $Na_2HPO_4 \cdot 7H_2O$ (MW = 268); 0.019 mole.

Converting the number of moles of each reagent to grams:

grams of D = 0.031 × 138 = 4.28 g
grams of A = 0.019 × 268 = 5.09 g

To prepare the solution, dissolve the appropriate amount of each reagent in about 975 mL of purified water in a 1-liter beaker. Check the pH of the solution and adjust to 7.0 by careful, dropwise addition (with stirring) of dilute NaOH or dilute phosphoric acid. Transfer the solution to a 1-liter volumetric flask, add water to the mark and do a final check of the pH.

Zwitterionic Buffers (Good's Buffers)

In the mid-1960s, N. E. Good and his colleagues recognized the need for a set of buffers specifically designed for biochemical studies (Good and Izawa, 1972). He and others noted major disadvantages of the established buffer systems. Good outlined several characteristics essential in a biological buffer system:

1. pK_a between 6 and 8.
2. Highly soluble in aqueous systems.
3. Exclusion or minimal transport by biological membranes.
4. Minimal salt effects.
5. Minimal effects on dissociation due to ionic composition, concentration, and temperature.
6. Buffer–metal ion complexes nonexistent or soluble and well defined.

7.   Chemically stable.

8.   Insignificant light absorption in the ultraviolet and visible regions.

9.   Readily available in purified form.

Good investigated a large number of synthetic zwitterionic buffers and found many of them to meet these criteria. Table 3.1 lists several of these buffers and their properties. Good's buffers are widely used, but their main disadvantage is high cost. Some zwitterionic buffers, such as Tris, HEPES, and PIPES, have been shown to produce radicals under a variety of experimental conditions, so they should be avoided if biological redox processes or radical-based reactions are being studied. Radicals are not produced from MES or MOPS buffers.

The use of the synthetic zwitterionic buffer Tris [tris(hydroxymethyl) aminomethane] is now probably greater than that of phosphate. It is useful in the pH range 7.5 to 8.5. Tris is available in a basic form as highly purified crystals, which makes buffer preparation especially convenient. Although Tris is a primary amine, it causes minimal interference with biochemical processes and does not precipitate calcium ions. However, Tris has several disadvantages, including (1) pH dependence on concentration, since the pH decreases 0.1 pH unit for each 10-fold dilution; (2) interference with some pH electrodes; and (3) a large $\Delta pK_a/C°$ compared to most other buffers. Most of these drawbacks can be minimized by (1) adjusting the pH *after* dilution to the appropriate concentration, (2) purchasing electrodes that are compatible with Tris, and (3) preparing the buffer at the temperature at which it will be used.

➤ | **Study Exercise 3.2**   We will illustrate the preparation of a Tris buffer using the titration method. (In Study Exercise 3.1, we used the calculation method to prepare the phosphate buffer.) Because Tris is commercially available in highly purified crystals, Tris Base (MW = 121.1), the appropriate amount may be weighed, dissolved in water, and titrated to the desired pH with an acid, usually HCl. For 1 liter of 0.1 M Tris-HCl buffer, pH 7.0, dissolve 12.11 g (0.1 mole) of Tris Base in about 975 mL of purified water in a 1-liter beaker. Adjust the pH (originally about 10) to 7.0 by careful, dropwise addition of concentrated hydrochloric acid. This should be done in a 1-liter beaker with gentle stirring. Transfer the solution to a 1-liter volumetric flask, add water to the mark, and make a final check of the pH.

As you can see by comparing the two procedures, the titration method for buffer preparation is faster and much more convenient than the calculation method. Is it possible to prepare the phosphate buffer in Study Exercise 3.1 with the use of the titration method? If so, describe the reagent(s) needed and the procedure you would use.

Carboxylic Acid Buffers

The most widely used buffers in this category are acetate, formate, citrate, and succinate. This group is useful in the pH range 3 to 6, a region that offers few

Table 3.1

Structures and Properties of Several Synthetic Zwitterionic Buffers

Buffer Name	Abbreviation	Structure (All structures shown in salt form)	pK_a (20°C)	Useful pH Range	$\Delta pK_a/C°$	Concentration of a Saturated Solution (M, 0°C)
N-2-Acetamido-2-amino-ethanesulfonic acid	ACES	$H_2NCOCH_2\overset{+}{N}H_2CH_2CH_2SO_3^-$	6.9	6.4–7.4	−0.020	0.22
N-2-Acetamidoiminodiacetic acid	ADA	$H_2NCOCH_2\overset{+}{N}\underset{H}{\Big\langle}\overset{CH_2COO^-}{\underset{CH_2COO^-}{}}$	6.6	6.2–7.2	−0.011	—
N,N-Bis(2-hydroxyethyl)-2-aminoethanesulfonic acid	BES	$(HOCH_2CH_2)_2\overset{+}{N}HCH_2CH_2SO_3^-$	7.15	6.6–7.6	−0.016	3.2
N,N-Bis(2-hydroxyethyl)-glycine	Bicine	$(HOCH_2CH_2)_2\overset{+}{N}HCH_2COO^-$	8.35	7.5–9.0	−0.018	1.1
3-(Cyclohexylamino)-propanesulfonic acid	CAPS	$\text{—}NHCH_2CH_2CH_2SO_3^-$ (cyclohexyl)	10.4	10.0–11.0	−0.009	0.85
Cyclohexylaminoethane-sulfonic acid	CHES	$\text{—}NHCH_2CH_2SO_3^-$ (cyclohexyl)	9.5	9.0–10.0	−0.009	0.85
Glycylglycine	Gly-Gly	$H_3\overset{+}{N}CH_2CONHCH_2COO^-$	8.4	7.5–9.5	−0.028	1.1
N-2-Hydroxyethyl-piperazine-N'-2-ethanesulfonic acid	HEPES	$HO(CH_2)_2N\overset{}{\underset{}{\bigcirc}}\overset{+}{N}CH_2CH_2SO_3^-$	7.55	7.0–8.0	−0.014	2.25
2-(N-Morpholino)ethane-sulfonic acid	MES	$\overset{+}{N}HCH_2CH_2SO_3^-$ (morpholino)	6.15	5.8–6.5	−0.011	0.65
3-(N-Morpholino)-propanesulfonic acid	MOPS	$\overset{+}{N}HCH_2CH_2CH_2SO_3^-$ (morpholino)	7.20	6.5–7.9	—	—
Piperazine-N,N'-bis-2-ethanesulfonic acid	PIPES	$^-O_3SCH_2CH_2\overset{+}{N}\overset{}{\underset{}{\bigcirc}}\overset{+}{N}HCH_2CH_2SO_3^-$	6.8	6.4–7.2	−0.0085	—
N-Tris(hydroxymethyl)-methyl-2-aminoethane-sulfonic acid	TES	$(HOCH_2)_3\overset{+}{N}HCH_2CH_2SO_3^-$	7.5	7.0–8.0	−0.020	2.6
N-Tris(hydroxymethyl)-methylglycine	Tricine	$(HOCH_2)_3\overset{+}{C}NH_2CH_2CH_2COO^-$	8.15	7.5–8.5	−0.021	0.8
Tris(hydroxymethyl)-aminomethane	Tris	$(HOCH_2)_3\overset{+}{C}NH_3$	8.3	7.5–9.0	−0.031	2.4

other buffer choices. All of these acids are natural metabolites, so they may interfere with the biological processes under investigation. Also, citrate and succinate may interfere by binding metal ions (Fe^{3+}, Ca^{2+}, Zn^{2+}, Mg^{2+}, etc.). Formate buffers are especially useful because they are volatile and can be removed by evaporation under reduced pressure.

Borate Buffers

Buffers of boric acid are useful in the pH range 8.5 to 10. Borate has the major disadvantage of complex formation with many metabolites, especially carbohydrates.

Amino Acid Buffers

The most commonly used amino acid buffers are glycine (pH 2 to 3 and 9.5 to 10.5), histidine (pH 5.5 to 6.5), glycine amide (pH 7.8 to 8.8), and glycylglycine (pH 8 to 9). These provide a more "natural" environment to cellular components and extracts; however, they may interfere with some biological processes, as do the carboxylic acid and phosphate buffers.

Buffer Dilutions

In biochemistry lab activities, buffers of many different concentrations and volumes are required. It is usually not practical, economical, or convenient to prepare all needed buffers from scratch. However, it is possible to have available a few concentrated stock buffer solutions that may be diluted to produce a new buffer solution of desired concentration and amount. It should be noted that the dilution of a concentrated stock buffer solution with purified water should not change its pH. Dilution decreases the actual concentrations of the acceptor and donor species, but their ratio does not change.

Dilutions are usually defined in the following way: A dilution of 1 mL to 10 mL (also 1:10; or 1 to 10) means that you take 1 mL of stock solution and add water to a final total volume of 10 mL. It is important to note that the ***amount*** of the buffer components taken from the stock solution is equal to the ***amount*** of buffer components in the diluted solution. However, the ***concentration*** of the buffer components has been changed. If the stock solution above had a buffer concentration of 0.5 *M*, then the 1:10 diluted solution has a buffer concentration of 0.05 *M*.

Two kinds of dilutions, linear and serial, are used most often in biochemistry.

- **Linear dilutions** Table 3.2 shows the results of a linear dilution process where a stock solution of 1 *M* is used to produce samples of different concentrations. Note that there is a linear (progressive) decrease in the concentration of the diluted samples, but the final volumes

Table 3.2				
Example of a Linear Dilution Process				
Stock (*M*)	Final dilution (*M*)	Stock (mL)	Water (mL)*	Final volume (mL)
1.0	0.80	8.0	2.0	10
1.0	0.60	6.0	4.0	10
1.0	0.40	4.0	6.0	10
1.0	0.20	2.0	8.0	10
1.0	0.00	0.0	10.0	10

*Because all solutions are not additive, this number is the approximate amount of water needed to bring the final volume to 10 mL. It may be necessary to add a few drops of water to bring the final volume to 10 mL.

remain the same (10 mL). The volume of stock solution to use in each case is calculated from the following relationship:

$$\text{volume (stock)} = \frac{\text{volume (diluted)} \times \text{concentration (diluted)}}{\text{concentration (stock)}}$$

Sample calculation for the first diluted solution in Table 3.2:

$$\text{volume (stock)} = \frac{10 \text{ mL} \times 0.80\,M}{1.0\,M} = 8.0 \text{ mL}$$

- **Serial dilutions**   Here dilution starts first with a stock buffer solution and then each diluted solution produced is used to prepare the next, etc. For example:

  2.0 mL of Solution I is diluted to a final volume of 100 mL (Solution II)

  2.0 mL of Solution II is diluted to a final volume of 100 mL (Solution III)

  2.0 mL of Solution III is diluted to a final volume of 100 mL (Solution IV)

  If Solution I has a buffer concentration of 10 *M* and water is used for dilution in each case, what is the buffer concentration of Solution II? The equation above for linear dilution may also be used here:

$$\text{For Solution II: concentration (diluted)} = \frac{2.0 \text{ mL} \times 10\,M}{100 \text{ mL}} = 0.2\,M$$

➤ **Study Exercise 3.3**   What are the concentrations of Solutions III and IV made from the above serial dilution?

The Oxygen Electrode

Second in popularity only to the pH electrode is the oxygen electrode. This device is a polarographic electrode system that can be used to measure oxygen

Figure 3.3

Biological oxygen monitor, YSI Model 5300A. *Courtesy of YSI Life Sciences; www.ysi.com*

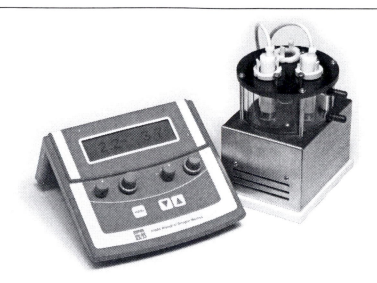

concentration in a liquid sample or to monitor oxygen uptake or evolution by a chemical or biological system.

The basic electrode was developed by L. C. Clark, Jr. in 1953. The popular Clark-type oxygen electrode with biological oxygen monitor is shown in Figure 3.3. The electrode system, which consists of a platinum cathode and silver anodes, is molded into an epoxy block and surrounded by a Lucite holder. A thin Teflon membrane is stretched over the electrode end of the probe and held in place with an O-ring. The membrane separates the electrode elements, which are bathed in an electrolyte solution (saturated KCl), from the test solution in the sample chamber. The membrane is permeable to oxygen and other gases, which then come into contact with the surface of the platinum cathode. If a suitable polarizing voltage (about 0.8 volt) is applied across the cell, oxygen is reduced at the cathode.

$$O_2 + 2e^- + 2H_2O \longrightarrow [H_2O_2] + 2OH^-$$

$$[H_2O_2] + 2e^- \longrightarrow 2OH^-$$

$$\text{Total} \quad O_2 + 4e^- + 2H_2O \longrightarrow 4OH^-$$

The reaction occurring at the anode is

$$4Ag^\circ + 4Cl^- \longrightarrow 4AgCl + 4e^-$$

The overall electrochemical process is

$$4Ag^\circ + O_2 + 4Cl^- + 2H_2O \longrightarrow 4AgCl + 4OH^-$$

The current flowing through the electrode system is, therefore, directly proportional to the amount of oxygen passing through the membrane. According to the laws of diffusion, the rate of oxygen flowing through the membrane depends on the concentration of dissolved oxygen in the sample. Therefore, the magnitude of the current flow in the electrode is directly related to the concentration of dissolved oxygen. The function of the electrode can also be explained in terms of oxygen pressure. The oxygen concentration at the cathode (inside the membrane) is virtually zero because oxygen is rapidly reduced. Oxygen pressure in the sample chamber (outside the membrane) will force oxygen to diffuse through the membrane at a rate that is directly proportional to the pressure (or concentration) of oxygen in the test solution. The current generated by the electrode system is electronically amplified and transmitted to a meter, which usually reads in units of % oxygen saturation.

Many biochemical processes consume or evolve oxygen. The oxygen electrode is especially useful for monitoring such changes in the concentration of dissolved oxygen. If oxygen is consumed, for example, by a suspension of mitochondria or by an oxygenase enzyme system, the rate of oxygen diffusion through the probe membrane will decrease, leading to less current flow. A linear relationship exists between the electrode current and the amount of oxygen consumed by the biological system.

The use of an oxygen electrode is not free of experimental and procedural problems. Probably the most significant source of trouble is the Teflon membrane. A torn, damaged, or dirty membrane is easily recognized by recorder noise or electronic "spiking." Temperature control of the sample chamber and electrode is essential because the rate of oxygen diffusion through the membrane is temperature dependent. The sample chamber is equipped with a bath assembly for temperature regulation and magnetic stirrer in order to maintain a constant oxygen concentration throughout the solution. Air bubbles in the sample chamber lead to considerable error in measurements and must be removed.

Biosensors

Electrodes or electrode-like devices are currently being developed for the specific measurement of physiologically important molecules such as urea, carbohydrates, enzymes, antibodies, and metabolic products. This type of device, now referred to as a **biosensor,** is an analytical tool or system consisting of an immobilized biological material (such as an enzyme, antibody, whole cell, organelle, or combinations thereof) in intimate contact with a suitable transducer device which will convert the biochemical signal into a quantifiable electric signal. The important components of a biosensor as shown in Figure 3.4 are (1) a reaction center consisting of a membrane or gel containing the biochemical system to be studied, (2) a transducer, (3) an amplifier, and (4) a computer system for data acquisition and processing. When biomolecules in the reaction center interact, a physicochemical change

Figure 3.4

Components of a biosensor system for determining the identity and concentration of biomolecule A. A is stoichiometrically converted to B by a specific enzyme in the reaction center. The concentration of B is measured by the sensor.

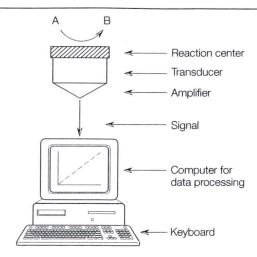

A   B

Reaction center

Transducer

Amplifier

Signal

Computer for data processing

Keyboard

occurs. This change in the molecular system, which may be a modification of concentration, absorbance, mass, conductance, or redox state, is converted into an electrical signal by the transducer. The signal is then amplified and displayed on a computer screen. Each biosensor is specifically designed for a type of molecule or biological interaction; therefore, details of construction and function vary. Some specific examples of biosensors include:

1. Electrodes based on enzyme activity. These are selective and sensitive devices that may be used to measure substrate concentrations. A biosensor based on glucose oxidase is used to measure the concentration of glucose by detecting the production of H_2O_2.

    Glucose + O_2 $\rightleftharpoons$ gluconic acid + H_2O_2

2. Optical biosensors that respond to the absorption of light from a laser. The optical response is transmitted to a computer by a fiber-optic system.

B.   MEASUREMENT OF PROTEIN SOLUTIONS

Biochemical research often requires the quantitative measurement of protein concentrations in solutions. Several techniques have been developed; however, most have limitations because either they are not sensitive enough or they are based on reactions with specific amino acids in the protein. Since the amino acid content varies from protein to protein, no single assay will be suitable for all proteins. In this section we discuss five assays: three older, classical methods that are occasionally used today and two newer methods that are widely used. In four of the methods, chemical reagents are added to protein solutions to develop a color whose intensity is measured in a spectrophotometer. A "standard protein" of known concentration is also treated with the same reagents and a calibration curve is constructed. The other assay relies on a direct spectrophotometric measurement. None of the methods is perfect because each is dependent on the amino acid content of the protein. However, each will provide a satisfactory result if the proper experimental conditions are used and/or a suitable standard protein is chosen. Other important factors in method selection include the sensitivity and accuracy desired, the presence of interfering substances, and the time available for the assay. The various methods are compared in Table 3.3.

The Biuret and Lowry Assays

When substances containing two or more peptide bonds react with the biuret reagent, alkaline copper sulfate, a purple complex is formed. The colored product is the result of coordination of peptide nitrogen atoms with Cu^{2+}. The amount of product formed depends on the concentration of protein.

In practice, a calibration curve must be prepared by using a standard protein solution. The best protein to use as a standard is a purified preparation of the protein to be assayed. Since this is rarely available, the experimenter must choose another protein as a relative standard. A relative standard must be selected that provides a similar color yield. An aqueous solution of bovine serum albumin is a commonly used standard. Various known amounts of this solution are treated with the biuret reagent, and the color is allowed to develop. Measurements of absorbance at 540 nm (A_{540}) are made against a blank containing biuret reagent and buffer or water. The A_{540} data are plotted versus protein concentration (mg/mL). Unknown protein samples are treated with biuret reagent and the A_{540} is measured after color development. The protein concentration is determined from the standard curve. The biuret assay has several advantages, including speed, similar color development with different proteins, and few interfering substances. Its primary disadvantage is its lack of sensitivity.

The **Lowry protein assay** is one of the more sensitive essays and has been widely used. The Lowry procedure can detect protein levels as low as 5 μg. The principle behind color development is identical to that of the biuret assay except that a second reagent (Folin-Ciocalteu) is added to increase the amount of color development. Two reactions account for the intense blue color that develops: (1) the coordination of peptide bonds with alkaline copper (biuret reaction) and (2) the reduction of the Folin-Ciocalteu reagent (phosphomolybdate-phosphotungstate) by tyrosine and tryptophan residues in the protein. The procedure is similar to that of the biuret assay. A standard curve is prepared with bovine serum albumin or other pure protein, and the concentration of unknown protein solutions is determined from the graph.

The obvious advantage of the Lowry assay is its sensitivity, which is up to 100 times greater than that of the biuret assay; however, more time is required for the Lowry assay. Since proteins have varying contents of tyrosine

Table 3.3

Methods for Protein Measurement

Method	Sensitivity	Time	Principle	Interferences	Comments
Biuret	Low 1–20 mg	Moderate 20–30 min	Peptide bonds + alkaline $Cu^{2+} \rightarrow$ purple complex	Zwitterionic buffers, some amino acids	Similar color with all proteins. Destructive to protein samples.
Lowry	High ~5 μg	Slow 40–60 min	(1) Biuret reaction (2) Reduction of phosphomolybdate-phosphotungstate by Tyr and Trp	Ammonium sulfate, glycine, zwitterionic buffers, mercaptans	Time-consuming. Intensity of color varies with proteins. Critical timing of procedure. Destructive to protein samples.
Bradford	High ~1 μg	Rapid 15 min	λ_{max} of Coomassie dye shifts from 465 nm to 595 nm when protein-bound	Strongly basic buffers, detergents Triton X-100, SDS	Stable color which varies with proteins. Reagents commercially available. Destructive to protein samples. Discoloration of glassware.
BCA	High 1 μg	Slow 60 min	(1) Biuret reaction (2) Copper complex with BCA; $\lambda_{max} =$ 562 nm	EDTA, DTT, ammonium sulfate	Compatible with detergents. Reagents commercially available. Destructive to protein samples.
Spectrophotometric (A_{280})	Moderate 50–1000 μg	Rapid 5–10 min	Absorption of 280-nm light by aromatic residues	Purines, pyrimidines, nucleic acids	Useful for monitoring column eluents. Nucleic acid absorption can be corrected. Nondestructive to protein samples. Varies with proteins.

Figure 3.5

Coomassie Brilliant Blue G-250 dye is used to measure protein concentration in the Bradford assay.

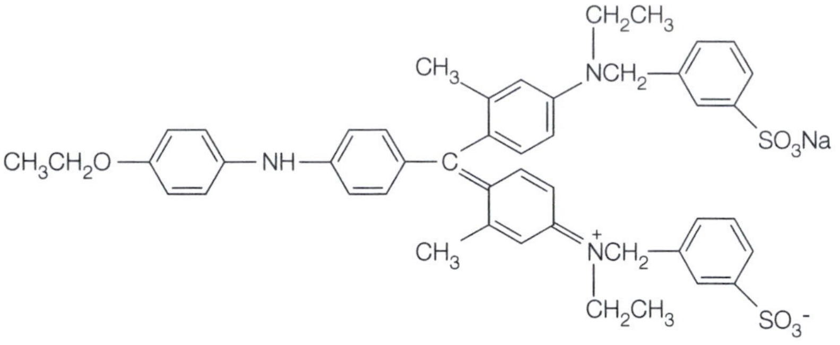

and tryptophan, the amount of color development changes with different proteins, including the bovine serum albumin standard. Because of this, the Lowry protein assay should be used only for measuring changes in protein concentration, not absolute values of protein concentration.

The Bradford Assay

The many limitations of the biuret and Lowry assays have encouraged researchers to seek better methods for quantitation of protein solutions. The **Bradford assay**, based on protein binding of a dye (perhaps to arginyl and lysyl residues), provides numerous advantages over other methods.

The binding of Coomassie Brilliant Blue dye (Figure 3.5) to protein in acidic solution causes a shift in wavelengh of maximum absorption (λ_{max}) of the dye from 465 nm to 595 nm. The absorption at 595 nm is directly related to the concentration of protein.

In practice, a calibration curve is prepared using bovine plasma gamma globulin or bovine serum albumin as a standard (Figure 3.6). The assay requires only a single reagent, an acidic solution of Coomassie Brilliant Blue G-250. After addition of dye solution to a protein sample, color development is complete in 2 minutes and the color remains stable for up to 1 hour. The sensitivity of the Bradford assay rivals and may surpass that of the Lowry assay. With a microassay procedure, the Bradford assay can be used to determine proteins in the range of 1 to 20 μg. The Bradford assay shows significant variation with different proteins, but this also occurs with the Lowry assay. The Bradford method not only is rapid but also has very few interferences by nonprotein components. The only known interfering substances are the detergents, Triton X-100 and sodium dodecyl sulfate. The many advantages of the Bradford assay have led to its wide adoption in biochemical research laboratories.

The BCA Assay

The **BCA protein assay** is based on chemical principles similar to those of the biuret and Lowry assays. The protein to be analyzed is reacted with Cu^{2+} (which produces Cu^+) and bicinchoninic acid (BCA). The Cu^+ is

Figure 3.6

Standard calibration curve for the Bradford protein assay using bovine gamma globulin as the protein standard.

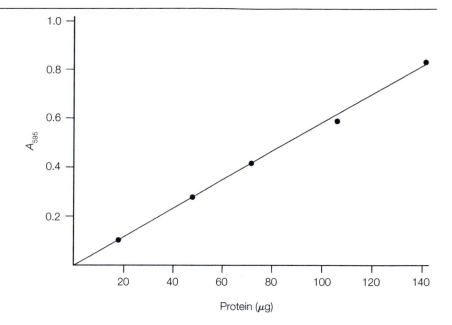

chelated by BCA, which converts the apple-green color of the free BCA to the purple color of the copper-BCA complex. Samples of unknown protein and relative standard are treated with the reagent, the absorbance is read in a spectrophotometer at 562 nm, and concentrations are determined from a standard calibration curve. This assay has the same sensitivity level as the Lowry and Bradford assays. Its main advantages are its simplicity and its usefulness in the presence of 1% detergents such as Triton or sodium dodecyl sulfate (SDS).

The Spectrophotometric Assay

Most proteins have relatively intense ultraviolet light absorption centered at 280 nm. This is due to the presence of tyrosine and tryptophan residues in the protein. However, the amount of these amino acid residues varies in different proteins, as was pointed out earlier. If certain precautions are taken, the A_{280} value of a protein solution is proportional to the protein concentration. The procedure is simple and rapid. A protein solution is transferred to a quartz cuvette and the A_{280} is read against a reference cuvette containing the protein solvent only (buffer, water, etc.).

Cellular extracts contain many other compounds that absorb in the vicinity of 280 nm. Nucleic acids, which can be common contaminants in a protein extract, absorb strongly at 280 nm ($\lambda_{max} = 260$). Early researchers developed a method to correct for this interference by nucleic acids. Mixtures of pure protein and pure nucleic acid were prepared, and the ratio

A_{280}/A_{260} was experimentally determined. The following empirical equation may be used for protein solutions containing up to 20% nucleic acids.

Protein concentration (mg/mL) = $1.55A_{280} - 0.76A_{260}$

Although the spectrophotometric assay of proteins is fast, relatively sensitive, and requires only a small sample size, it is still only an estimate of protein concentration. It has certain advantages over the colorimetric assays in that most buffers and ammonium sulfate do not interfere and the procedure is nondestructive to protein samples. The spectrophotometric assay is particularly suited to the rapid measurement of protein elution from a chromatography column, where only protein concentration changes are required.

C. MEASUREMENT OF NUCLEIC ACID SOLUTIONS

The Spectrophotometric Assay

Solutions of nucleic acids strongly absorb ultraviolet light with a λ_{max} of 260 nm. The intense absorption is due primarily to the presence of aromatic rings in the purine and pyrimidine bases. The concentration of nucleic acid in a solution can be calculated if one knows the value of A_{260} of the solution. *A solution of double-stranded DNA at a concentration of 50 μg/mL in a 1-cm quartz cuvette will give an A_{260} reading of 1.0. A solution of single-stranded DNA or RNA that has an A_{260} of 1.0 in a cuvette with a 1-cm path length has a concentration of 40 μg/mL.*
The spectrophotometric assay for nucleic acids is simple, rapid, nondestructive, and very sensitive; however, it measures both DNA and RNA. The minimum concentration of nucleic acid that can be accurately determined is 2.5–5.0 μg/mL.

➤ | **Study Exercise 3.4** Two solutions, one of double-stranded DNA and one of single-stranded RNA, were transferred to 1-cm quartz cuvettes and the absorbance (A) at 260 nm of each of the solutions was found to be 0.15. What are the concentrations of the DNA and RNA solutions? Hint: A standard solution of double-stranded DNA at a concentration of 50 μg/mL in a 1-cm quartz cuvette will yield an A_{260} reading of 1.0. A solution of single-stranded DNA or RNA that has an A_{260} of 1.0 in a cuvette with a 1-cm path length has a concentration of 40 μg/mL.

Other Assays for Nucleic Acids

The spectrophotometric assay for nucleic acids, DNA and RNA, is simple and rapid, but it does not have a sufficient level of sensitivity for many applications. Other assays for DNA and RNA depend on the enhanced fluorescence of dyes when bound to nucleic acids. Features of fluorescence assays are compared to the spectrophotometric assay in Table 3.4.

Table 3.4

Measurement of Nucleic Acid Solutions

Method	Sensitivity	Time	Principle	Interferences	Comments
Spectrophotometric (A_{260})	High 2.5 μg	Rapid 5–10 min	Absorption of 260 nm light by aromatic bases	Proteins	Protein absorption may be corrected. A_{280}/A_{260} indicates purity. Nondestructive. Measures both DNA and RNA.
Bisbenzimidazole	Very high 10 ng	Rapid 5–10 min	Enhancement of fluorescence of dye when bound to DNA	None known	Not reliable with plasmids or small DNA. Does not measure RNA. May be used with crude extracts. Nondestructive.
DNA Dipstick	Very high 0.1–10 ng	Moderate 30 min	Proprietary	SDS, agarose	Gives no indication of purity. Measures single and double-stranded DNA and RNA.
EtBr-Agarose	Very high 1–10 ng	Moderate 30–45 min	Fluorescence of bound EtBr	None known	Measures DNA and RNA.

The fluorescence dye bisbenzimidazole, when bound to DNA, emits light at 458 nm that is about five times greater in intensity than the emission by free, unbound dye. The increase in fluorescence is linear with increasing amounts of DNA in the concentration range of 0.01 μg/mL to 5 μg/mL. One special advantage of this assay is that it can be used to quantify DNA in cell suspensions and crude homogenates. It does not respond to RNA or other biomolecules in the extracts. The assay is fast and requires only a fluorimeter, an instrument found in most laboratories.

Another simple and relatively inexpensive fluorescence method uses ethidium bromide–agarose plates. The fluorescence emission of ethidium bromide at 590 nm is enhanced about 25-fold when the dye interacts with DNA. DNA samples for measurement are spotted onto 1% agarose containing ethidium bromide in a Petri dish or spot plates. The samples are exposed to UV light and photographed, and the light intensity is compared to standard samples of DNA. This method is sensitive (in the 1–10-ng range), but it detects both DNA and RNA.

One new and interesting method for nucleic acid analysis is the DNA Dipstick Kit developed by Invitrogen. The kit is a colorimetric-based assay where the nucleic acid sample interacts with a proprietary dye solution. The method can detect nucleic acids and oligonucleotides in the range of 0.1–10 ng/μL and requires only about 30 minutes.

D. TECHNIQUES FOR SAMPLE PREPARATION

Dialysis

One of the oldest procedures applied to the purification and characterization of biomolecules is **dialysis,** an operation used to separate dissolved molecules on the basis of their molecular size. The technique involves sealing an aqueous solution containing both macromolecules and small molecules in a porous membrane. The sealed membrane is placed in a large container of low-ionic-strength buffer. The membrane pores are too small to allow diffusion of macromolecules of molecular weight greater than about 10,000. Smaller molecules diffuse freely through the openings (Figure 3.7). The passage of smaller molecules continues until their concentrations inside the dialysis tubing and outside in the large volume of buffer are equal. Thus, the concentration of small molecules inside the membrane is reduced. Equilibrium is volume dependent and is reached after 4 to 6 hours. Of course, if the outside solution (dialysate) is replaced with fresh buffer after equilibrium is reached, the concentration of small molecules inside the membrane will be further reduced by continued dialysis.

Dialysis membranes are available in a variety of materials and sizes. The most common materials are collodion, cellophane, and cellulose. Recent modifications in membrane construction make a range of pore sizes available. Spectrum Laboratories offers Spectra/Por membrane tubing with complete molecular weight cutoffs ranging from 100 to 300,000.

Figure 3.7

Diffusion of smaller molecules through dialysis membrane pores.

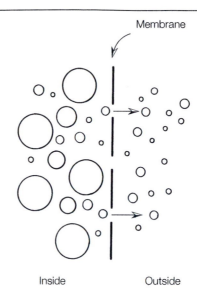

Membrane

Inside   Outside

Dialysis is most commonly used to remove salts and other small molecules from solutions of macromolecules. During the separation and purification of biomolecules, small molecules are added to selectively precipitate or dissolve the desired molecule. For example, proteins are often precipitated by addition of organic solvents or salts such as ammonium or sodium sulfate. Since the presence of organics or salts usually interferes with further purification and characterization of the molecule, they must be removed. Dialysis is a simple, inexpensive, and effective method for removing all small molecules, ionic or nonionic.

Dialysis is also useful for removing small ions and molecules that are weakly bound to biomolecules. Protein cofactors such as NAD, FAD, and metal ions can often be dissociated by dialysis. The removal of metal ions is facilitated by the addition of a chelating agent (EDTA) to the dialysate.

Dialysis of small-volume samples (less than 3 mL) is often tedious, inconvenient, and inefficient. Commercial vials for dialysis are now available for sample sizes from 0.05 to 3.0 mL, with molecular weight cutoffs ranging from 3500 to 14,000 daltons. For example, GeBAflex tubes, offered by Gene Bio-Applications (www.geba.org), may be used for dialysis (the removal of small molecules like urea, detergents, and ethidium bromide), buffer exchange, sample concentration, and electroelution of protein and nucleic acid samples (see Chapter 6).

Ultrafiltration

Although dialysis is still used occasionally as a purification tool, it has been largely replaced by ultrafiltration and gel filtration (see Chapter 5). The major disadvantage of dialysis that is overcome by the newer methods is that it may take several days of dialysis to attain a suitable separation. The other methods require 1 to 2 hours or less.

Ultrafiltration involves the separation of molecular species on the basis of size, shape, and/or charge. The solution to be separated is forced through a membrane by an external force. Membranes may be chosen for optimum flow rate, molecular specificity, and molecular weight cutoff. Two applications of membrane filtration are obvious: (1) desalting buffers or other solutions and (2) clarification of turbid solutions by removal of micron- or submicron-sized particles. Other applications are discussed below.

Ultrafiltration membranes have molecular weight cutoffs in the range of 100 to 1,000,000. They are usually composed of two layers: (1) a thin (0.1–0.5 μm), surface, semipermeable membrane made from a variety of materials including cellulose acetate, nylon, and polyvinylidene, and (2) a thicker, inert, support base (Figure 3.8). These filters function by retaining particles on the surfaces, not within the base matrix.

Membrane filters of these materials can be manufactured with a predetermined and accurately controlled pore size. Filters are available with a mean pore size ranging from 0.025 to 15 μm. These filters require suction, pressure, or centrifugal force for liquid flow. A typical flow rate for the commonly used 0.45-μm membrane is 57 mL min^{-1} cm^{-2} at 10 psi.

Ultrafiltration devices are available for macroseparations (up to 50 L) or for microseparations (milli- to microliters). For solutions larger than a few milliliters, gas-pressurized cells or suction-filter devices are used. For concentration and purification of samples in the milli- to microliter range, disposable filters are available. These devices, often called microconcentrators, offer the user simplicity, time saving, and high recovery. The sample is placed in a

Figure 3.8

Electron micrograph of an ultrafiltration membrane showing the two layers. Particles greater than 0.1 μm in diameter are retained on the surface or within pores. *Courtesy of the Millipore Corporation; www.millipore.com/*

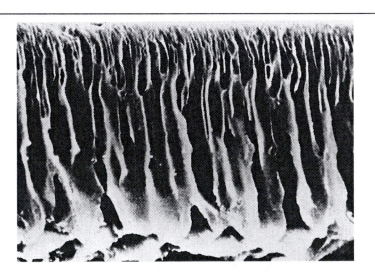

Figure 3.9

Use of a centrifuge
microfilter. *Courtesy of the
Millipore Corporation;
www.millipore.com/*

1. Pipet up to 2 ml of sample into the top reservoir.

2. Cover with the cone-shaped cap and spin in a centrifuge
   with a fixed-angle rotor.

3. Detach and invert top half.

4. Centrifuge 1 to 2 minutes to spin the concentrate into the
   cone-shaped cap.

5. Concentrate is now completely accessible for further
   analysis. Both concentrate and filtrate can be stored in collection cups.

reservoir above the membrane and centrifuged. The time and centrifugal
force required depend on the membrane, with spin times varying from
30 minutes to 2 hours and forces from $1000 \times g$ to $7500 \times g$. Figure 3.9 outlines the use of a centrifuge microfilter.

The principles behind ultrafiltration are sometimes misunderstood. The nomenclature implies that separations are the result of physical trapping of the particles and molecules by the filter. With polycarbonate and fiberglass filters, separations are made primarily on the basis of physical size. Other filters (cellulose nitrate, polyvinylidene fluoride, and to a lesser extent cellulose acetate) trap particles that cannot pass through the pores, but also retain macromolecules by adsorption. In particular, these materials have protein and nucleic acid binding properties. Each type of membrane displays a different affinity for various molecules. For protein, the relative binding affinity is polyvinylidene fluoride > cellulose nitrate > cellulose acetate. We can expect to see many applications of the "affinity membranes" in the future as the various membrane surface chemistries are altered and made more specific. Some applications are described in the following pages.

Clarification of Solutions

Because of low solubility and denaturation, solutions of biomolecules or cellular extracts are often turbid. This is a particular disadvantage if spectrophotometric analysis is desired. The transmittance of turbid solutions can be greatly increased by passage through a membrane filter system.

This simple technique may also be applied to the sterilization of nonautoclavable materials such as protein and nucleic acid solutions or heat-labile reagents. Bacterial contamination can be removed from these solutions by passing them through filter systems that have been sterilized by autoclaving.

Collection of Precipitates for Analysis

The collection of small amounts of very fine precipitates is the basis for many chemical and biochemical analytical procedures. Membrane filtration is an ideal method for sample collection. This is of great advantage in the collection of radioactive precipitates. Cellulose nitrate and fiberglass filters are often used to collect radioactive samples because they can be analyzed by direct suspension in an appropriate scintillation cocktail.

Harvesting of Bacterial Cells from Fermentation Broths

The collection of bacterial cells from nutrient broths is typically done by batch centrifugation. This time-consuming operation can be replaced by membrane filtration. Filtration is faster than centrifugation, and it allows extensive cell washing.

Concentration of Biomolecule Solutions

Protein or nucleic acid solutions obtained from extraction or various purification steps are often too dilute for further investigation. Since they cannot be concentrated by high-temperature evaporation of solvent, gentler methods have been developed. One of the most effective is the use of ultrafiltration pressure cells, as shown in Figure 3.10. A membrane filter is placed in

Figure 3.10

Schematic of an ultrafiltration cell. *Courtesy of Millipore/Amicon Corporation; www.millipore.com/*

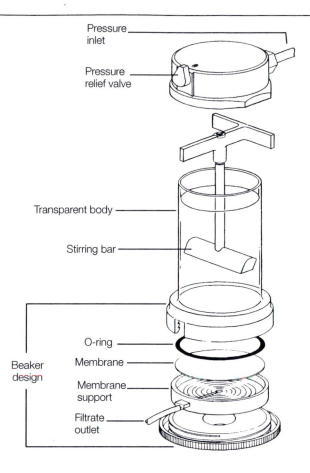

the bottom and the solution is poured into the cell. High pressure, exerted by compressed nitrogen (air could cause oxidation and denaturation of bio-molecules), forces the flow of small molecules, including solvent, through the filter. Membranes are available in a number of sizes, allowing a large variety of molecular weight cutoffs. Larger molecules that cannot pass through the pores are concentrated in the sample chamber. This method of concentration is rapid and gentle and can be performed at cold temperatures to ensure minimal inactivation of the molecules. One major disadvantage is clogging of the pores, which reduces the flow rate through the filter; this is lessened by constant but gentle stirring of the solution.

Lyophilization and Centrifugal Vacuum Concentration

Although ultrafiltration is being used more and more for the concentration of biological solutions, the older technique of **lyophilization** (freeze-drying) is still used. There are some situations (storing or transporting biological materials) in which lyophilization is preferred. Lyophilization is a drying technique that uses the process of sublimation to change a solvent (water) in the frozen state directly to the vapor state under vacuum. The product after lyophilization is a fluffy matrix that may be reconstituted by the addition of liquid. This is one of the most effective methods for drying or concentrating heat-sensitive materials. In practice, a biological solution to be concentrated is "shell-frozen" on the walls of a round-bottom or freeze-drying flask. Freezing of the solution is accomplished by placing the flask (half full with sample) in a dry ice–acetone bath and slowly rotating it as it is held at a 45° angle. The aqueous solution freezes in layers on the wall of the flask. This provides a large surface area for evaporation of water. The flask is then connected to the lyophilizer, which consists of a refrigeration unit and a vacuum pump (see Figure 3.11). The combined unit maintains the sample at −40°C for stability of the biological materials and applies a vacuum of approximately 5 to 25 mm Hg on the sample. Ice formed from the aqueous solution sublimes and is pumped from the sample vial. In fact, all materials that are volatile under these conditions (−40°C, 5 to 25 mm Hg) will be removed, and nonvolatile materials (proteins, buffer salts, nucleic acids, etc.) will be concentrated into a light, sometimes fluffy precipitate. Most freeze-dried biological materials are stable for long periods of time and some remain viable for many years.

As with any laboratory method, there are precautions and limitations of lyophilization that must be understood. Only aqueous solutions should be lyophilized. Organic solvents lower the melting point of aqueous solutions and increase the chances that the sample will melt and become denatured during

Figure 3.11

A typical freeze dryer.
*Courtesy of Thermo
Electron Corporation;
www.thermoelectron.com/*

The SpeedVac centrifugal
vacuum concentrator. *Courtesy
of Thermo Electron Corporation;
www.thermoelectron.com/*

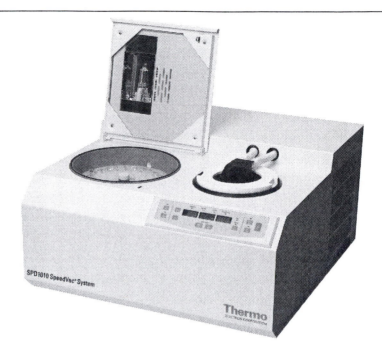

freeze-drying. There is also the possibility that organic vapors will pass through
the cold trap into the vacuum pump, where they may cause damage.

A new and increasingly popular technique for sample concentration
and drying is **centrifugal vacuum concentration.** The method may be
used to dry a wide variety of biological samples. The process starts with a
sample dissolved in a solvent (water or organic). The solvent is evaporated under vacuum in a centrifuge, thus producing a pellet in the bottom of
the container. The most widely used instrument is the SpeedVac, available
through Savant (Figure 3.12). This method is better than freeze-drying
because it is faster, it does not require a prefreezing step, it provides
100% sample recovery, and it may be used with solvents other than
water. The SpeedVac is usually used for relatively small volumes—1–2-mL
samples.

E.   RADIOISOTOPES IN BIOCHEMISTRY

The use of radioactive isotopes in experimental biochemistry has provided us with a wealth of information about biological processes. In the earlier days of biochemistry, radioisotopes were primarily used to elucidate
metabolic pathways. In modern biochemistry, there are few experimental
techniques that offer such a diverse range of applications as that provided
by radioactive isotopes. The measurement of radioactivity is a tool that

has the potential for use in all areas of experimental biochemistry, including enzyme assays, biochemical pathways of synthesis and degradation, analysis of biomolecules, measurement of antibodies, binding and transport studies, clinical biochemistry, and many others. Radioisotope applications have two major advantages over other analytical methods: (1) sensitive instrumentation allows the detection of minute quantities of radioactive material (some radioactive substances can be measured in the picomole, 10^{-12}, range), and (2) radioisotope techniques offer the ability to differentiate physically between substances that are chemically indistinguishable. The major disadvantages of radioisotope use are, of course, the safety and environmental concerns that must be considered when working with radioactive materials. These concerns have greatly slowed the development of new procedures using radioisotopes, as some states require that all who work with radioactive materials, including students, must have completed an intensive training course. Because of the potential hazards and problems associated with using and disposing radioactive waste materials, radioisotopes are generally not used in undergraduate biochemistry teaching labs. They are used in research, in industrial biotechnology applications, and in medicine only when there are no other techniques that can be applied to the problem and then are handled only by experienced personnel.

Origin and Properties of Radioactivity

Introduction

Radioactivity results from the spontaneous nuclear disintegration of unstable isotopes. The hydrogen nucleus, consisting of a proton, is represented as ^1_1H. Two additional forms of the hydrogen nucleus contain one and two neutrons; they are represented by ^2_1H and ^3_1H. These **isotopes** of hydrogen are commonly called deuterium and tritium, respectively. All isotopes of hydrogen have an identical number of protons (constant charge, $+1$) but they differ in the number of neutrons. Only tritium is radioactive.

The stability of a nucleus depends on the ratio of neutrons to protons. Some nuclei are unstable and undergo spontaneous nuclear disintegration accompanied by emission of particles. Unstable isotopes of this type are called **radioisotopes.** Three main types of radiation are emitted during nuclear decay: α particles, β particles, and γ rays. The α particle, a helium nucleus, is emitted only by elements of mass number greater than 140. These elements are seldom used in biochemical research.

Most of the commonly used radioisotopes in biochemistry are β emitters. The β particles exist in two forms (β^+, positrons, and β^-, electrons) and are emitted from a given radioisotope with a continuous range of energies. However, an average energy (called the *mean* energy) of β particles from an element can be determined. The mean energy is a characteristic of that isotope and can be used to identify one β emitter in the presence of a second

Figure 3.13

Energy spectra for the β emitters ^{3}H and ^{14}C. The dashed lines indicate the upper and lower limits for discrimination counting.

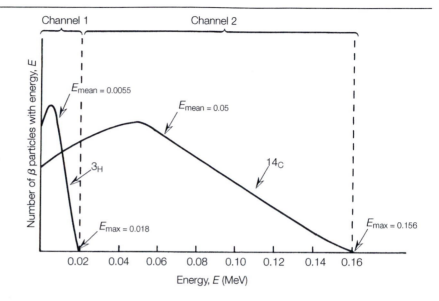

(see Figure 3.13). Equations 3.10 and 3.11 illustrate the disintegration of two β emitters. Here $\overline{v}_e$ represents the antineutrino and v_e the neutrino.

$$^{32}_{15}\text{P} \longrightarrow \, ^{32}_{16}\text{S} + \beta^- + \overline{v}_e \qquad \text{Equation 3.10}$$

$$^{65}_{30}\text{Zn} \longrightarrow \, ^{65}_{29}\text{Cu} + \beta^+ + v_e \qquad \text{Equation 3.11}$$

A few radioisotopes of biochemical significance are γ emitters. Emission of a γ ray (a photon of electromagnetic radiation) is often a secondary process occurring after the initial decay by β emission. The disintegration of the isotope ^{131}I is an example of this multistep process.

$$^{131}_{53}\text{I} \longrightarrow \beta^- + \, ^{131}_{54}\text{Xe} \longrightarrow \, ^{131}_{54}\text{Xe} + \gamma \qquad \text{Equation 3.12}$$

Each radioisotope emits γ rays of a distinct energy, which can be measured for identification of the isotope.

The spontaneous disintegration of a nucleus is a first-order kinetic process. That is, the rate of radioactive decay of N atoms ($-dN/dt$, the change of N with time, t) is proportional to the number of radioactive atoms present (Equation 3.13).

$$-\frac{dN}{dt} = \lambda N \qquad \text{Equation 3.13}$$

The proportionality constant, λ, is called the **disintegration** or **decay constant**. Equation 3.13 can be transformed to a more useful equation by integration within the limits of $t = 0$ and t. The result is shown in Equation 3.14.

>>
$$N = N_0 e^{-\lambda t}$$
Equation 3.14

where

N_0 = the number of radioactive atoms present at $t = 0$

N = the number of radioactive atoms present at time $= t$

Equation 3.14 can be expressed in the natural logarithmic form as in Equation 3.15.

>>
$$\ln \frac{N}{N_0} = -\lambda t$$
Equation 3.15

Combining Equations 3.14 and 3.15 leads to a relationship that defines the **half-life**. Half-life, $t_{1/2}$, is a term used to describe the time necessary for one-half of the radioactive atoms initially present in a sample to decay. At the end of the first half-life, N in Equation 3.15 becomes $\frac{1}{2}N_0$, and the result is shown in Equations 3.16 and 3.17.

>>
$$\ln \frac{\frac{1}{2}N_0}{N_0} = -\lambda t_{1/2}$$
Equation 3.16

$$\ln \frac{1}{2} = -\lambda t_{1/2}$$

>>
$$t_{1/2} = \frac{0.693}{\lambda}$$
Equation 3.17

Equations 3.15 and 3.17 allow calculation of the ratio N/N_0 at any time during an experiment. This calculation is especially critical when radioisotopes with short half-lives are used.

Isotopes in Biochemistry

The properties of several radioisotopes that are important in biochemical research are listed in Table 3.5. note that many of the isotopes are β emitters; however, a few are γ emitters.

The half-life, defined in the previous section and listed for each isotope in Table 3.5, is an important property to consider when one is designing experiments using radioisotopes. Using an isotope with a short half-life (for example, ^{24}Na with $t_{1/2} = 15$ hr) is difficult because the radioactivity lost during the course of the experiment is significant. Quantitative measurements made before and after the experiment must be corrected for this loss of activity. Radioactive phosphorus, ^{32}P, an isotope of significant value in biochemical research, has a relatively short half-life (14 days), so if quantitative measurements are made they must be corrected as described in Equations 3.16 and 3.17.

Table 3.5			
Properties of Biochemically Important Radioisotopes			
Isotope	Particle Emitted	Energy of Particle (MeV)	Half-Life, $t_{1/2}$
^{3}H	β^-	0.018	12.3 yr
^{14}C	β^-	0.155	5570 yr
^{22}Na	β^+	0.55	2.6 yr
	γ	1.28	
^{24}Na	β^-	1.39	15 hr
	γ	1.7, 2.75	
^{32}P	β^-	1.71	14.2 days
^{35}S	β^-	0.167	87.1 days
^{36}Cl	β^-	0.714	3×10^5 days
^{40}K	β^-	1.33	1.3×10^9 yr
^{45}Ca	β^-	0.25	165 days
^{59}Fe	β^-	0.46	45 days
	γ	1.1	
^{60}Co	β^-	0.318	5.3 yr
	γ	1.3	
^{65}Zn	β^+	0.33	245 days
	γ	1.14	
^{90}Sr	β^-	0.54	29 yr
^{125}I	γ	0.035	60 days
	γ	0.027	
^{131}I	β^-	0.61, 0.33	8.1 days
	γ	0.64, 0.36, 0.28	
^{137}Cs	β^-	0.51	30 yr
	γ	0.66	
^{226}Ra	α	4.78	1600 yr
	γ	0.19	

Mention should also be made of short-lived isotopes that are important in biotechnology and medical biochemistry. The isotopes ^{11}C, ^{13}N, ^{15}O, and ^{18}F, which are positron emitters, are crucial for use in positron emission tomography (PET).

Units of Radioactivity

The basic unit of radioactivity is the **curie,** Ci, named in honor of Marie and Pierre Curie. One curie is the amount of radioactive material that emits particles at a rate of 3.7×10^{10} disintegrations per second (dps), or 2.2×10^{12} min^{-1} (dpm). Amounts that large are seldom used in experimentation, so subdivisions are convenient. The **millicurie** (mCi, 2.2×10^9 min^{-1}) and **microcurie** (μCi, 2.2×10^6 min^{-1}) are standard units for radioactive measurements (see Table 3.6). The radioactivity unit of the meter-kilogram-seconds (MKS) system is the **becquerel** (Bq). A becquerel, named in honor of Antoine Becquerel, who studied uranium radiation, represents one disintegration per second. The two systems of measurement are related by the definition 1 curie = 3.70×10^{10} becquerels. Since the becquerel is such a small unit, radioactive units are sometimes reported in MBq

Table 3.6		
Units of Radioactivity		
Unit's Name	Multiplication Factor (relative to curie)	Activity (dps)
Curie (Ci)	1.0	3.70×10^{10}
Millicurie (mCi)	10^{-3}	3.70×10^{7}
Microcurie (μCi)	10^{-6}	3.70×10^{4}
Nanocurie (nCi)	10^{-9}	3.70×10
Becquerel (Bq)	2.7×10^{-11}	1.0
Mega becquerel (MBq)	2.7×10^{-5}	1.0×10^{6}

(mega, 10^6) or TBq (tera, 10^{12}). Both unit systems are in common use today, and radioisotopes received through commercial sources are labeled in curies and becquerels.

The number of disintegrations emitted by a radioactive sample depends on the purity of the sample (number of radioactive atoms present) and the decay constant, λ. Therefore, radioactive decay is also expressed in terms of **specific activity,** the disintegration rate per unit mass of radioactive atoms. Typical units for specific activity are mCi/mmole and μCi/μmole.

Although radioactivity is defined in terms of nuclear disintegrations per unit of time, rarely does one measure this absolute number in the laboratory. Instruments that detect and count emitted particles respond to only a small fraction of the particles. Data from a radiation counter are in counts per minute (cpm), which can be converted to actual disintegrations per minute if the counting efficiency of the instrument is known. The percent efficiency of an instrument is determined by counting a standard compound of known radioactivity and using the ratio of detected activity (observed cpm) to actual activity (disintegrations per minute). Equation 3.18 illustrates the calculation using a standard of 1 μCi.

>>
$$\% \text{ efficiency} = \frac{\text{observed cpm of standard}}{\text{dpm of } 1\mu\text{Ci of standard}} \times 100$$

$$= \frac{\text{observed cpm}}{2.2 \times 10^6 \text{ dpm}} \times 100$$

Equation 3.18

Detection and Measurement of Radioactivity

Since most of the radioisotopes used in biochemical research are β emitters, only methods that detect and measure β particles will be emphasized.

Liquid Scintillation Counting

Samples for liquid scintillation counting consist of three components: (1) the radioactive material; (2) a solvent, usually aromatic, in which the radioactive substance is dissolved or suspended; and (3) one or more organic fluorescent

Figure 3.14

Diagram of a typical scintillation counter, showing coincidence circuitry.

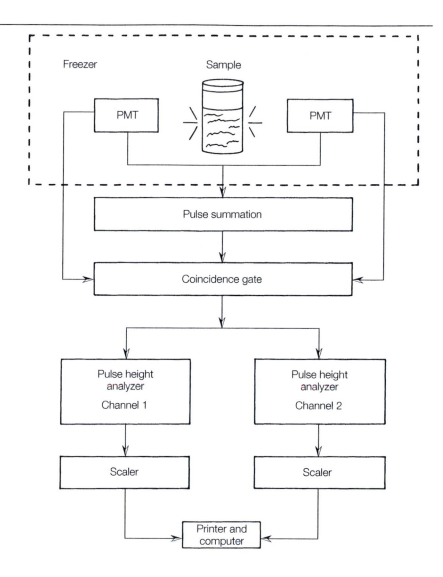

substances. Components 2 and 3 make up the **liquid scintillation system** or cocktail. β particles emitted from the radioactive sample interact with the scintillation system, producing small flashes of light or scintillations. The light flashes are detected by a photomultiplier tube (PMT). Electronic pulses from the PMT are amplified and registered by a counting device called a scaler. A schematic diagram of a typical scintillation counter is shown in Figure 3.14.

The scintillation process, in detail, begins with the collision of emitted β particles with solvent molecules, S (Equation 3.19).

>> $\beta + S \longrightarrow S^* + \beta$ (less energetic)   *Equation 3.19*

Contact between the energetic β particles and S in the ground state results in transfer of energy and conversion of an S molecule into an excited state, S*. Aromatic solvents are most often used because their electrons are easily promoted to an excited state orbital (see discussion of fluorescence, Chapter 6). The β particle after one collision still has sufficient energy to excite several more solvent molecules. The excited solvent molecules normally return to the ground state by emission of a photon, $S^* \rightarrow S + h\upsilon$. Photons from the typical aromatic solvent are of short wavelength and are not efficiently detected by photocells. A convenient way to resolve this technical problem is to add one or more fluorescent substances (fluors) to the scintillation mixture. Excited solvent molecules interact with a **primary fluor,** F_1, as shown in Equation 3.20.

>> $$S^* + F_1 \longrightarrow S + F_1^*$$ *Equation 3.20*

Energy is transferred from S* to F_1, resulting in ground state S molecules and excited F_1 molecules, F_1^*. F_1^* molecules are fluorescent and emit light of a longer wavelength than S* (Equation 3.21).

>> $$F_1^* \longrightarrow F_1 + h\upsilon_1$$ *Equation 3.21*

If the light emitted during the decay of F_1^* is still of a wavelength too short for efficient measurement by a PMT, a **secondary fluor,** F_2, that accepts energy from F_1^* may be added to the scintillation system. Equations 3.22 and 3.23 outline the continued energy transfer process and fluorescence of F_2^*.

>> $$F_1^* + F_2 \longrightarrow F_1 + F_2^*$$ *Equation 3.22*

>> $$F_2^* \longrightarrow F_2 + h\upsilon_2$$ *Equation 3.23*

The light, $h\upsilon_2$, from F_2^* is of longer wavelength than $h\upsilon_1$ from F_1^* and is more efficiently detected by a PMT. Two widely used primary and secondary fluors are 2,5-diphenyloxazole (PPO) with an emission maximum of 380 nm and 1,4-bis-2-(5-phenyloxazolyl)benzene (POPOP) with an emission maximum of 420 nm. A more efficient but more expensive fluor is 2-(4'-*t*-butylphenyl)-5-(4''-biphenyl)-1,3,4-oxadiazole (Butyl-PBD).

The most basic elements in a liquid scintillation counter are the PMT, a pulse amplifier, and a counter, called a **scaler.** This simple assembly may be used for counting; however, there are many problems and disadvantages with this setup. Many of the difficulties can be alleviated by more sophisticated instrumental features. Some of the problems and practical solutions are outlined below.

Thermal Noise in Photomultiplier Tubes

The energies of the β particles from most β emitters are very low. This, of course, leads to low-energy photons emitted from the fluors and relatively low-energy electrical pulses in the PMT. In addition, photomultiplier tubes produce thermal background noise with 25 to 30% of the energy associated with the fluorescence-emitted photons. This difficulty cannot be completely eliminated, but its effects can be lessened by placing the samples and the PMT in a freezer at -5 to $-8°C$ in order to decrease thermal noise.

A second way to help resolve the thermal noise problem is to use two photomultiplier tubes for detection of scintillations. Each flash of light that is detected by the photomultiplier tubes is fed into a **coincidence circuit.** A coincidence circuit counts only the flashes that arrive simultaneously at the two photodetectors. Electrical pulses that are the result of simultaneous random emission (thermal noise) in the individual tubes are very unlikely. A schematic diagram of a typical scintillation counter with coincidence circuitry is shown in Figure 3.14.

Counting More Than One Isotope in a Sample

The basic liquid scintillation counter with coincidence circuitry can be used only to count samples containing one type of isotope. Many experiments in biochemistry require the counting of just one isotope; however, more valuable experiments can be performed if two radioisotopes can be simultaneously counted in a single sample (double-labeling experiments). The basic scintillation counter just described has no means of discriminating between electrical pulses of different energies.

The size of the current generated in a photocell is nearly proportional to the energy of the β particle initiating the pulse. Recall that β particles from different isotopes have characteristic energy spectra with an average energy (see Figure 3.13). Modern scintillation counters are equipped with **pulse height analyzers** that measure the size of the electrical pulse and count only the pulses within preselected energy limits set by **discriminators.** The circuitry required for pulse height analysis and energy discrimination of β particles is shown in Figure 3.14. Discriminators are electronic "windows" that can be adjusted to count β particles within certain energy or voltage ranges called **channels.** The channels are set to a **lower limit** and an **upper limit,** and all voltages within those limits are counted. Figure 3.13 illustrates the function of discriminators for the counting of ^3H and ^{14}C in a single sample. Discriminator channel 1 is adjusted to accept typical β particles emitted from ^3H and channel 2 is adjusted to receive β particles of the energy characteristic of ^{14}C.

Quenching

Any chemical agent or experimental condition that reduces the efficiency of the scintillation and detection process leads to a reduced level of counting, or **quenching.** Quenching may be caused by a decrease in the amount of fluorescence from the primary and secondary scintillators (fluors) or a decrease in light activating the PMT. There are four common origins of quenching.

Color quenching is a problem if chemical substances that absorb photons from the secondary fluors are present in the scintillation mixture. Since the secondary fluors emit light in the visible region between 410 and 420 nm, colored substances may absorb the emitted light before it is detected by the photocells. Radioactive samples may be treated to remove colored impurities before mixing with the scintillation solvent.

Chemical quenching occurs when chemical substances in the scintillation solution interact with excited solvent and fluor molecules and decrease the efficiency of the scintillation process. To avoid this type of quenching the sample can be purified or the fluors can be increased in concentration. Modern scintillation counters have computer programs to correct for color and chemical quenching.

Point quenching occurs if the radioactive sample is not completely dissolved in the solvent. The emitted β particles may be absorbed before they have a chance to interact with solvent molecules. The addition of solubilizing agents such as Cab-O-Sil or Thixin decreases point quenching by converting the liquid scintillator to a gel.

Dilution quenching results when a large volume of liquid radioactive sample is added to the scintillation solution. In most cases, this type of quenching cannot be eliminated, but it can be corrected by one of the techniques discussed below.

Since quenching can occur during all experimental counting of radioisotopes, it is important to be able to determine the extent of the reduced counting efficiency. Two methods for quench correction are in common use.

In the **internal standard ratio** method, the sample, X, under study is first counted; then a known amount of a standard radioactive solution is added to the sample, and it is counted again. The absolute activity of the original sample A_X is represented by Equation 3.24.

>>
$$A_X = A_s \frac{C_X}{C_s}$$
Equation 3.24

where

A_X = activity of the sample, X
C_X = radioactive counts from sample
A_s = activity of the standard
C_s = radioactive counts from standard

The absolute value of C_s is determined from Equation 3.25.

>>
$$C_T = C_X + C_s$$
Equation 3.25

where

C_T = total radioactive counts from sample plus standard

Equation 3.24 can then be modified to Equation 3.26.

>> $$A_X = \frac{A_s C_X}{C_T - C_X}$$ *Equation 3.26*

The internal standard ratio method for quench correction is tedious and time-consuming and it destroys the sample, so it is not an ideal method. Scintillation counters are equipped with a standard radiation source inside the instrument, but outside the scintillation solution. The radiation source, usually a gamma emitter, is mechanically moved into a position next to the vial containing the sample, and the combined system of standard and sample is counted. Gamma rays from the standard excite solvent molecules in the sample, and the scintillation process occurs as previously described. However, the instrument is adjusted to register only scintillations due to γ particle collisions with solvent molecules. This method for quench correction, called the **external standard method,** is fast and precise.

Scintillation Cocktails and Sample Preparation

Many of the quenching problems discussed earlier can be lessened if an experiment is carefully planned. Many sample preparation methods and scintillation liquids are available for use. Only a few of the more common techniques will be mentioned here. Most liquid or solid radioactive samples can be counted by mixing with a **scintillation cocktail,** a mixture of solvent and fluor(s). The radioactive sample and scintillation solution are placed in a glass, polyethylene, polypropylene, polyester, or polycarbonate vial for counting. Traditional counting vials include a 20-mL-standard vial, a 6-mL miniature vial, a 1-mL Eppendorf tube, or a 200-μL microfuge tube. If glass is used, it must contain a low potassium content because naturally occurring ^{40}K is a β emitter. Two major types of solvent systems are available: (1) those that are immiscible with water, in which only organic samples may be used, and (2) those that dissolve aqueous samples. During the early years of liquid scintillation counting the most common solvents used were toluene and xylene for organic samples and dioxane for aqueous samples. These solvents provided high counting efficiency. However, on the downside, they were flammable (flash points between 4 and 25°C), highly toxic, and presented major disposal costs and problems. In addition, they penetrated plastic counting vials, releasing vapors and liquids into the laboratory. A new environmental awareness has prompted the development of biosafe liquid scintillation cocktails. The use of highly alkylated aromatic solvents has produced cocktails that are fully biodegradable (converted to CO_2 and H_2O), have higher flash points (above 120°C) and offer high counting efficiency. The newer solvents may be disposed of after use by incineration, which greatly reduces costs compared to road transport to a disposal site. Occupational risks are also lessened because the solutions are defined by the Environmental Protection Agency as nontoxic.

Solid radioactive samples or those that are insoluble in either type of solvent may be quantified by collection onto small pieces of a solid support (filter paper or cellulose membrane) and added directly to the cocktail for counting. The efficiency of counting these samples depends on the support but is usually less than that of counting a homogeneous sample.

Applications of Radioisotopes

The applications of radioisotopes in biochemical measurements are too numerous to outline here. Excellent reference books and Web sites, some of which are listed at the end of this chapter, provide experimental procedures for radioisotope utilization. The technique of **autoradiography,** however, should be mentioned. This procedure allows the detection and localization of a radioactive substance (molecule or atom) in a tissue, cell, or cell organelle. Briefly, the technique involves placing a radioactive material directly onto or close to a photographic emulsion. Radioisotopes in the material emit radiation that impinges on the photographic plate, activating silver halide crystals in the emulsion. Upon development of the plate, a pattern or image is displayed that yields information about the location and amount of radioactive material in the sample. A densitometer may be used to scan the autoradiograph and quantify the amount of radioactivity in each region. One of the most common uses of autoradiography in biochemistry and molecular biology is in the detection and quantification of ^{32}P-labeled nucleic acids on polyacrylamide or agarose gels after electrophoresis (Figure 3.15). Autoradiography has also been useful in concentration and localization studies of biomolecules in cells and cell organelles.

Statistical Analysis of Radioactivity Measurements

Radioactive decay is a random process. It is impossible to predict when a radioactive event will occur. Therefore, counting measurements provide

---

Figure 3.15

Autoradiograph showing the transfer of ^{32}P from γ-^{32}P-labeled ATP to the 5′ end of a polynucleotide, (dT)$_8$, catalyzed by polynucleotide kinase. Reagents were incubated for 30 min and electrophoresed on a 20% polyacrylamide, 7 M urea gel and autoradiographed. Note that as more units of kinase are added, more ^{32}P-labeled polynucleotide is made. The dark regions indicate the presence of ^{32}P. Source: *Copyright ©1992 New England Biolabs Catalog. Reprinted with permission.*

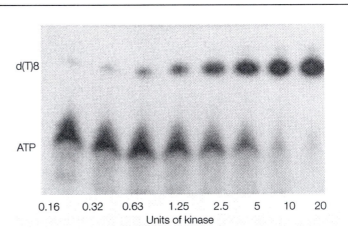

only average rates of decay. However, counting measurements may be treated by Gaussian distribution analysis to determine an average counting rate, standard deviation, percent confidence level, and other statistical parameters. An introduction to statistical analysis of radioactivity counting data and other experimental measurements is given in Chapter 1.

Radioisotopes and Safety

Safety should be of major concern in performing any chemistry experiment, but when radioisotopes are involved, special precautions must be taken. Many chemistry and biochemistry departments have specially equipped laboratories for radioisotope work. Alternatively, your instructor should set aside a specific area of your laboratory for using radioactive materials. In either situation, specific guidelines should be mandated and enforced by all institutions.

Study Problems

1. You need to prepare a buffer for biochemistry lab. The required solution is 0.5 M sodium phosphate, pH 7.0. Use the Henderson-Hasselbalch equation to calculate the number of moles and grams of monobasic sodium phosphate (NaH_2PO_4) and dibasic sodium phosphate (Na_2HPO_4) necessary to make 1 liter of the solution.

2. Design a "shortcut" method for preparing the phosphate buffer in Study Problem 1. Hint: You need only NaH_2PO_4, a solution of NaOH, and a pH meter.

3. Describe how you would prepare a 0.1 M glycine buffer of pH 10.0. You have available isoelectric glycine and sodium glycinate.

4. Describe how you would prepare a 0.20 M Tris-HCl buffer of pH 8.0. The only Tris reagent you have available is Tris base. What other reagent do you need and how would you use it to prepare the solution?

5. Below is a table prepared by a biochemistry student to construct a standard curve for protein analysis. The Bradford assay was used with bovine serum albumin (BSA, 0.1 mg/mL) as standard protein. Complete the table by filling in the weight of BSA in each tube and the approximate A_{595} that will be obtained for each tube. Assume the procedure was conducted correctly.

	Tube No.					
Reagents	1	2	3	4	5	6
H_2O (mL)	1.0	0.9	0.8	0.6	0.2	—
BSA volume (mL)	—	0.1	0.2	0.4	0.8	1.0
BSA weight (μg)	0					
Bradford reagent (mL)	5.0					→
A_{595}	0.00	0.08				

> 6. Assume that you use the standard curve produced in Study Problem 5 to measure the concentration of an unknown protein. The A_{595} for 1.0 mL of the unknown was 0.52. Prepare a standard curve from the data in Problem 5 and estimate the concentration of unknown protein in the sample in $\mu g/mL$.

> 7. A solution of purified DNA gave in the spectrophotometric assay an A_{260} of 0.35 when measured in a 1-cm quartz cuvette. What is the concentration of the DNA in $\mu g/mL$?

8. Compare the techniques of lyophilization and centrifugal vacuum concentration. Give advantages and disadvantages of each.

> 9. What amino acid residues are detected when the spectrophotometric assay is used to quantify proteins? Are those amino acids present in the same quantity in all proteins? Explain how this may affect measurement of proteins by this method.

> 10. What biomolecules would interfere with the measurement of nucleic acids using the spectrophotometric (A_{260}) assay?

> 11. Select a buffer system that can be used at each of the following pH values.
> (a) pH 3.5
> (b) pH 6.0
> (c) pH 9.5

> 12. Prepare a linear dilution table similar to Table 3.2. Assume that the stock solution is 2.0 M and you require 20-mL diluted samples of final concentrations: 1.66 M, 1.33 M, 1.0 M, 0.66 M, 0.33 M, and 0.00 M.

> 13. The following serial dilution is carried out in the lab:

1.0 mL of Solution A is diluted to a final volume of 10 mL (Solution B).

1.0 mL of Solution B is diluted to a final volume of 10 mL (Solution C).

1.0 mL of Solution C is diluted to a final volume of 10 mL (Solution D).

Assume that Solution A has a concentration of 2.0 M. Calculate the concentrations of Solutions B, C, and D.

14. You need to prepare a buffer solution at each of the following concentrations:

1 M, 0.1 M, 0.01 M, and 0.001 M

Describe how you would use a single stock solution of the buffer (5.0 M) to prepare 10 mL of each diluted buffer solution.

15. Describe how you would prepare 1 liter of the following buffer: 0.025 M Tris-chloride, pH 8.0, containing 0.05 M glucose, and 0.01 M EDTA (MW EDTA disodium salt, dihydrate = 372.2).

> 16. Calculate the half-life of an isotope from the following experimental measurements. At time = 0, the activity is 300 disintegrations

per minute. After 1 hour, the activity is 200 disintegrations per minute.

➤ *17.* Calculate λ, the decay constant for ^{32}P.

➤ *18.* How long will it take a sample of ^{32}P that contains 65,000 disintegrations per minute to decay to 1500 disintegrations per minute?

➤ *19.* List five radioisotopes that may be measured using a liquid scintillation counter.

Further Reading

General Procedures

K. Stewart and R. Ebel, *Chemical Measurements in Biological Systems* (2000), Wiley-Interscience (New York).

J. Walker, Editor, *The Protein Protocols Handbook,* 2nd ed. (2002), Humana Press (Totowa, NJ).

K. Wilson and J. Walker, Editors, *Principles and Techniques of Practical Biochemistry,* 5th ed. (2000), Cambridge University Press (Cambridge).

Measurement of Proteins and Nucleic Acids

M. Bradford, *Anal. Biochem.* **72,** 248–254 (1976). "A Rapid and Sensitive Method for the Quantitation of Microgram Quantities of Protein Utilizing the Principles of Ligand-Dye Binding."

P. Hengen, *Trends Biochem. Sci.* **19,** 93–94 (1994). "Methods and Reagents: Determining DNA Concentrations and Rescuing PCR Primers."

P. Smith et al., *Anal. Biochem.* **150,** 76–85 (1985). "Measurement of Protein Using Bicinchoninic Acid."

J. Walker, Editor, *The Protein Protocols Handbook,* 2nd ed. (2002), Humana Press, Inc. (Totowa, NJ), pp. 3–21.

http://www.turnerdesigns.com/
    Measurement of DNA solution using bisbenzimidazole dye.

pH, Buffers, Electrodes, and Biosensors

K. Barker, *At the Bench: A Laboratory Navigator* (1998), Cold Spring Harbor Laboratory Press (Cold Spring Harbor, NY).

R. Boyer, *Concepts in Biochemistry,* 2nd ed. (2002), John Wiley & Sons (New York), pp. 48–64.

R. Boyer, *Modern Experimental Biochemistry,* 3rd ed. (2000), Benjamin Cummings (San Francisco), pp. 29–41.

R. Curtright et al., *Biochem. Mol. Biol. Educ.* **32,** 71–77 (2004). "Facilitating Student Understanding of Buffering by an Integration of Mathematics and Chemical Concepts."

R. Garrett and C. Grisham, *Biochemistry,* 3rd ed. (2005), Brooks/Cole (Belmont, CA), pp. 31–50.

N. Good and S. Izawa, in *Methods in Enzymology,* Vol. XXIV, A. San Pietro, Editor (1972), Academic Press (New York), pp. 53–68. "Hydrogen Ion Buffers for Photosynthesis Research."

D. E. Gueffroy, Editor, *Buffers—A Guide for the Preparation and Use of Buffers in Biological Systems* (1993), Calbiochem-Novabiochem Corp., P.O. Box 12087, La Jolla, CA 92039–2087.

D. Nelson and M. Cox, *Lehninger Principles of Biochemistry* (2005), Freeman (New York), pp. 47–74.

J. Risley, *J. Chem. Educ.* **68,** 1054 (1991). "Preparing Solutions in the Biochemistry Lab."

K. Stewart and R. Ebel, *Chemical Measurements in Biological Systems* (2000), Wiley-Interscience (New York), pp. 15–38.

D. Voet and J. Voet, *Biochemistry* (2004), John Wiley & Sons (Hoboken, NJ), pp. 39–50.

J. Wang and C. Macca, *J. Chem. Educ.* **73,** 797 (1996). "Use of Blood-Glucose Strips for Introducing Enzyme Electrodes and Modern Biosensors."

K. Wilson and J. Walker, Editors, *"Principles and Techniques of Practical Biochemistry,"* 5th ed. (2000), Cambridge University Press (Cambridge), pp. 37–42.

http://lab3d.chem.virginia.edu
   Virtual lab on amino acid titration.

http://www.ysi.com
   Information on the biological oxygen monitor (oxygen electrode).

Radioisotopes

G. Chase and J. Rabinowitz, *Principles of Radioisotope Methodology,* 3rd ed. (1967), Burgess (Minneapolis, MN), pp. 75–108.

G. Choppin, J. Liljenzin, and J. Rydberg, *Radiochemistry and Nuclear Chemistry,* 3rd ed. (2002), Butterworth-Heinemann (Burlington, MA).

R. Garrett and C. Grisham, *Biochemistry,* 3rd ed. (2005), Brooks/Cole (Belmont, CA), pp. 550–551.

K. Leiser, *Nuclear and Radiochemistry: Fundamentals and Applications,* 2nd ed. (2002), John Wiley & Sons (New York).

R. Slater, Editor, *Radioisotopes in Biology,* 2nd ed. (2002), Oxford University Press (Oxford).

D. Voet and J. Voet, *Biochemistry,* 3rd ed. (2004), John Wiley & Sons (Hoboken, NJ), pp. 562–565.

K. Wilson and J. Walker, Editors, *Principles and Techniques of Practical Biochemistry,* 5th ed. (2000), Cambridge University Press (Cambridge), pp. 687–728.

http://nucleus.wpi.edu/Reactor/labs.html
  Radioisotopes in the lab.

http://beckman.com
  Click on LS 6500 liquid scintillation counters.

CENTRIFUGATION TECHNIQUES IN BIOCHEMISTRY

A centrifuge of some kind is found in every biochemistry laboratory. Centrifuges have many applications, but they are used primarily for the preparation of biological samples and for the analysis of the physical properties of biomolecules, organelles, and cells. Centrifugation is carried out by spinning a biological sample at a high rate of speed, thus subjecting it to an intense force (artificial gravitational field). Most centrifuge techniques fit into one of two categories—**preparative centrifugation** or **analytical centrifugation.** A preparative procedure is one that can be applied to the separation or purification of biological samples (cells, organelles, macromolecules, etc.). Analytical procedures are used to measure physical characteristics of biological samples. For example, the purity, size, shape, and density of macromolecules may be defined by centrifugation.

In this chapter, we will first explore the principles and theory underlying centrifugation techniques in order to provide a fundamental background, and then turn to a discussion of the application of these technique to the isolation and characterization of biological molecules and cellular components.

A.   BASIC PRINCIPLES OF CENTRIFUGATION

A particle, whether it is a floating solid, a precipitate, a macromolecule, or a cell organelle, is subjected to a centrifugal force when it is rotated at a high rate of speed. The **centrifugal force,** F, is defined by Equation 4.1.

$$F = m\omega^2 r$$

Equation 4.1

where

> F = intensity of the centrifugal force
>
> m = effective mass of the sedimenting particle
>
> ω = angular velocity of rotation in rad/sec
>
> r = distance of the migrating particles from the central axis of rotation

The force on a sedimenting particle increases with the velocity of the rotation and the distance of the particle from the axis of rotation. A more common measurement of F, in terms of the earth's gravitation force, g, is **relative centrifugal force,** RCF, defined by Equation 4.2.

>> $$RCF = (1.119 \times 10^{-5})(rpm)^2(r)$$ *Equation 4.2*

This equation relates RCF to revolutions per minute of the sample. Equation 4.2 dictates that the RCF on a sample will vary with r, the distance of the sedimenting particles from the axis of rotation (see Figure 4.1). It is convenient to determine RCF by use of the nomogram in Figure 4.2. It should be clear from Figures 4.1 and 4.2 that, since RCF varies with r, it is important to define r for an experimental run. Often an average RCF is determined using a value for r midway between the top and bottom of the sample container. The RCF value is reported as "a number times gravity, g."

Figure 4.1

A diagram illustrating the variation of RCF with r, the distance of the sedimenting particles from the axis of rotation. *Courtesy of Beckman Coulter; www.beckman.com*

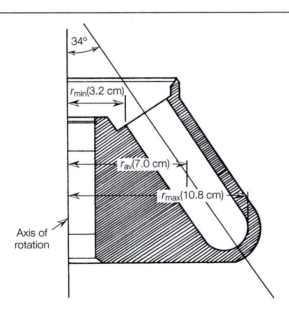

Figure 4.2

A nomogram for estimating RCF. To use, place a straight edge connecting the instrumental revolutions (right column) with the distance from the center of rotation (left column). The RCF is read where the straight edge crosses the middle column. *Courtesy of Beckman Coulter; www.beckman.com.*

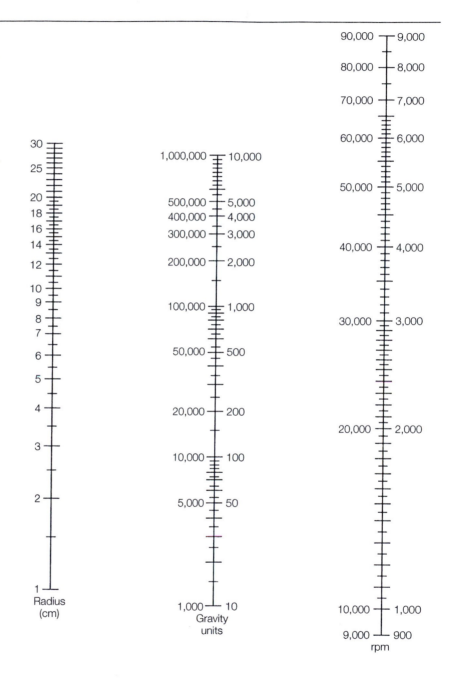

➤ | **Study Exercise 4.1**   To illustrate the use of the nomogram in Figure 4.2, consider a centrifuge tube with average r of 7 cm (see Figure 4.1), being rotated at 20,000 rpm. Use the nomogram to determine the relative centrifugal force (RCF) in gravity (g) units.

Solution: Follow the instructions in the caption for Figure 4.2 to estimate the RCF to be about 32,000 $\times$ g.

Although this introduction outlines the basic principles of centrifugation, it does not take into account other factors that influence the rate of particle sedimentation. Centrifuged particles migrate at a rate that depends on the mass, shape, and density of the particle and the density of the medium. The centrifugal force felt by the particle is defined by Equation 4.1. The term m is the **effective mass** of the particle, that is, the actual mass, m_0, minus a correction factor for the weight of water displaced (buoyancy factor) (Equation 4.3).

>> $\qquad$ $m = m_0 - m_0 \bar{v}\rho$ $\qquad\qquad$ **Equation 4.3**

where

$\qquad$ m = effective mass of the sedimenting particle

$\qquad$ m_0 = actual mass of the particle

$\qquad$ $\bar{v}$ = partial specific volume, the volume change occurring when a particle is placed in a large excess of solvent

$\qquad$ ρ = density of the solvent or medium

Equations 4.1 and 4.3 may be combined to describe the centrifugal force on a particle (Equation 4.4).

>> $\qquad$ $F = m_0(1 - \bar{v}\rho)\omega^2 r$ $\qquad\qquad$ **Equation 4.4**

As particles sediment under the influence of the centrifugal field, their movement is countered by a resistance force, the **frictional force.** The frictional force is defined by Equation 4.5.

>> $\qquad$ Frictional force = fv $\qquad\qquad$ **Equation 4.5**

where

$\qquad$ f = frictional coefficient

$\qquad$ v = velocity of the sedimenting particle (sedimentation velocity)

The **frictional coefficient,** f, depends on the size and shape of the particle, as well as the viscosity of the solvent. The frictional force increases with the velocity of the particle until a constant velocity is reached. At this point, the two forces are balanced (Equation 4.6).

>> $\qquad$ $m_0(1 - \bar{v}\rho)\omega^2 r = fv = f\left(\dfrac{dr}{dt}\right)$ $\qquad\qquad$ **Equation 4.6**

The rate of sedimentation, sometimes called **sedimentation velocity, v,** is defined by Equation 4.7.

>>

$$v = \frac{dr}{dt} = \frac{m_0(1 - \bar{v}\rho)\omega^2 r}{f}$$

Equation 4.7

It is cumbersome and sometimes impractical to express sedimentation velocity in terms of ρ, $\bar{v}$, and f, since these factors are difficult to measure. A new term, **sedimentation coefficient,** s (the ratio of sedimentation velocity to centrifugal force) is introduced by rearranging Equation 4.7 to Equation 4.8.

>>

$$s = \frac{v}{\omega^2 r} = \frac{m_0(1 - \bar{v}\rho)}{f}$$

Equation 4.8

The term s is most often defined under standard conditions, 20°C and water as the medium, and denoted by $s_{20, \text{w}}$. The s value is a physical characteristic used to classify biological macromolecules and cell organelles. Sedimentation coefficients are in the range 1×10^{-13} to $10,000 \times 10^{-13}$ second. For numerical convenience, sedimentation coefficients are expressed in Svedberg units, S, where $1\,\text{S} = 1 \times 10^{-13}$ second. Human hemoglobin has an s value of 4.5×10^{-13} second or 4.5 S. The value of S for several biomolecules, bacterial cells, and cell organelles is shown in Figure 4.3. Note in the figure that there appears to be a direct relationship between the S value and the molecular weight or particle size. This, however, is not always true, as in the case of nonspherical molecules.

➤ | **Study Exercise 4.2**   The goal of many centrifugation experiments is the measurement of s. This value is important because it can be used to calculate the size (molecular weight, kilo base pairs, etc.) of a molecule or cell organelle. The units of s are not obvious from Equation 4.8. Dimensional analysis shows the following: v in cm/sec, ω in radians/sec, r in cm, m_0 in grams, $\bar{v}$ in cm³/g, ρ in g/cm³, and f in g/sec. Determine the units for s. (Ans. = second)

B.   INSTRUMENTATION FOR CENTRIFUGATION

The basic centrifuge consists of two components, an electric motor with drive shaft to spin the sample and a **rotor** to hold tubes or other containers of the sample. A wide variety of centrifuges is available, ranging from a low-speed centrifuge used for routine pelleting of relatively heavy particles to sophisticated instruments that include accessories for making analytical measurements during centrifugation. Here we will describe three types, the low-speed or clinical centrifuge; the high-speed centrifuge, including the "microfuge;" and the ultracentrifuge. Major characteristics and applications of each type are compared in Table 4.1.

Figure 4.3

Range of S values for biomolecules, cell organelles, and cells. kb = kilobase pairs. Molecular weight or kb is shown in parentheses.

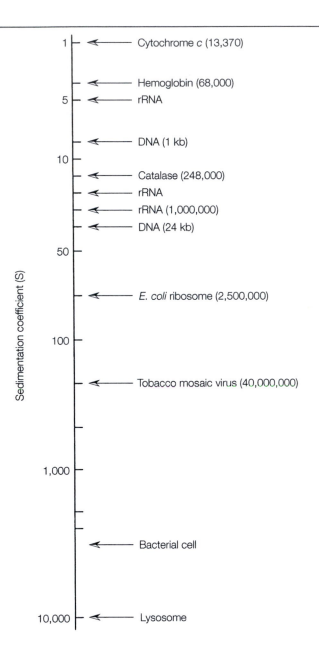

Low-Speed Centrifuges

Most laboratories have a standard low-speed centrifuge used for routine sedimentation of relatively heavy particles. The common centrifuge has a maximum speed in the range of 4000 to 5000 rpm, with RCF values up to 3000 × *g*. These instruments usually operate at room temperature with no means of temperature control of the samples. Two types of rotors, **fixed**

Table 4.1

Types of Centrifuges and Applications

	Type of Centrifuge			
Characteristic	Low Speed	Microfuge (Medium Speed)	High Speed	Ultracentrifuge
Range of speed (rpm)	1–6000	1000–15,000	1000–25,000	20–100,000
Maximum RCF (g)	6000	12,000	50,000	1,000,000
Refrigeration	Some	Some	Yes	Yes
Applications				
Pelleting of cells	Yes	Yes	Yes	Yes
Pelleting of nuclei	Yes	Yes	Yes	Yes
Pelleting of organelles	No	No	Yes	Yes
Pelleting of ribosomes	No	No	No	Yes
Pelleting of macromolecules	No	No	No	Yes
Analytical techniques	No	No	No	Yes

angle and **swinging bucket,** may be used in the instrument. Centrifuge tubes or bottles that contain 12 or 50 mL of sample are commonly used. Low-speed centrifuges are especially useful for the rapid sedimentation of coarse precipitates or red blood cells. The sample is centrifuged until the particles are tightly packed into a **pellet** at the bottom of the tube. (This technique is sometimes called **pelleting.**) The upper, liquid portion, the **supernatant,** is then separated by decantation.

High-Speed Centrifuges

For more sophisticated biochemical applications, higher speeds and temperature control of the rotor chamber are essential. A typical high-speed centrifuge is shown in Figure 4.4. The operator of this instrument can carefully control speed and temperature, which is especially important for carrying out reproducible centrifugations of temperature-sensitive biological samples. Rotor chambers in most instruments are maintained at or near 4°C.

Three types of rotors are available for high-speed centrifugation, the fixed-angle, the swinging-bucket, and the vertical rotor (Figure 4.5A–C). Fixed-angle rotors are especially useful for differential pelleting of particles (Figure 4.6A). In swinging-bucket rotors (Figure 4.5B), the sample tubes move to a position perpendicular to the axis of rotation during centrifugation, as shown in Figure 4.7. These are used most often for density gradient centrifugation (see below). In the vertical rotor (Figure 4.5C), the sample tubes remain in an upright position (Figure 4.8). These rotors are used often for gradient centrifugation. Prior to the early 1990s, rotors were constructed from metals such as aluminum and titanium. Although metal rotors have great strength, they do have several disadvantages: they are very heavy to handle, they are not corrosion resistant, and they become fatigued with use. Rotors are now available

Figure 4.4

A typical high-speed, refrigerated centrifuge. *Courtesy of Sorvall/Kendro; www.kendro.spx.com.*

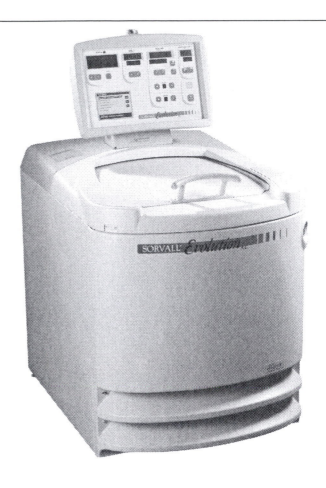

Figure 4.5

Rotors for a high-speed centrifuge. **A** Fixed angle; **B** swinging bucket; **C** vertical. *Courtesy of Beckman Coulter.*

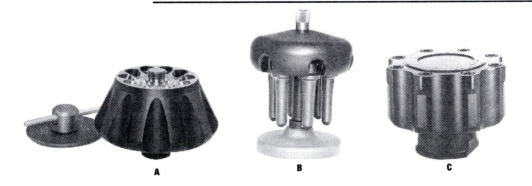

A   B   C

Figure 4.6

Operation of a fixed angle rotor. **A** Loading of sample. **B** Sample at start of centrifugation. **C** Bands form as molecules sediment. **D** Rotor at rest showing separation of two components.

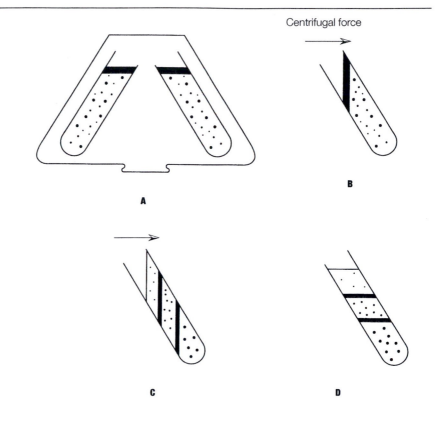

Figure 4.7

Operation of a swinging-bucket rotor. **A** Loading of sample. **B** Sample at start of centrifugation. **C** Sample during centrifugation separates into two components. **D** Rotor at rest.

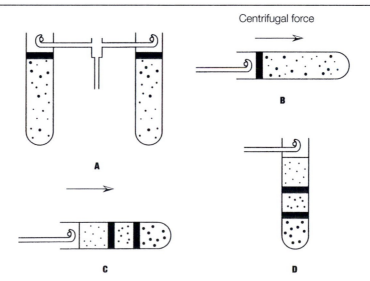

Figure 4.8

Operation of a vertical rotor.
A Loading of sample. **B** Beginning
of centrifugation. **C, D** During
centrifugation. **E** Deceleration
of sample. **F** Rotor at rest.

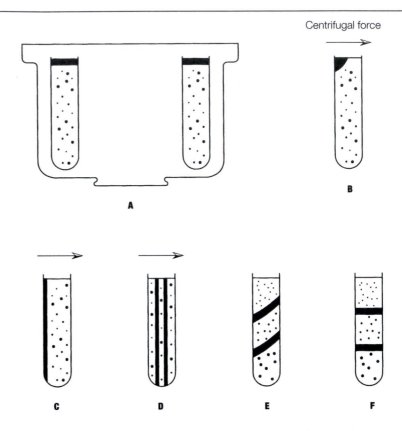

that are fabricated from carbon-fiber composite materials. They have several advantages over heavy metal rotors. These new rotors are 60% lighter than comparable aluminum and titanium rotors. Because of the lighter weight, acceleration and deceleration times are reduced, thus centrifuge run times are shorter. This also results in lower service and maintenance costs. Instruments are equipped with a brake to slow the rotor rapidly after centrifugation.

Widely used in the category of medium-speed centrifuges is the **microfuge** (Table 4.1 and Figure 4.9). These instruments, which are designed for the bench-top, are used for rapid pelleting of small samples. Fixed-angle rotors are available to hold up to eighteen 1.5- or 0.5-mL tubes. The maximum speed of most commercial microfuges is between 12,000 and 15,000 rpm, which delivers a force of $11,000-12,000 \times g$. Some instruments can accelerate to full speed in 6 seconds and decelerate within 18 seconds. Most instruments have a variable speed control and a momentary pulse button for minispins.

The preparation of biological samples almost always requires the use of a high-speed centrifuge. Specific examples will be described later, but high-speed centrifuges may be used to sediment (1) cell debris after cell homogenization, (2) ammonium sulfate precipitates of proteins, (3) microorganisms, and (4) cellular organelles such as chloroplasts, mitochondria, and nuclei.

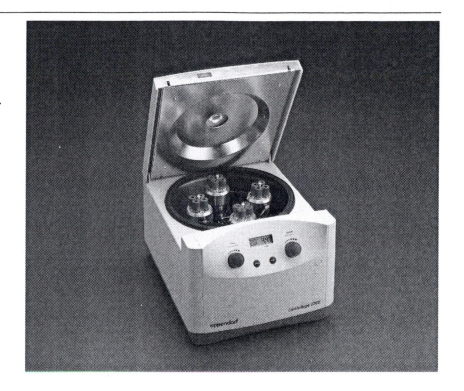

Ultracentrifuges

The most sophisticated of the centrifuges are the **ultracentrifuges.** Because
of the high speeds attainable (see Table 4.1), intense heat is generated in
the rotor, so the spin chamber must be refrigerated and placed under a high
vacuum to reduce friction. The sample in a cell or tube is placed in a rotor,
which is then driven by an electric motor. Although it is rare, metal rotors
when placed under high stress sometimes break into fragments. The rotor
chamber on all ultracentrifuges is covered with protective steel armor plate.
The drive shaft of the ultracentrifuge is constructed of a flexible material to
accommodate any "wobble" of the rotor due to imbalance of the samples.
It is still important to counterbalance samples as carefully as possible.

The previously discussed centrifuges—the low, medium, and high
speed—are of value only for preparative work, that is, for the isolation and
separation of precipitates and biological samples. Ultracentrifuges can be
used both for preparative work and for analytical measurements. Thus, two
types of ultracentrifuges are available, **preparative models,** primarily used
for separation and purification of samples for further analysis, and
analytical models, which are designed for performing physical measurements on the sample during sedimentation. Two of the most versatile models are Beckman Optima MAX and TLX microprocessor-controlled tabletop

Figure 4.10

Outside **A** and inside **B** of the Beckman Optima TLX ultracentrifuge. *Courtesy of Beckman Coulter; www.beckman.com/*

A

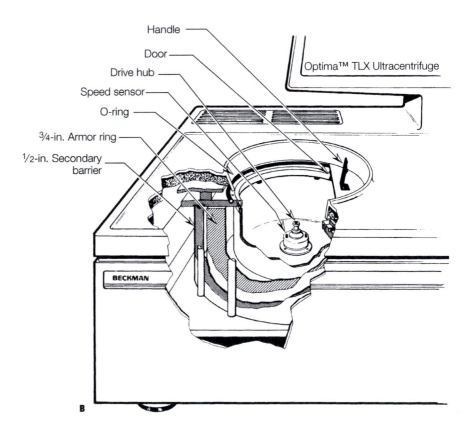

Handle

Door

Drive hub

Speed sensor

O-ring

¾-in. Armor ring

½-in. Secondary barrier

Optima™ TLX Ultracentrifuge

BECKMAN

B

ultracentrifuges (Figure 4.10). With a typical fixed-angle rotor, which holds six 0.2- to 2.2-mL samples, the instruments can generate 100,000 rpm and an RCF of over $1,000,000 \times g$.

Analytical ultracentrifuges have the same basic design as preparative models except that they are equipped with optical systems to monitor directly the sedimentation of the sample during centrifugation. The first commercial instrument of this type was the Beckman Model E, introduced in 1947.

For analysis, a sample of nucleic acid or protein (0.1 to 1.0 mL) is sealed in a special analytical cell and rotated. Light is directed through the sample parallel to the axis of rotation, and measurements of absorbance by sample molecules are made. (The Beckman instrument can scan the sample over the wavelength range 190 to 800 nm.) If sample molecules have no significant absorption bands in the wavelength range, then optical systems that measure changes in the refractive index may be used. Optical systems aided by computers are capable of relating absorbance changes or index of refraction changes to the rate of movement of particles in the sample. The optical system actually detects and measures the front edge or moving boundary of the sedimenting molecules. These measurements can lead to an analysis of concentration distributions within the centrifuge cell. Applications of these measurements will be discussed in the next section.

C.   APPLICATIONS OF CENTRIFUGATION

Preparative Techniques

Centrifuges in undergraduate biochemistry laboratories are used most often for preparative-scale separation procedures. This technique is quite straightforward, consisting of placing the sample in a tube or similar container, inserting the tube in the rotor, and spinning the sample for a fixed period. The sample is removed and the two phases, pellet and supernatant (which should be readily apparent in the tube), may be separated by careful decantation. Further characterization or analysis is usually carried out on the individual phases. This technique, called **velocity sedimentation centrifugation,** separates particles ranging in size from coarse precipitates to cellular organelles. Relatively heavy precipitates are sedimented in low-speed centrifuges, whereas lighter organelles such as ribosomes require the high centrifugal forces of an ultracentrifuge.

Much of our current understanding of cell structure and function depends on separation of subcellular components by centrifugation. The specific method of separation, called **fractional centrifugation,** consists of successive centrifugations at increasing rotor speeds. Figure 4.11 illustrates the fractional centrifugation of a cell homogenate, leading to the separation and isolation of the common cell organelles. For most biochemical applications, the rotor chamber must be kept at low temperatures to maintain the native structure and function of each cellular organelle and its component biomolecules. A high-speed centrifuge equipped with a fixed-angle rotor is most

Figure 4.11

Fractional centrifugation of a cell homogenate. See text for details.

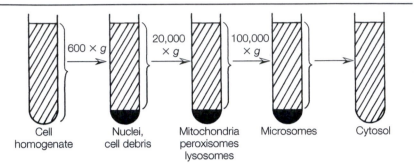

| Cell homogenate | Nuclei, cell debris | Mitochondria peroxisomes lysosomes | Microsomes | Cytosol |

appropriate for the first two centrifugations at $600 \times g$ and $20,000 \times g$. After each centrifuge run, the supernatant is poured into another centrifuge tube, which is then rotated at the next higher speed. The final centrifugation at $100,000 \times g$ to sediment microsomes and ribosomes must be done in an ultracentrifuge. The $100,000 \times g$ supernatant, the **cytosol,** is the soluble portion of the cell and consists of soluble proteins and smaller molecules.

➤ | **Study Exercise 4.3**   The enzymes required for the metabolic process of glycolysis are located in the cytoplasm of the cell. Describe how you could use centrifugation to prepare a cell extract containing these enzymes separated from cell organelles.

Analytical Measurements

A variety of analytical measurements can be made on biological samples during and after a centrifuge run. Most often, the measurements are taken to determine molecular weight, density, and purity of biological samples. All analytical techniques require the use of an ultracentrifuge and can be classified as **differential** or **density gradient.**

Differential Centrifugation

Differential methods involve sedimentation of particles in a medium of homogeneous density. Although the technique is similar to preparative differential centrifugation as previously discussed, the goal of an experiment is to measure the sedimentation coefficient of a particle. The underlying principles of this technique are illustrated in Figure 4.12A. During centrifugation, a moving boundary is generated between pure solvent and sedimenting particles. An analytical ultracentrifuge is capable of detecting and measuring the rate of movement of the boundary. Hence, the sedimentation velocity, v, can be experimentally determined. By using Equation 4.8 the sedimentation coefficient, s, can be calculated. The value of s for a sedimenting particle is related to

Figure 4.12

A comparison of differential and density gradient measurements. **A** Differential centrifugation in a medium of unchanging density. **B** Zonal centrifugation in a prepared density gradient. **C** Isopycnic centrifugation; the density gradient forms during centrifugation.

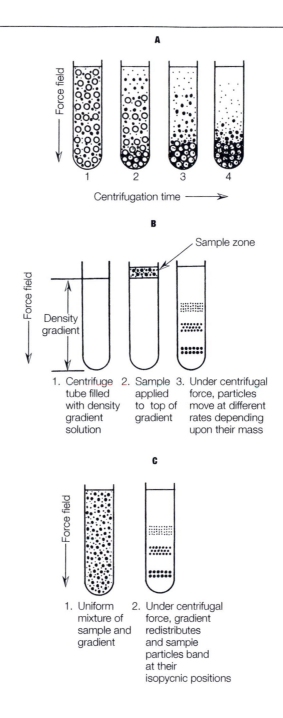

A

Force field

Centrifugation time ⟶

B

Sample zone

Force field

Density gradient

1. Centrifuge tube filled with density gradient solution
2. Sample applied to top of gradient
3. Under centrifugal force, particles move at different rates depending upon their mass

C

Force field

1. Uniform mixture of sample and gradient
2. Under centrifugal force, gradient redistributes and sampie particles band at their isopycnic positions

the molecular weight of that particle by the Svedberg equation. The Svedberg equation is derived from Equation 4.8 by recognizing that the frictional force, f, may be defined by Equation 4.9.

>>
$$f = \frac{RT}{ND}$$
Equation 4.9

where

R = the gas constant, 8.3×10^7 g cm^2/sec/deg/mole

T = the absolute temperature

D = the diffusion coefficient of the solute in units of cm^2/sec

N = Avogadro's number

Thus, Equation 4.8 may be transformed into Equation 4.10.

>>
$$s = \frac{m_0(1 - \bar{v}\rho)}{RT/ND}$$
Equation 4.10

$$RTs = m_0ND(1 - \bar{v}\rho)$$

Since molecular weight, MW, is equal to m_0N, Equation 4.10 is converted to Equation 4.11, the Svedberg equation. In this book two terms are used to define the molecular masses of biomolecules:

Molecular weight (MW): defined as the ratio of the particle mass to 1/12th the mass of a C-12 atom. MW has no units because it is a ratio.

Molecular mass: defined as the mass of one molecule, or the molar mass divided by Avogadro's number. Molecular masses are expressed in the units of daltons (D, 1/12th the mass of a C-12 atom), or kilodaltons (kD).

>>
$$MW = \frac{RTs}{D(1 - \bar{v}\rho)}$$
Equation 4.11

This equation provides an accurate calculation of molecular weight and is applicable to macromolecules such as proteins and nucleic acids. However, its usefulness is limited because diffusion coefficients are difficult to measure and are not readily available in the literature.

An alternative method sometimes used to determine molecular weights of macromolecules is **sedimentation equilibrium.** In the previous example, using the Svedberg equation, the sample is rotated at a rate sufficient to sediment the particles. Here, the sample is rotated at a lower rate, and the particles sediment until they reach an equilibrium position at the point where the centrifugal force is equal to the frictional component opposing their movement (see Equation 4.6). The molecular weight is then calculated using Equation 4.12.

>>
$$MW = \frac{RT}{(1 - \bar{v}\rho)\omega^2}\left(\frac{1}{rc}\right)\left(\frac{dc}{dr}\right)$$
Equation 4.12

where

$$\frac{dc}{dr} = \begin{array}{l}\text{change in particle concentration as a function of distance} \\ \text{from the rotation center, as measured in the ultracentrifuge}\end{array}$$

and MW, R, T, $\bar{v}$, ρ, ω, and r have the definitions previously given. This technique is time consuming, as the system may require 1 to 2 days to reach equilibrium. Also, the use of Equation 4.12 is complicated by the difficulty of making concentration measurements in the ultracentrifuge.

Differential ultracentrifugation methods may also be applied to analysis of the purity of macromolecular samples. If one sharp moving boundary is observed in a rotating centrifuge cell, it indicates that the sample has one component and therefore is pure. In an impure sample, each component would be expected to form a separate moving boundary upon sedimentation.

Density Gradient Centrifugation

In differential procedures, the sample is uniformly distributed in a cell before centrifugation, and the initial concentration of the sample is the same throughout the length of the centrifuge cell. Although useful analytical measurements can be made with this technique, it has disadvantages when applied to impure samples or samples with more than one component. Large particles that sediment faster pass through a medium consisting of solvent and particles of smaller size. Therefore, clear-cut separations of macromolecules are seldom obtained. This can be avoided if the sample is centrifuged in a fluid medium that gradually increases in density from top to bottom. This technique, called **density gradient centrifugation,** permits the separation of multicomponent mixtures of macromolecules and the measurement of sedimentation coefficients. Two methods are used, **zonal centrifugation,** in which the sample is centrifuged in a preformed gradient, and **isopycnic centrifugation,** in which a self-generating gradient forms during centrifugation.

Zonal Centrifugation

Figure 4.12B outlines the procedure for zonal centrifugation of a mixture of macromolecules. A density gradient is prepared in a tube prior to centrifugation. This is accomplished with the use of an automatic gradient mixer. Solutions of low-molecular-weight solutes such as sucrose or glycerol are allowed to flow into the centrifuge cell. The sample under study is layered on top of the gradient and placed in a swinging-bucket rotor. Sedimentation in an ultracentrifuge results in movement of the sample particles at a rate dependent on their individual s values. As shown in Figure 4.12B, the various types of particles sediment as *zones* and remain separated from the other components. The centrifuge run is terminated before any particles reach the bottom of the gradient. The various zones in the centrifuge tubes are then isolated by collecting fractions from the bottom of the tube and analyzing them for the presence of macromolecules. The zones of separated macromolecules are relatively stable in the gradient because it slows diffusion and convection. The gradient conditions can be varied by using different ranges of sucrose concentration. Sucrose concentrations up to 60% can be used, with a density limit of 1.28 g/cm^3. The zonal method can be applied to the separation and isolation of macromolecules (preparative ultracentrifuge) and to the determination of s (analytical ultracentrifuge).

Isopycnic Centrifugation

In the isopycnic technique, the density gradient is formed during the centrifugation. Figure 4.12C outlines the operation of isopycnic centrifugation. The sample under study is dissolved in a solution of a dense salt such as cesium chloride or cesium sulfate. The cesium salts may be used to establish gradients to an upper density limit of 1.8 g/cm^3. The solution of biological sample and cesium salt is uniformly distributed in a centrifuge tube and rotated in an ultracentrifuge. Under the influence of the centrifugal force, the cesium salt redistributes to form a continuously increasing density gradient from the top to the bottom. The macromolecules of the biological sample seek an area in the tube where the density is equal to their respective densities. That is, the macromolecules move to a region where the sum of the forces (centrifugal and frictional) is zero (Equation 4.6). The macromolecules either sediment or float to this region of equal density. Stable zones or bands of the individual components are formed in the gradient (see Figure 4.12C). These bands can be isolated as previously described. Cesium salt gradients are especially valuable for separation of nucleic acids.

Density gradients are widely used in separating and purifying biological samples. In addition to this preparative application, measurements of s can be made. Gradient techniques have been used to isolate and purify the subcellular components, microsomes, ribosomes, lysosomes, mitochondria, peroxisomes, chloroplasts, and others. After isolation, they have been biochemically characterized as to their protein, lipid, nucleic acid, and enzyme contents.

Nucleic acids, in particular, have been extensively studied by density gradient techniques. Both RNA and DNA are routinely classified according to their s values. The different structural forms of DNA discussed in Chapter 9 can be determined by density gradient centrifugation.

A summary of the centrifugation techniques described in this chapter is given in Table 4.2. Space does not allow an exhaustive review of centrifuge applications. Interested students should consult the references at the end of the chapter for recent developments.

Care of Centrifuges and Rotors

Centrifuge equipment represents a sizable investment for a laboratory, so proper maintenance is essential. In addition, poorly maintained equipment is especially dangerous. Since many instruments are now available, specific instructions will not be given here, but general guidelines are outlined.

1. Carefully read the operating manual or receive proper instructions before you use any centrifuge.

Table 4.2

A Summary of Centrifuge Techniques and Applications

Centrifuge Method	Applications
Preparative	
Velocity sedimentation centrifugation (pelleting)	Separation and isolation of particles in a solution. May be applied to precipitates, cell organelles, cells, or biomolecules.
Fractional centrifugation	Isolation of particles, based on size, by successive centrifugation at increasing rotor speeds. For example, may be used to separate cell organelles (Figure 4.11).
Analytical	
Differential centrifugation	Sedimentation of particles in a medium of homogeneous density. Used to measure the sedimentation coefficient, s, and MW of a particle (Figure 4.12A).
Sedimentation equilibrium centrifugation	Used to determine MW of macromolecule or other particle.
Density gradient centrifugation	
Zonal	Gradient is present in the tube before centrifugation and sample is layered on top. Used to isolate purified molecules and determine s.
Isopycnic	Gradient is formed during centrifugation. Used to isolate purified molecules and determine s.

2. Select the proper operating conditions on the instrument. If refrigeration is necessary, set the temperature to the appropriate level and allow 1 to 2 hours for temperature equilibration.

3. Check the rotor chamber for cleanliness and for damage. Clean with soap and warm water and rinse with distilled water.

4. Select the proper rotor. Many sizes and types are available. Follow guidelines already stated in this chapter or consult your instructor.

5. Be sure the rotor is clean and undamaged. Observe any nicks, scratches, or other damage that may cause imbalance. If dirty, the rotor should be cleaned with warm water and a mild, nonbasic detergent. A soft brush can be used inside the cavities. Rinse well with distilled water and dry. Scratches should not be made on the surface coating, as corrosion may result.

6. Filled centrifuge tubes or bottles should be weighed carefully and balanced before centrifugation.

7. Rotor manufacturers provide a maximum allowable speed limit for each rotor. Do not exceed that limit.

8. Keep an accurate record of centrifuge and rotor use. Just as your automobile needs service after a certain number of miles, the centrifuge should be serviced after certain intervals of use. Centrifuge maintenance is usually determined by hours of use and total revolutions of the rotor.

It is also essential to maintain a record of the use for each rotor. Metal rotors weaken with use, and the maximum allowable speed limit decreases. Rotor manufacturers usually provide guidelines for decreasing the allowable speed for a rotor (derating a rotor).

9. If an unusual noise or vibration develops during centrifugation, immediately turn the centrifuge off.

10. Carefully clean the rotor chamber and rotor after each centrifugation.

Study Problems

> 1. Which of the following factors will have an effect on the sedimentation rate of a particle during centrifugation?
> (a) Mass of the sedimenting particle
> (b) Angular velocity of rotation
> (c) Atmospheric pressure
> (d) Density of the solvent

> 2. You wish to centrifuge a biological sample so that it experiences an RCF of $100,000 \times g$. At what rpm must you set the centrifuge assuming an average r value of 4?

> 3. Cytochrome c has an s value of 1×10^{-13} second and hemoglobin an s value of about 4.5×10^{-13} second. Which protein has the larger molecular weight?

> 4. An enzyme has a sedimentation coefficient of 3.5 S. When a substrate molecule is bound into the active site of the enzyme, the sedimentation coefficient decreases to 3.0 S. Explain this change.

> 5. A protein with molecular weight of 100,000 shows a single boundary when centrifuged in aqueous buffer. If the protein is centrifuged in a medium of the same buffer plus 6 M urea, two boundaries are observed, one corresponding to a molecular weight of 10,000, the other 30,000. The area of the slower peak is two-thirds that of the faster. Describe the subunit structure of the protein.

> 6. Describe how you would design a centrifuge experiment to isolate sediments containing cell nuclei.

> 7. Explain the following observation. The density of DNA in 7 M CsCl containing 0.15 M MgCl$_2$ is less than the density of the same DNA in 7 M CsCl.

> 8. Assume that you have centrifuged in a density gradient a sample of DNA that contained both closed, circular DNA and supercoiled DNA. Would you expect to see two bands in the sedimentation pattern? Explain.

> 9. Assume that a centrifuge is operating at 43,000 rpm. What is the relative centrifugal force at a distance from the central axis of 6 cm?

> 10. Could you use a low-speed centrifuge to sediment mitochondria? Explain.

Further Reading

J. Berg, J. Tymoczko, and L. Stryer, *Biochemistry, 5th ed.* (2002), W. H. Freeman (New York), pp. 79–80; 87–89. Applications of centrifugation.

D. Cantor and P. Schimmel, *Biophysical Chemistry,* Part II (1980), W. H. Freeman (San Francisco), pp. 591–641. Excellent, but advanced treatment of theory and applications of analytical centrifugation.

R. Garrett and C. Grisham, *Biochemistry, 3rd ed.* (2005), Brooks-Cole (Belmont, CA), p. 152. Techniques of centrifugation.

J. Graham and D. Rickwood, *Biological Centrifugation* (2001), Springer Verlag (New York). Applications of centrifugation in biology.

E. Holme and H. Peck, *Analytical Biochemistry, 3rd ed.* (1998), Addison-Wesley Longman (Essex, U.K.), pp. 153–163. Principles of centrifugation.

N. Price, R. Dwek, and M. Wormald, *Principles and Problems in Physical Chemistry for Biochemists, 3rd ed.* (2002), Oxford University Press (Oxford).

I. Tinoco, K. Sauer, J. Wang, and J. Puglisi, *Physical Chemistry: Principles and Applications in Biological Sciences, 4th ed.* (2002), Prentice Hall (Upper Saddle River, NJ). Chapter 6 covers centrifugation and sedimentation.

D. Voet and J. Voet, *Biochemistry, 3rd ed.* (2004), John Wiley & Sons (Hoboken, NJ), pp. 151–154. "Centrifugation in Protein Purification."

K. Wilson and J. Walker, Editors, *Principles and Techniques of Practical Biochemistry, 5th ed.* (2000), Cambridge University Press (Cambridge), pp. 263–311. "Techniques in Centrifugation."

http://www.beckman.com
  Click on Centrifuges and follow links to Applications, Literature, Rotor Calculations, etc.

http://www.brinkmann.com
  Eppendorf centrifuges.

http://www.sorvall.com
  Superspeed centrifuges.

http://www.piramoon.com
  New carbon-fiber rotors by Fiberlite Centrifuge, Inc.

CHAPTER 5

PURIFICATION AND IDENTIFICATION OF BIOMOLECULES BY CHROMATOGRAPHY

The primary goal of biochemical research is to understand the molecular nature of life processes. The molecular details of a biological process cannot be fully elucidated until the interacting molecules have been isolated and characterized. Therefore, our understanding of the mechanisms of life processes has increased at about the same pace as the development of techniques for the separation and characterization of biomolecules.

Chromatography, the most important technique for isolating and purifying biomolecules, was developed by Mikhail Tswett, an Italian-born, Russian botanist. In 1902, Tswett began his studies on the isolation and characterization of the colorful pigments in plant chloroplasts. He prepared separating columns by packing fine powders like sucrose and chalk (calcium carbonate) into long glass tubes. He then poured petroleum ether–derived plant extracts through the columns. As he continued eluting the columns with solvent, he noted the formation of yellow and green zones. Tswett had invented "chromatography," which he defined in his 1906 publication as "a method in which the components of a mixture are separated on an adsorbent in a flowing solvent." In addition to introducing a new technique, Tswett also showed by these experiments that chlorophyll exists in different forms. From such humble beginnings, chromatography has been developed into the ultimate tool, not only for the isolation and purification of biomolecules, but also for their characterization. Chromatography, which now has been expanded into multiple forms, continues to be the most effective technique for separating and purifying all types of biomolecules. In addition, it is widely used as an analytical tool to measure biophysical and other quantitative properties of molecules.

A.   INTRODUCTION TO CHROMATOGRAPHY

All types of chromatography are based on a very simple concept: The sample to be examined is allowed to interact with two physically distinct entities—a **mobile phase** and a **stationary phase** (see Figure 5.1). The sample most often contains a mixture of several components to be separated. The molecules targeted for analysis are called **analytes.** The mobile phase, which may be a liquid or gas, moves the sample components through a region containing the solid or liquid stationary phase called the **sorbent.** The stationary phase will not be described in detail at this time, because it varies from one chromatographic method to another. However, it may be considered as having the ability to "bind" some types of analytes. The molecular components in the sample distribute themselves between the mobile phase and sorbent and thus have the opportunity to interact intimately with the stationary phase. If some of the sample molecules (analytes) are preferentially bound by the sorbent, they spend more time in the sorbent and are retarded in their movement through the chromatographic system. Molecules that show weak affinity for the sorbent spend more time with the mobile phase and are more easily removed or **eluted** from the system. The many interactions that occur between analytes and the stationary phase sorbent bring about a separation of molecules because of different affinities for the stationary phase. The general process of moving a sample mixture through a chromatographic system is called **development.**

    The mobile phase can be collected as a function of time at the end of the chromatographic system. The mobile phase, now called the **effluent,** contains the purified analytes. If the chromatographic process has been effective, fractions or "cuts" that are collected at different times will contain

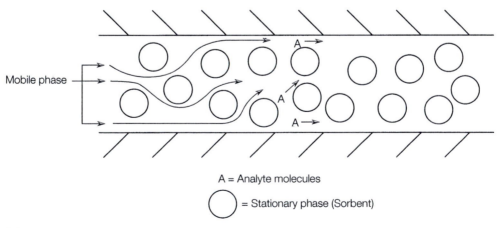

A = Analyte molecules

◯ = Stationary phase (Sorbent)

Figure 5.1

A representation of the principles of chromatography.

the different components of the original sample. *In summary, molecules are separated because they differ in the extent to which they are distributed between the mobile phase and the stationary phase.*

Throughout this chapter and others, biochemical techniques will be designated as **preparative** or **analytical,** or both. A preparative procedure is one that can be applied to the purification of a relatively large amount of a biological material (mg or g). The purpose of such an experiment would be to obtain purified material for further characterization and study. Analytical procedures are used most often to determine the purity of a biological sample; however, they may be used to evaluate any physical, chemical, or biological characteristic of a biomolecule or biological system.

Partition versus Adsorption Chromatography

Chromatographic methods are divided into two types according to how analytes bind to or interact with the stationary phase. **Partition** chromatography is the distribution of an analyte between two liquid phases. This may involve direct extraction using two liquids, or it may use a liquid immobilized on a solid support as in the case of paper, thin-layer, and gas–liquid chromatography. For partition chromatography, the sorbent in Figure 5.1 consists of inert solid particles coated with liquid adsorbent. The distribution of analytes between the two phases is based primarily on solubility differences. The distribution may be quantified by using the **partition coefficient,** K_D (Equation 5.1).

$$K_D = \frac{\text{concentration of analyte in sorbent}}{\text{concentration of analyte in mobile phase}}$$

Equation 5.1

Adsorption chromatography refers to the use of a stationary phase or support, such as an ion-exchange resin, that has a finite number of relatively specific binding sites for analytes. There is not a clear distinction between the processes of partition and adsorption. All chromatographic separations rely, to some extent, on adsorptive processes. However, in some methods (paper, thin-layer, and gas chromatography) these specific adsorptive effects are minimal and the separation is based primarily on nonspecific solubility factors. Adsorption chromatography relies on relatively specific interactions between the analytes and binding sites on the surface of the sorbent. The attractive forces between analyte and support may be ionic, hydrogen bonding, or hydrophobic interactions. Binding of analyte is, of course, reversible.

Because of the different interactions involved in partition and adsorption processes, they may be applied to different separation problems. Partition processes are the most effective for the separation of small molecules, especially those in homologous series. Partition chromatography has been widely used for the separation and identification of amino acids, carbohydrates, and fatty acids. Adsorption techniques, represented by ion-exchange chromatography, are most effective when applied to the separation of macromolecules including proteins and nucleic acids.

In the rest of the chapter, various chromatographic methods will be discussed. You should recognize that no single chromatographic technique relies solely on adsorption or partition effects. Therefore, little emphasis will be placed on classification of the techniques; instead, theoretical and practical aspects will be discussed.

B. PAPER AND THIN-LAYER CHROMATOGRAPHY (PLANAR CHROMATOGRAPHY)

Because of the similarities in the theory and practice of these two procedures, they will be considered together. Both are examples of partition chromatography. In paper chromatography, the cellulose support is extensively hydrated, so distribution of the analyte occurs between the immobilized water (sorbent) and the mobile developing solvent. The initial stationary liquid phase in thin-layer chromatography (TLC) is the solvent used to prepare the thin layer of adsorbent. However, as developing solvent molecules move through the sorbent, polar solvent molecules may bind to the immobilized support and become the sorbent.

Preparation of the Sorbent

The support medium may be a sheet of cellulose or a glass or plastic plate covered with a thin coating of silica gel, alumina, or cellulose. Large sheets of cellulose chromatography paper are available in different porosities. These may be cut to the appropriate size and used without further treatment. The paper should never be handled with bare fingers. Although thin-layer plates can easily be prepared, it is much more convenient to purchase ready-made plates. These are available in a variety of sizes, materials, and thicknesses of stationary support. They are relatively inexpensive and have a more uniform support thickness than handmade plates.

Figure 5.2 outlines the application procedure. The sample to be analyzed is usually dissolved in a volatile solvent. A very small drop of solution is spotted onto the plate with a disposable microcapillary pipet and allowed to dry; then the spotting process is repeated by superimposing more drops on the original spot. The exact amount of sample applied is critical. There must be enough sample so the developed spots can be detected, but overloading will lead to "tailing" and lack of resolution. Finding the proper sample size is a matter of trial and error. It is usually recommended that two or three spots of different concentrations be applied for each sample tested. Spots should be applied along a very faint line drawn with a pencil and ruler. TLC plates should not be heavily scratched or marked. Identifying marks may be made on the top of the chromatogram, where solvent does not reach.

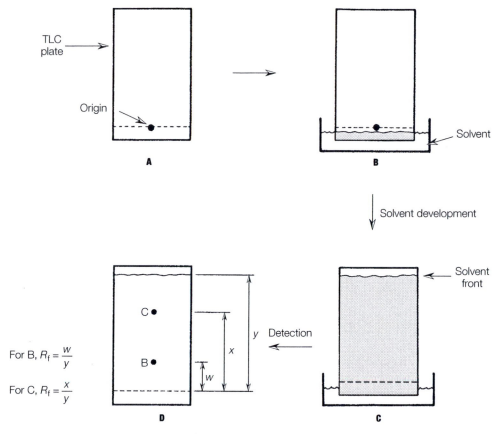

For B, $R_f = \dfrac{w}{y}$

For C, $R_f = \dfrac{x}{y}$

Figure 5.2

The procedure of paper and thin-layer chromatography. **A** Application of the sample. **B** Setting plate in solvent chamber. **C** Movement of solvent by capillary action. **D** Detection of separated components and calculation of R_f.

Solvent Development

A wide selection of solvent systems is available in the biochemical literature. If a new solvent system must be developed, a preliminary analysis must be done on the sample with a series of solvents. Solvents can be rapidly screened by developing several small chromatograms (2×6 cm) in small sealed bottles containing the solvents. For the actual analysis, the sample should be run on a larger plate with appropriate standards in a development chamber (Figure 5.3). The chamber must be airtight and saturated with solvent vapors. Filter paper on two sides of the chamber, as shown in Figure 5.3, enhances vaporization of the solvent.

Paper chromatograms may be developed in either of two types of arrangements–ascending or descending solvent flow. Descending solvent flow leads to faster development because of assistance by gravity,

Figure 5.3

A typical chamber for paper and thin-layer chromatography.

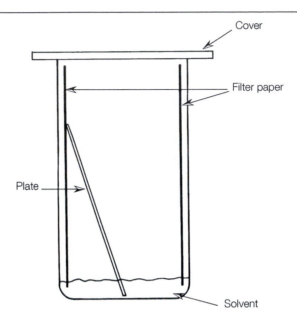

Cover

Filter paper

Plate

Solvent

and it can offer better resolution for compounds with small R_f values because the solvent can be allowed to run off the paper. R_f values cannot be determined under these conditions, but it is useful for qualitative separations.

Two-dimensional chromatography is used for especially difficult separations. The chromatogram is developed in one direction by a solvent system, air dried, turned 90°, and developed in a second solvent system.

Detection and Measurement of Components

Unless the components in the sample are colored, their location on a chromatogram will not be obvious after solvent development. Several methods can be used to locate the spots, including fluorescence, radioactivity, and treatment with chemicals that develop colors. Substances that are highly conjugated may be detected by fluorescence under a UV lamp. Chromatograms may be treated with different types of reagents to develop a color. **Universal reagents** produce a colored spot with any organic compound. When a solvent-developed plate is sprayed with concentrated H_2SO_4 and heated at 100°C for a few minutes, all organic substances appear as black spots. A more convenient universal reagent is I_2. The solvent-developed chromatogram is placed in an enclosed chamber containing a few crystals of I_2. The I_2 vapor reacts with most organic substances on the plate to produce brown spots. The spots are more intense with unsaturated compounds.

Specific reagents react with a particular class of compound. For example, rhodamine B is often used for visualization of lipids, ninhydrin for amino acids, and aniline phthalate for carbohydrates.

The position of each component of a mixture is quantified by calculating the distance traveled by the component relative to the distance traveled by the solvent. This is called **relative mobility** and symbolized by R_f. In Figure 5.2D, the R_f values for components B and C are calculated. The R_f for a substance is a constant for a certain set of experimental conditions. However, it varies with solvent, type of stationary support (paper, alumina, silica gel), temperature, humidity, and other environmental factors. R_f values are always reported along with solvent and temperature.

Applications of Planar Chromatography

Thin-layer chromatography is now more widely used than paper chromatography. In addition to its greater resolving power, TLC is faster and plates are available with several sorbents (cellulose, alumina, silica gel).

Partition chromatography as described in this section may be applied to two major types of problems: (1) identification of unknown samples and (2) isolation of the components of a mixture. The first application is, by far, the more widely used. Paper chromatography and TLC require only a minute sample size, the analysis is fast and inexpensive, and detection is straightforward. Unknown samples are applied to a plate along with appropriate standards, and the chromatogram is developed as a single experiment. In this way any changes in experimental conditions (temperature, humidity, etc.) affect standards and unknowns to the same extent. It is then possible to compare the R_f values directly.

Purified substances can be isolated from developed chromatograms; however, only tiny amounts are present. In paper chromatography, the spot may be cut out with a scissors and the piece of paper extracted with an appropriate solvent. Isolation of a substance from a TLC plate is accomplished by scraping the solid support from the region of the spot with a knife edge or razor blade and extracting the sorbent with a solvent. "Preparative" thin-layer plates with a thick coating of sorbent (up to 2 mm) are especially useful because they have higher sample capacity.

➤ **Study Exercise 5.1**   A mixture containing five amino acids (Ala, Asp, Gly, Phe, Pro) was analyzed using two methods of planar chromatography, paper and cellulose-coated thin layer. The solvent system was *n*-propanol/water (70/30 v/v). Predict the order of the mobility of the amino acids (low R_f to high R_f) on the chromatograms.

Solution: In cellulose planar chromatography, the amino acids interact with two phases, the sorbent (extensively hydrated cellulose, which is very polar) and the mobile phase (*n*-propanol/water, which is less polar than the sorbent). The more polar the amino acid, the stronger it will interact with the hydrated cellulose, thus the slower it will move with solvent during development (lower R_f). The order may be predicted by looking at the polarity of each amino acid side chain and arranging the amino acids in order of decreasing polarity. The correct order of migration (low to high R_f) is: Asp, Gly, Ala, Pro, Phe.

C. COLUMN CHROMATOGRAPHY

Adsorption chromatography in biochemical applications usually consists of a solid stationary phase and a liquid mobile phase. The most useful technique is column chromatography, in which the stationary phase is confined to a glass or plastic tube and the mobile phase (a solvent or buffer) is allowed to flow through the solid adsorbent. A small amount of the sample to be analyzed is layered on top of the column. The sample mixture enters the column of adsorbing material and the molecules present are distributed between the mobile phase and the stationary phase. The various components in the sample have different affinities for the two phases and move through the column at different rates. Collection of the liquid phase emerging from the column yields separate fractions containing the individual components in the sample.

Specific terminology is used to describe various aspects of column chromatography. When the actual adsorbing material is made into a column, it is said to be **poured** or **packed.** Application of the sample to the top of the column is **loading** the column. Movement of solvent through the loaded column is called **developing** or **eluting** the column. The **bed volume** is the total volume of solvent and adsorbing material taken up by the column. The volume taken up by the liquid phase in the column is the **void volume.** The **elution volume** is the amount of solvent required to remove a particular analyte from the column. This is analogous to R_f values in thin-layer or paper chromatography.

In adsorption chromatography, solute molecules take part in specific interactions with the stationary phase. Herein lies the great versatility of adsorption chromatography. Many varieties of adsorbing materials are available, so a specific sorbent can be chosen that will effectively separate a mixture. There is still an element of trial and error in the selection of an effective stationary phase. However, experiences of many investigators are recorded in the literature and are of great help in choosing the proper system. Table 5.1 lists the most common stationary phases employed in adsorption column chromatography.

Adsorbing materials come in various forms and sizes. The most suitable forms are dry powders or a slurry form of the material in an aqueous buffer or organic solvent. Alumina, silica gel, and fluorisil do not normally need special pretreatment. The size of particles in an adsorbing material is defined by **mesh size.** This refers to a standard sieve

Table 5.1

Adsorbents Useful in Biochemical Applications

Adsorbing Material	Uses
Alumina	Small organics, lipids
Silica gel	Amino acids, lipids, carbohydrates
Fluorisil (magnesium silicate)	Neutral lipids
Calcium phosphate (hydroxyapatite)	Proteins, polynucleotides, nucleic acids
Cellulose	Proteins

Table 5.2	
Mesh Sizes of Adsorbents and Typical Applications	
Mesh Size	Applications
20–50	Crude preparative work, very high flow rate
50–100	Preparative applications, high flow rate
100–200	Analytical separations, medium flow rate
200–400	High-resolution analytical separations, low flow rate

through which the particles can pass. A 100-mesh sieve has 100 small openings per square inch. Adsorbing material with high mesh size (400 and greater) is extremely fine and is most useful for very high resolution chromatography. Table 5.2 lists standard mesh sizes and the most appropriate application. For most biochemical applications, 100 to 200 mesh size is suitable.

Operation of a Chromatographic Column

A typical column setup is shown in Figure 5.4. The heart of the system is, of course, the column of adsorbent. In general, the longer the column, the better the resolution of components. However, a compromise must be made because flow rate decreases with increasing column length. The actual size of a column depends on the nature of the adsorbing material and the amount of chemical sample to be separated. For preparative purposes, column heights of 20 to 50 cm are usually sufficient to achieve acceptable resolution. Column inside diameters may vary from 0.5 to 5 cm.

Packing the Column

Once the adsorbing material and column size have been selected, the column is poured. If the tube does not have a fritted disc in the bottom, a small piece of glass wool or cotton should be used to support the column. Most columns are packed by pouring a slurry of the sorbent into the tube and allowing it to settle by gravity into a tight bed. The slurry is prepared with the solvent or buffer that will be used as the initial developing solvent. Pouring of the slurry must be continuous to avoid formation of sorbent layers. Excess solvent is eluted from the bottom of the column while the sorbent is settling. The column must never run dry. Additional slurry is added until the column bed reaches the desired height. The top of the settled adsorbent is then covered with a small circle of filter paper or glass wool to protect the surface while the column is loaded with sample or the eluting solvent is changed.

Sometimes it is necessary to pack a column under pressure (5 to 10 psi). This leads to a tightly packed bed that yields more reproducible results, especially with gradient elution (see the following subsections).

Figure 5.4

Setup for the operation of a chromatography column.

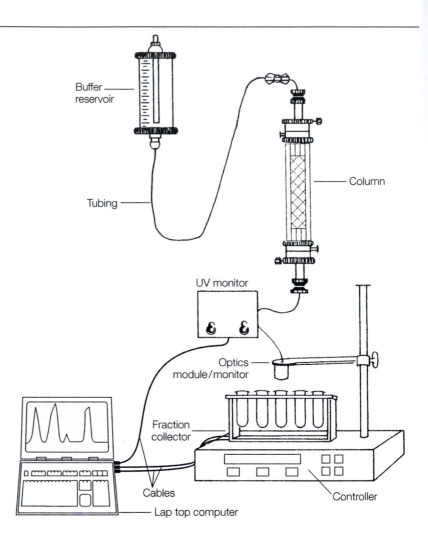

Loading the Column

The sample to be analyzed by chromatography should be applied to the top of the column in a concentrated form. If the sample is solid, it is dissolved in a minimum amount of solvent; if already in solution, it may be concentrated by ultrafiltration as described in Chapter 3. After the sample is loaded onto the column with a graduated or disposable pipet, it is allowed to percolate into the adsorbent. A few milliliters of solvent are then carefully added to wash the sample into the column material. The column is then filled with eluting solvent.

Eluting the Column

The chromatography column is developed by continuous flow of a solvent. Maintaining the appropriate flow rate is important for effective separation.

If the flow rate is set too high, there is not sufficient time for complete equilibration of the analytes with the two phases. Too low a flow rate allows diffusion of analytes, which leads to poor resolution and broad elution peaks. It is difficult to give guidelines for the proper flow rate of a column, but, in general, a column should be adjusted to a rate slightly less than "free flow." Sometimes it is necessary to find the proper flow rate by trial and error. One problem encountered during column development is a changing flow rate. As the solvent height above the column bed is reduced, there is less of a "pressure head" on the column, so the flow rate decreases. This can be avoided by storing the developing solvent in a large reservoir and allowing it to enter the column at the same rate as it is emerging from the column (see Figure 5.4).

Adsorption columns are eluted in one of three ways. All components may be eluted by a single solvent or buffer. This is referred to as **continual elution.** In contrast, **stepwise elution** refers to an incremental change of solvent to aid development. The column is first eluted with a volume of one solvent and then with a second solvent. This may continue with as many solvents or solvent mixtures as desired. In general, the first solvent should be the least polar of any used in the analysis, and each additional solvent should be of greater polarity or ionic strength. Finally, adsorption columns may be developed by **gradient elution** brought about by a gradual change in solvent composition. The composition of the eluting solvent can be changed by the continuous mixing of two different solvents to gradually change the ratio of the two solvents. Alternatively, the concentration of a component in the solvent can be gradually increased. This is most often done by addition of a salt (KCl, NaCl, etc.). Devices are commercially available to prepare predetermined, reproducible gradients.

Collecting the Eluent

The separated components emerging from the column in the eluent are usually collected as discrete fractions. This may be done manually by collecting specified volumes of eluent in Erlenmeyer flasks or test tubes. Alternatively, if many fractions are to be collected, a mechanical fraction collector is convenient and even essential. An automatic fraction collector (see Figure 5.4) directs the eluent into a single tube until a predetermined volume has been collected or until a preselected time period has elapsed; then the collector advances another tube for collection. Specified volumes are collected by a drop counter activated by a photocell, or a timer can be set to collect a fraction over a specific period.

Detection of Eluting Components

The completion of a chromatographic experiment calls for a means to detect the presence of analytes in the collected fractions. The detection method used will depend on the nature of the analytes. Smaller molecules such as lipids, amino acids, and carbohydrates can be detected by spotting fractions on a thin-layer plate or a piece of filter paper and treating them with a chemical reagent that produces a color. The same reagents that are used to visualize spots on a thin-layer or paper chromatogram are useful for this. Proteins and nucleic acids are conveniently detected by spectroscopic absorption measurements at 280 and 260 nm, respectively. Enzymes can be detected by measurements of catalytic activity associated with each fraction. Research-grade chromatographic systems are equipped with detectors that continuously monitor some physical property of the eluent and display the separation results on a computer screen (see Figure 5.4). The newest advance in detectors is the diode array (see Chapter 7). Most often the eluent is directed through a flow cell where absorbance or fluorescence characteristics can be measured. The detector is connected to a recorder or computer for a permanent record of spectroscopic changes. When the location of the various analytes is determined, adjacent fractions containing identical components are pooled and stored for later use.

D.   ION-EXCHANGE CHROMATOGRAPHY

Ion-exchange chromatography is a form of adsorption chromatography in which ionic analytes display reversible electrostatic interactions with a charged stationary phase. The chromatographic setup is identical to that described in the last section and Figure 5.4. The column is packed with a stationary phase consisting of a synthetic resin that is tagged with ionic functional groups. The steps involved in ion-exchange chromatography are outlined in Figure 5.5. In stage 1, the insoluble resin material (positively charged) in the column is surrounded by buffer counterions. Loading of

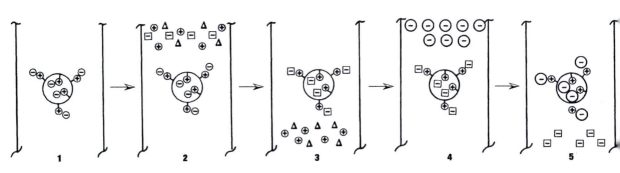

Figure 5.5

Illustration of the principles of ion-exchange chromatography. See text for explanation.

the column in stage 2 brings analytes of different charge into the ion-exchange medium. Solutes entering the column may be negatively charged, positively charged, or neutral under the experimental conditions. Analytes that have a charge opposite to that of the resin bind tightly but reversibly to the stationary phase (stage 3). The strength of binding depends on the size of the charge and the charge density (amount of charge per unit volume of molecule) of the analyte. The greater the charge or the charge density, the stronger the interaction. Neutral analytes (Δ) or those with a charge identical to that of the resin show little or no affinity for the stationary phase and move with the eluting buffer. The bound analytes can be released by eluting the column with a buffer of increased ionic strength or pH (stage 4). An increase in buffer ionic strength releases bound analytes by displacement. Increasing the buffer pH decreases the strength of the interaction by reducing the charge on the analyte or on the resin (stage 5).

The following sections will focus on the properties of ion-exchange resins, selection of experimental conditions, and applications of ion-exchange chromatography.

Ion-Exchange Resins

Ion exchangers are made up of two parts—an insoluble, three-dimensional matrix and chemically bonded charged groups within and on the surface of the matrix. The resins are prepared from a variety of materials, including polystyrene, acrylic resins, polysaccharides (dextrans), agarose, and celluloses. An ion exchanger is classified as **cationic** or **anionic** depending on whether it exchanges cations or anions. A resin that has negatively charged functional groups exchanges positive ions and is a **cation exchanger.** Each type of exchanger is also classified as **strong** or **weak** according to the ionizing strength of the functional group. An exchanger with a quaternary amino group is, therefore, a **strongly basic anion exchanger,** whereas primary or secondary aromatic or aliphatic amino groups would lead to a **weakly basic anion exchanger.** A **strongly acidic cation exchanger** contains the sulfonic acid group. Table 5.3 lists the more common ion exchangers according to each of these classifications.

The ability of an ion exchanger to adsorb counterions is defined quantitatively by **capacity.** The **total capacity** of an ion exchanger is the quantity of charged and potentially charged groups per unit weight of dry exchanger. It is usually expressed as milliequivalents of ionizable groups per milligram of dry weight, and it can be experimentally determined by titration. The capacity of an ion exchanger is a function of the porosity of the resin. The resin matrix contains covalent cross-linking that creates a "molecular sieve." Ionized functional groups within the matrix are not readily accessible to large molecules that cannot fit into the pores. Only surface charges would be available to these molecules for exchange. The purely synthetic resins (polystyrene and acrylic) have cross-linking ranging from 2 to 16%, with 8% being the best for general purposes.

Table 5.3

Ion-Exchange Resins

Name	Functional Group	Matrix	Class
Anion Exchangers			
AG 1	Tetramethylammonium	Polystyrene	Strong
AG 3	Tertiary amine	Polystyrene	Weak
DEAE-Sephacel	Diethylaminoethyl	Sephacel	Weak
PEI-cellulose	Polyethyleneimine	Cellulose	Weak
DEAE-Sephadex	Diethylaminoethyl	Dextran	Weak
QAE-Sephadex	Diethyl-(2-hydroxyl-propyl)-aminoethyl	Dextran	Strong
DEAE-Sepharose	Diethylaminoethyl	Agarose	Weak
Cation Exchangers			
AG 50	Sulfonic acid	Polystyrene	Strong
Bio-Rex 70	Carboxylic acid	Acrylic	Weak
CM-Sephacel	Carboxymethyl	Sephacel	Weak
P-Cellulose	Phosphate	Cellulose	Intermediate
CM-Sephadex	Carboxymethyl	Dextran	Weak
SP-Sephadex	Sulfopropyl	Dextran	Strong
CM-Sepharose	Carboxymethyl	Agarose	Weak
SP-Sepharose	Sulfonic acid	Agarose	Strong

With so many different experimental options and resin properties to consider, it is difficult to select the proper conditions for a particular separation. The next section will outline the choices and offer guidelines for proper experimental design.

Selection of the Ion Exchanger

Before a proper choice of ion exchanger can be made, the nature of the molecules to be separated must be considered. For relatively small, stable molecules (amino acids, lipids, nucleotides, carbohydrates, pigments, etc.) the synthetic resins based on polystyrene are most effective. They have relatively high capacity for small molecules because the extensive cross-linking still allows access to the interior of the resin beads. For separations of peptides, proteins, nucleic acids, polysaccharides, and other large biomolecules, one must consider the use of fibrous cellulosic ion exchangers and low-percent cross-linked dextran or acrylic exchangers. The immobilized functional groups in these resins are readily available for exchange even to larger molecules.

The choice of ion exchanger has now been narrowed considerably. The next decision is whether to use a cationic or anionic exchanger. If the analyte has only one type of charged group, the choice is simple. A molecule that has a positive charge will bind to a cationic exchanger and vice versa. However, many biomolecules have more than one type of ionizing group and may have both negatively and positively charged groups (they are amphoteric). The net charge on such molecules depends on pH. At the isoelectric point,

Figure 5.6

The effect of pH on the net charge of a protein.

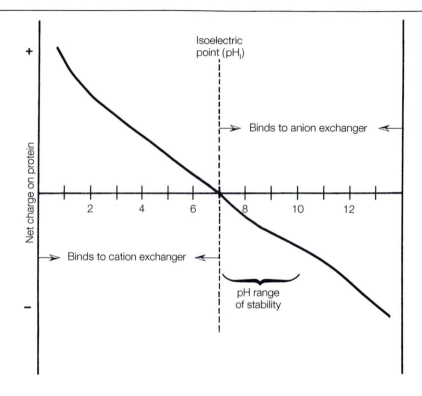

the substance has no net charge and would not bind to any type of ion exchanger.

In principle, amphoteric molecules should bind to both anionic and cationic exchangers. However, when one is dealing with large biomolecules, the pH **range of stability** must also be evaluated. The range of stability refers to the pH range in which the biomolecule is not denatured. Figure 5.6 shows how the net charge of a hypothetical protein changes as a function of pH. Below the isoelectric point (pH$_I$) the molecule has a net positive charge and would be bound to a cation exchanger. Above the isoelectric point, the net charge is negative, and the protein would bind to an anion exchanger. Superimposed on this graph is the pH range of stability for the hypothetical protein. Because it is stable in the range of pH 7.0–9.0, the ion exchanger of choice is an anionic exchanger. In most cases, the isoelectric point of the protein is not known. The type of ion exchanger must be chosen by trial and error as follows. Small samples of the protein mixture in buffer are equilibrated for 10 to 15 minutes in separate test tubes, one with each type of ion exchanger. The tubes are then centrifuged or let stand to sediment the ion exchanger. Check each supernatant for the presence of the desired analyte (A_{260} for nucleic acids, A_{280} for proteins, catalytic activity for enzymes, etc.). If a supernatant has a relatively low level of added protein, that ion exchanger would be suitable for use. This simple test can also be extended to find conditions for elution of the desired macromolecule from the ion exchanger. The ion exchanger charged with the macromolecule is treated with buffers of increasing ionic strength or changing pH. The supernatant after each treatment is analyzed as before for release of the macromolecule.

Choice of Buffer

This decision includes not just the buffer substance but also the pH and the ionic strength. Buffer ions will, of course, interact with ion-exchange resins. Buffer ions with a charge opposite to that on the ion exchanger compete with analyte for binding sites and greatly reduce the capacity of the column. Cationic buffers should be used with anionic exchangers; anionic buffers should be used with cationic exchangers.

The pH chosen for the buffer depends first of all on the range of stability of the macromolecule to be separated (see Figure 5.6). Second, the buffer pH should be chosen so that the desired macromolecule will bind to the ion exchanger. In addition, the ionic strength should be relatively low to avoid "damping" of the interaction between analyte and ion exchanger. Buffer concentrations in the range 0.05 to 0.1 M are recommended.

Preparation of the Ion Exchanger

The commercial suppliers of ion exchangers provide detailed instructions for the preparation of the adsorbents. Failure to pretreat ion exchangers will greatly reduce the capacity and resolution of a column. Most new ion-exchange resins are commercially available in slurry form and are ready to use with a minimum number of pretreatment steps.

Sometimes pretreatment steps do not remove the small particles that are present in most ion-exchange materials. If left in suspension, these particles, called fines, result in decreased resolution and low column flow rates. The fines are removed from an exchanger by suspending the swollen adsorbent in a large volume of water in a graduated cylinder and allowing at least 90% of the exchanger to settle. The cloudy supernatant containing the fines is decanted. This process is repeated until the supernatant is completely clear. The number of washings necessary to remove most fines is variable, but for a typical ion exchanger 8 to 10 times is probably sufficient.

Using the Ion-Exchange Resin

Ion exchangers are most commonly used in a column form. The column method discussed earlier in this chapter can be directly applied to ion-exchange chromatography.

An alternative method of ion exchange is **batch separation.** This involves mixing and stirring equilibrated exchanger directly with the analyte mixture to be separated. After an equilibration time of approximately 1 hour, the slurry is filtered and washed with buffer. The ion exchanger can be chosen so that the desired analyte is adsorbed onto the exchanger or remains unbound in solution. If the latter is the case, the desired material is in the filtrate. If the desired analyte is bound to the exchanger, it can be removed by suspending the exchanger in a buffer of greater ionic strength or different pH. Batch processes have some advantages over column methods. They are rapid, and the problems of packing, channeling, and dry columns are avoided.

Another development in ion-exchange column chromatography allows the separation of proteins according to their isoelectric points. This technique, **chromatofocusing,** involves the formation of a pH gradient on an ion-exchange column. If a buffer of a specified pH is passed through an ion-exchange column that was equilibrated at a second pH, a pH gradient is formed on the column. Proteins bound to the ion exchanger are eluted in the order of their isoelectric points. In addition, protein band concentration (focusing) takes place during elution. Chromatofocusing is similar to isoelectric focusing, introduced in Chapter 6, in which a column pH gradient is produced by an electric current.

Storage of Resins

Most ion exchangers in the dry form are stable for many years. Aqueous slurried ion exchangers are still useful after several months. One major storage problem with a wet exchanger is microbial growth. This is especially true for the cellulose and dextran exchangers. If it is necessary to store pretreated exchangers, an antimicrobial agent must be added to the slurry. Sodium azide (0.02%) is suitable for cation exchangers and phenylmercuric salts (0.001%) are effective for anion exchangers. Since these preservative reagents are toxic, they must be used with caution.

➤   **Study Exercise 5.2**   A mixture of amino acids (Lys, Ala, Asp) is subjected to cation exchange chromatography at pH 3.0. Predict the order of elution of the three amino acids and explain your answer. **Hint:** Draw the structures of the three amino acids as they would exist at pH 3. Study the structures and determine how each would interact with the negatively charged ion-exchange resin. Determine the relative strength of binding for each amino acid.

➤   **Study Exercise 5.3**   A mixture of three proteins was subjected to ion-exchange chromatography using CM-cellulose as the stationary phase. Predict the order of elution of the proteins assuming that the mixture was applied at low ionic strength and eluted with buffers of increasing ionic strength. The proteins in the mixture are listed below with isoelectric pH (pH_I).

Pepsinogen, 1.0
Cytochrome c, 10.6
Myoglobin, 6.8

Hint: The column separates proteins on the basis of net charge. The more negatively charged a protein, the more weakly it will bind to an anion exchange resin like CM-cellulose, and the faster it will elute. At any pH, the protein pepsinogen will have the most negative net charge of any of the three proteins.

➤   **Study Exercise 5.4**   What kind of ion-exchange resin would be most effective for purifying RNA molecules, an anion exchanger or cation exchanger?

E.   GEL EXCLUSION CHROMATOGRAPHY

The chromatographic methods discussed up to this point allow the separation of molecules according to polarity and charge. The method of **gel exclusion chromatography** (also called gel filtration, molecular sieve chromatography, or gel permeation chromatography) exploits the physical property of molecular size to achieve separation. The molecules of nature range in molecular weight from less than 100 to as large as several million. It should be obvious that a technique capable of separating molecules of molecular weight 10,000 from those of 100,000 would be very popular among research biochemists. Gel filtration chromatography has been of major importance in the purification of thousands of proteins, nucleic acids, enzymes, polysaccharides, and other biomolecules. In addition, the technique may be applied to molecular weight determination and quantitative analysis of molecular interactions. In this section the theory and practice of gel filtration will be introduced and applied to several biochemical problems.

Theory of Gel Filtration

The operation of a gel filtration column is illustrated in Figure 5.7. The stationary phase consists of inert particles that contain small pores of a controlled size. Microscopic examination of a particle reveals an interior resembling a sponge. A solution containing analytes of various molecular sizes is allowed to pass through the column under the influence of continuous solvent flow. Analytes larger than the pores cannot enter the interior of the gel beads, so they are limited to the space between the beads. The volume of the column accessible to very large molecules is, therefore, greatly reduced. As a

Figure 5.7

Separation of molecules by gel filtration. **A** Application of sample containing large and small molecules. **B** Large molecules cannot enter gel matrix, so they move rapidly through the column. **C** Elution of the large molecules first and then smaller molecules.

A   B   C

result, they are not slowed in their progress through the column and elute rapidly in a single zone. Small molecules capable of diffusing in and out of the beads have a much larger volume available to them. Therefore, they are delayed in their journey through the column bed. Molecules of intermediate size migrate through the column at a rate somewhere between those for large and small molecules. Therefore, the order of elution of the various analytes is directly related to their molecular dimensions.

Physical Characterization of Gel Chromatography

Several physical properties must be introduced to define the performance of a gel and solute behavior. Some important properties are:

1. **Exclusion Limit**   This is defined as the molecular mass of the smallest molecule that cannot diffuse into the inner volume of the gel matrix. All molecules above this limit elute rapidly in a single zone. The exclusion limit of a typical gel, Sephadex G-50, is 30,000 daltons. All analytes having a molecular size greater than this value would pass directly through the column bed without entering the gel pores.

2. **Fractionation Range**   Sephadex G-50 has a fractionation range of 1500 to 30,000 daltons. Analytes within this range would be separated in a somewhat linear fashion.

3. **Water Regain and Bed Volume**   Gel chromatography media are often supplied in dehydrated form and must be swollen in a solvent, usually water, before use. The weight of water taken up by 1 g of dry gel is known as the water regain. For G-50, this value is 5.0 ± 0.3 g. This value does not include the water surrounding the gel particles, so it cannot be used as an estimate of the final volume of a packed gel column. Most commercial suppliers of gel materials provide, in addition to water regain, a bed volume value. This is the final volume taken up by 1 g of dry gel when swollen in water. For G-50, bed volume is 9 to 11 mL/g dry gel.

4. **Gel Particle Shape and Size**   Ideally, gel particles should be spherical to provide a uniform bed with a high density of pores. Particle size is defined either by mesh size or bead diameter (μm). The degree of resolution afforded by a column and the flow rate both depend on particle size. Larger particle sizes (50 to 100 mesh, 100 to 300 μm) offer high flow rates but poor chromatographic separation. The opposite is true for very small particle sizes ("superfine," 400 mesh, 10 to μm). The most useful particle size, which represents a compromise between resolution and flow rate, is 100 to 200 mesh (50 to 150 μm).

5. **Void Volume**   This is the total space surrounding the gel particles in a packed column. This value is determined by measuring the volume of solvent required to elute a solute that is completely excluded from the gel matrix. Most columns can be calibrated for void volume with a

dye, blue dextran, which has an average molecular mass of 2,000,000 daltons.

6. **Elution Volume** This is the volume of eluting buffer necessary to remove a particular analyte from a packed column.

Chemical Properties of Gels

Four basic types of gels are available: **dextran, polyacrylamide, agarose, and combined polyacrylamide-dextran.** The first gels to be developed were those based on a natural polysaccharide, dextran. These are supplied by Pharmacia under the trade name Sephadex. Table 5.4 gives the physical properties of the various sizes of Sephadex. The number given each gel refers to the water regain multiplied by 10. Sephadex is available in various particle sizes labeled coarse, medium, fine, and superfine. Dextran-based gels cannot be manufactured with an exclusion limit greater than 600,000 daltons because the small extent of cross-linking is not sufficient to prevent collapse of the particles. If the dextran is cross-linked with N,N'-methylenebisacrylamide, gels for use in higher fractionation ranges are possible. Table 5.4 lists these gels, called Sephacryl.

Polyacrylamide gels are produced by the copolymerization of acrylamide and the cross-linking agent N,N'-methylenebisacrylamide. These are supplied by Bio-Rad Laboratories (Bio-Gel P). The Bio-Gel media are available in 10 sizes with exclusion limits ranging from 1800 to 400,000 daltons. Table 5.4 lists the acrylamide gels and their physical properties.

The agarose gels have the advantage of very high exclusion limits. Agarose, the neutral polysaccharide component of agar, is composed of alternating galactose and anhydrogalactose units. The gel structure is stabilized by hydrogen bonds rather than by covalent cross-linking. Agarose gels, supplied by Bio-Rad Laboratories (Bio-Gel A) and by Pharmacia (Sepharose and Superose), are listed in Table 5.4.

The combined polyacrylamide-agarose gels are commercially available under the trade name Ultragel. These consist of cross-linked polyacrylamide with agarose trapped within the gel network. The polyacrylamide gel allows a high degree of separation and the agarose maintains gel rigidity, so high flow rates may be used.

Selecting a Gel

The selection of the proper gel is a critical stage in successful gel chromatography. Most gel chromatographic experiments can be classified as either **group separations** or **fractionations.** Group separations involve dividing the components of a sample into two groups, a fraction of relatively low-molecular-weight analytes and a fraction of relatively high-molecular-weight analytes. Specific examples of this are desalting a protein solution or removing small contaminating molecules from protein or nucleic acid extracts. For group separations, a gel should be chosen that allows complete exclusion of the high-molecular-weight molecules in the void volume. Sephadex G-25,

Table 5.4

Properties of Gel Filtration Media

Name	Fractionation Range for Proteins (daltons)	Water Regain (mL/g dry gel)	Bed Volume (mL/g dry gel)
Dextran (Sephadex)[1]			
G-10	0–700	1.0 ± 0.1	2–3
G-15	0–1500	1.5 ± 0.2	2.5–3.5
G-25	1000–5000	2.5 ± 0.2	4–6
G-50	1500–30,000	5.0 ± 0.3	9–11
G-75	3000–80,000	7.5 ± 0.5	12–15
G-100	4000–150,000	10 ± 1.0	15–20
G-150	5000–300,000	15 ± 1.5	20–30
G-200	5000–600,000	20 ± 2.0	30–40
Polyacrylamide (Bio-Gels)[2]			
P-2	100–1800	1.5	3.0
P-4	800–4000	2.4	4.8
P-6	1000–6000	3.7	7.4
P-10	1500–20,000	4.5	9.0
P-30	2500–40,000	5.7	11.4
P-60	3000–60,000	7.2	14.4
P-100	5000–100,000	7.5	15.0
P-150	15,000–150,000	9.2	18.4
P-200	30,000–200,000	14.7	29.4
P-300	60,000–400,000	18.0	36.0
Dextran-polyacrylamide (Sephacryl)[1]			
S-100 HR	1000–100,000	—	—
S-200 HR	5000–250,000	—	—
S-300 HR	10,000–1,500,000	—	—
S-400 HR	20,000–8,000,000	—	—
Agarose			
Sepharose[1] 6B	10,000–4,000,000	—	—
Sepharose 4B	60,000–20,000,000	—	—
Sepharose 2B	70,000–40,000,000	—	—
Superose[1] 12 HR	1000–300,000	—	—
Superose 6 HR	5000–5,000,000	—	—
Bio-Gel[2] A-0.5	10,000–500,000	—	—
Bio-Gel A-1.5	10,000–1,500,000	—	—
Bio-Gel A-5	10,000–5,000,000	—	—
Bio-Gel A-15	40,000–15,000,000	—	—
Bio-Gel A-50	100,000–50,000,000	—	—
Bio-Gel A-150	1,000,000–150,000,000	—	—
Vinyl (Fractogel TSK)[3]			
HW-40	100–10,000	—	—
HW-55	1000–700,000	—	—
HW-65	50,000–5,000,000	—	—
HW-75	500,000–50,000,000	—	—

[1]*Amersham – Pharmacia – LKB – Hoefer.*

[2]*Bio-Rad Laboratories.*

[3]*Pierce Biotechnology.*

Bio-Gel P-6, and Sephacryl S-100HR are recommended for most group separations. The particle size recommended is 100 to 200 mesh or 50 to 150 μm diameter.

Gel fractionation involves separation of groups of molecules of similar molecular weights in a multicomponent mixture. In this case, the gel should be chosen so that the fractionation range includes the molecular weights of the analytes. If the mixture contains macromolecules up to 120,000 in molecular weight, then Bio-Gel P-150, Sephacryl S-200HR, or Sephadex G-150 would be most appropriate. If P-100, G-100, or Sephacryl S-100HR were used, some of the higher-molecular-weight proteins in the sample would elute in the void volume. On the other hand, if P-200, P-300, or G-200 were used, there would be a decrease in both resolution and flow rate. If the molecular weight range of the mixture is unknown, empirical selection is necessary. The recommended gel grade for most fractionations is 100–200 or 200–400 mesh (20–80 μm or 10–40 μm). The finest grade that allows a suitable flow rate should be selected. For very critical separations, superfine grades offer the best resolution but with very low flow rates.

Gel Preparation and Storage

The dextran and acrylamide gel products are sometimes supplied in dehydrated form and must be allowed to swell in water before use. The swelling time required differs for each gel, but the extremes are 3 to 4 hours at 20°C for highly cross-linked gels and up to 72 hours at 20°C for P-300 or G-200. The swelling time can be shortened if a boiling-water bath is used. Agarose gels and combined polyacrylamide-agarose gels are supplied in a hydrated state, so there is no need for swelling.

Before a gel slurry is packed into the column, it should be defined and deaerated. Defining is necessary to remove very fine particles, which would reduce flow rates. To define, pour the gel slurry into a graduated cylinder and add water equivalent to two times the gel volume. Invert the cylinder several times and allow the gel to settle. After 90 to 95% of the gel has settled, decant the supernatant, add water, and repeat the settling process. Two or three defining operations are usually sufficient to remove most small particles.

Deaerating (removing dissolved gases) should be done on the gel slurry and all eluting buffers. Gel particles that have not been deaerated tend to float and form bubbles in the column bed. Dissolved gases are removed by placing the gel slurry in a side-arm vacuum flask and applying a vacuum from a water aspirator. The degassing process is complete when no more small air bubbles are released from the gel (usually 1 to 2 hours).

Antimicrobial agents must be added to stored, hydrated gels. One of the best agents is sodium azide (0.02%).

Operation of a Gel Column

The procedure for gel column chromatography is very similar to the general description given earlier. The same precautions must be considered in packing, loading, and eluting the column. A brief outline of important considerations follows.

Column Size

For fractionation purposes, it is usually not necessary to use columns greater than 100 cm in length. The ratio of bed length to width should be between 25 and 100. For group separations, columns less than 50 cm long are sufficient, and appropriate ratios of bed length to width are between 5 and 10.

Eluting Buffer

There are fewer restrictions on buffer choice in gel chromatography than in ion-exchange chromatography. Dextran and polyacrylamide gels are stable in the pH range 1 to 10, whereas agarose gels are limited to pH 4 to 10. Since there is such a wide range of stability of the gels, the buffer pH should be chosen on the basis of the range of stability of the macromolecules to be separated.

Sample Volume

The sample volume is a critical factor in planning a gel chromatography experiment. If too much sample is applied to a column, resolution is decreased; if the sample size is too small, the analytes are greatly diluted. For group separations, a sample volume of 10 to 25% of the column total volume is suitable. The sample volume for fractionation procedures should be between 1 and 5% of the total volume. Column total volume is determined by measuring the volume of water in the glass column that is equivalent to the height of the packed bed.

Column Flow Rate

The flow rate of a gel column depends on many factors, including length of column and type and size of the gel. It is generally safe to elute a gel column at a rate slightly less than free flow. A high flow rate reduces sample diffusion or zone broadening but may not allow complete equilibration of analyte molecules with the gel matrix.

A specific flow rate cannot be recommended, since each type of gel requires a different range. The average flow rate given in literature references for small-pore-size gels is 8 to 12 mL/cm^2 of cross-sectional bed area per hour (15 to 25 mL/hr). For large-pore-size gels, a value of 2 to 5 mL/cm^2 of cross-sectional bed area per hour (5 to 10 mL/hr) is average.

Eluent can be made to flow through a column by either of two methods, gravity or pump elution. Gravity elution is most often used because no special equipment is required. It is quite acceptable for developing a column used for group separations and fractionations when small-pore-sized gels are used. However, if the flow rate must be maintained at a constant value throughout an experiment or if large-pore gels are used, pump elution is recommended.

One variation of gel chromatography is ascending eluent flow. Some investigators report more reproducible results, better resolution, and a more constant flow rate if the eluting buffer is pumped backward through the gel. This type of experiment requires special equipment, including a specialized column, and a peristaltic pump.

Applications of Gel Exclusion Chromatography

Several experimental applications of gel chromatography have already been mentioned, but more detail will be given here.

Desalting

Inorganic salts, organic solvents, and other small molecules are used extensively for the purification of macromolecules. Gel chromatography provides an inexpensive, simple, and rapid method for removal of these small molecules. One especially attractive method for desalting very small samples (0.1 mL or less) of proteins or nucleic acid solutions is to use spin columns. These are prepacked columns of polyacrylamide exclusion gels. Spin columns are used in a similar fashion to microfiltration centrifuge tubes (Chapter 3, pp. 77–81). The sample is placed on top of the gel column and spun in a centrifuge. Large molecules are eluted from the column and collected in a reservoir. The small molecules to be removed remain in the gel.

Purification of Biomolecules

This is probably the most popular use of gel chromatography. Because of the ability of a gel to fractionate molecules on the basis of size, gel filtration complements other purification techniques that separate molecules on the basis of polarity and charge.

Estimation of Molecular Weight

The elution volume for a particular analyte is proportional to its molecular size. This indicates that it is possible to estimate the molecular weight of a molecule on the basis of its elution characteristics on a gel column. An elution curve for several standard proteins on Sephadex G-100 is shown in Figure 5.8. This curve, a plot of protein concentration A_{280} vs. volume collected, is representative of data obtained from a gel filtration experiment. The elution volume, V_e, for each protein can be estimated as shown in the figure. A plot of log molecular mass vs. elution volume for the proteins is shown in Figure 5.9. Note that the linear portion of the curve in Figure 5.9 covers the molecular mass range of 10,000 to 100,000 daltons. Molecules below 10,000 daltons are not eluted in an elution volume proportional to size. Since they readily diffuse into the gel particles, they are retarded. Molecules larger than 100,000 daltons are all excluded from the gel in the void volume. A solution of the unknown protein is chromatographed through the calibrated column under conditions identical to those for the standards and the elution volume is measured. The unknown molecular size is then read directly from the graph. This method of molecular weight estimation is widely used because it is simple, inexpensive, and fast. It

Figure 5.8

Elution curve for a mixture of several proteins using gel filtration chromatography.
A = hemoglobin;
B = egg albumin;
C = chymotrypsinogen;
D = myoglobin;
E = cytochrome *c*.

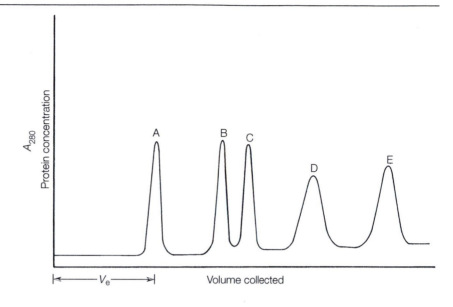

Figure 5.9

A plot of log molecular mass vs. elution volume for the proteins in Figure 5.8.

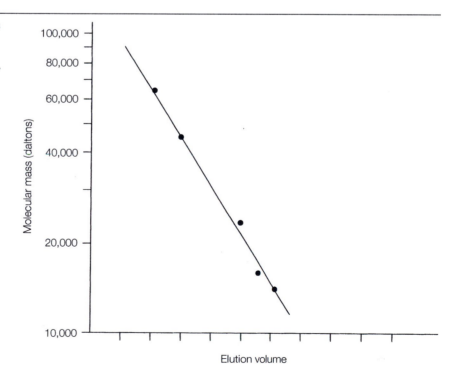

can be used with highly purified or impure samples. There are, of course, some limitations to consider. The gel must be chosen so that the molecular weight of the unknown is within the linear section of the curve. The method assumes that only steric and partition effects influence the elution of the standards and unknown. If a protein interacts with the gel by adsorptive or ionic processes, the estimate of molecular weight is lower than the true value. The assumption is also made that the unknown molecules have a general spherical shape, not an elongated or rod shape. Although this is a convenient and rapid method for molecular weight determination, the final number is only an estimate.

Gel Chromatography in Organic Solvents

The gels discussed so far in this section are hydrophilic, and the inner matrix retains its integrity only in aqueous solvents. Because there is a need for gel chromatography of nonhydrophilic molecules, gels have been produced that can be used with organic solvents. The trade names of some of these products are Sepharose CL and Sephadex LH, Bio Beads S, and Styragel. Organic solvents that are of value in gel chromatography are ethanol, acetone, dimethylsulfoxide, dimethylformamide, tetrahydrofuran, chlorinated hydrocarbons, and acetonitrile.

➤ **Study Exercise 5.5** Your first experiment in biochemistry lab is to analyze a mixture with the use of gel filtration. You are instructed to run a Sephadex G-50 column on the mixture containing the components listed below. All three components are colored, so you are able to observe the separation. The components are dissolved in water and the column is eluted with water.

Blue dextran, a blue dye (MW = about 2,000,000)

Cytochrome *c*, a red protein (MW = 12,400)

Flavin adenine dinucleotide, a yellow coenzyme (FAD, MW = 830)

Predict the results of your experiment.

Solution: According to Table 5.4, the fractionation range for G-50 is 1500–30,000. The Blue dextran is above the fractionation range, so it will elute very rapidly in the void volume. The FAD is below the fractionation range, so it will take its time and elute much later than Blue dextran and cyt *c*. The red-colored cyt *c*, within the fractionation range, will elute somewhere between the Blue dextran and FAD.

➤ **Study Exercise 5.6** Your biochemistry lab instructor gives you a mixture of three proteins and asks you to separate the proteins on a Sephadex G-100 gel. The three proteins are serum albumin, myoglobin, and chymotrypsinogen. The column is eluted with 0.1 M Tris buffer, pH 7.0. Predict the order of elution of the proteins from the Sephadex column. Use Appendix V to find the molecular weights of the proteins.

F.   HIGH-PERFORMANCE LIQUID CHROMATOGRAPHY (HPLC)

The previous discussions on the theory and practice of the various chromatographic methods should convince you of the tremendous influence chromatography has had on our biochemical understanding. It is tempting, but unfair, to make comparisons about the relative importance of the methods; because each serves a specific purpose. There will, for example, always be a need for fast, inexpensive, and qualitative analyses as afforded by planar chromatography. Traditional column chromatography will probably always be preferred in large-scale protein purification.

However, during the past three decades, an analytical method has been developed that now surpasses the traditional liquid chromatographic techniques in importance for analytical separations. This technique, **high-performance liquid chromatography (HPLC),** is ideally suited for the separation and identification of amino acids, carbohydrates, lipids, nucleic acids, proteins, pigments, steroids, pharmaceuticals, and many other biologically active molecules.

The future promise of HPLC is indicated by its classification as "modern liquid chromatography" when compared to other forms of column-liquid chromatography, now referred to as "classical" or "traditional." Compared to the classical forms of liquid chromatography (paper, TLC, column), HPLC has several advantages:

1.   Resolution and speed of analysis far exceed the classical methods.

2.   HPLC columns can be reused without repacking or regeneration.

3.   Reproducibility is greatly improved because the parameters affecting the efficiency of the separation can be closely controlled.

4.   Instrument operation and data analysis are easily automated.

5.   HPLC is adaptable to very small sample sizes or large-scale, preparative procedures.

The advantages of HPLC are the result of two major advances: (1) the development of stationary supports with very small particle sizes and large surface areas, and (2) the improvement of elution rates by applying high pressure to the solvent flow.

The great versatility of HPLC is evidenced by the fact that all chromatographic modes, including partition, adsorption, ion exchange, chromatofocusing, and gel exclusion, are possible. In a sense, HPLC can be considered as automated liquid chromatography. The theory of each of these chromatographic modes has been discussed and needs no modification for application to HPLC. However, there are unique theoretical and practical characteristics of HPLC that should be introduced.

The **retention time** of an analyte in HPLC (t_R) is defined as the time necessary for maximum elution of the particular analyte. This is analogous to retention time measurements in GC. **Retention volume** (V_R) of an analyte is the solvent volume required to elute the analyte and is defined by Equation 5.2, where F is the flow rate of the solvent.

>> $V_R = Ft_R$ **Equation 5.2**

In all forms of chromatography, a measure of column efficiency is **resolution**, R. Resolution indicates how well analytes are separated; it is defined by Equation 5.3, where t_R and t'_R are the retention times of two analytes and w and w' are the base peak widths of the same two analytes.

>> $$R = 2\frac{t_R - t'_R}{w + w'}$$ **Equation 5.3**

Instrumentation

The increased resolution achieved in HPLC compared to classical column chromatography is primarily the result of adsorbents of very small particle sizes (less than 20 μm) and large surface areas. The smallest gel beads used in gel exclusion chromatography are "superfine" grade with diameters of 20 to 50 μm. Recall that the smaller the particle size, the lower the flow rate; therefore, it is not feasible to use very small gel beads in liquid column chromatography because low flow rates lead to solute diffusion and the time necessary for completion of an analysis would be impractical. In HPLC, increased flow rates are obtained by applying a pressure differential across the column. A combination of high pressure and adsorbents of small particle size leads to the high resolving power and short analysis times characteristic of HPLC.

A schematic diagram of a typical high-pressure liquid chromatograph is shown in Figure 5.10. The basic components are a solvent reservoir, high-pressure pump, packed column, detector, and recorder. A computer is used to control the process and to collect and analyze data.

Solvent Reservoir

The solvent chamber should have a capacity of at least 500 mL for analytical applications, but larger reservoirs are required for preparative work. In order to avoid bubbles in the column and detector, the solvent must be degassed. Several methods may be used to remove unwanted gases, including refluxing, filtration through a vacuum filter, ultrasonic vibration, and purging with an inert gas. The solvent should also be filtered to remove particulate matter that would be drawn into the pump and column.

Pumping Systems

The purpose of the pump is to provide a constant, reproducible flow of solvent through the column. Two types of pumps are available–constant pressure and constant volume. Typical requirements for a pump are:

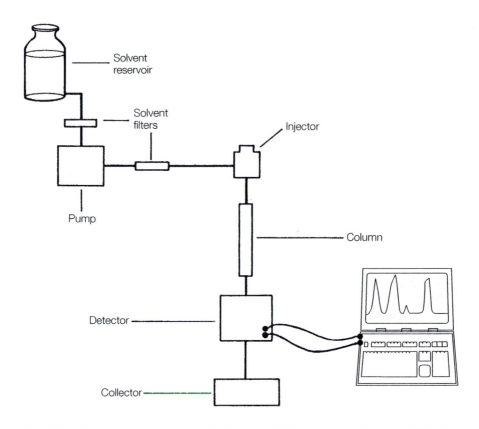

Figure 5.10

A schematic diagram of a high-performance liquid chromatograph.

1. It must be capable of pressure outputs of at least 500 psi and preferably up to 5000 psi.
2. It should have a controlled, reproducible flow delivery of about 1 mL/min for analytical applications and up to 100 mL/min for preparative applications.
3. It should yield pulse-free solvent flow.
4. It should have a small holdup volume.

Although neither type of pump meets all these criteria, constant-volume pumps maintain a more accurate flow rate and a more precise analysis is obtained.

Injection Port

A sample must be introduced onto the column in an efficient and reproducible manner. One of the most popular injectors is the syringe injector. The sample, in a microliter syringe, is injected through a neoprene/Teflon septum. This type of injection can be used at pressures up to 3000 psi.

Columns

HPLC columns are prepared from stainless steel or glass-Teflon tubing. Typical column inside diameters are 2.1, 3.2, or 4.5 mm for analytical separations and up to 30 mm for preparative applications. The length of the column can range from 5 to 100 cm, but 10 to 20 cm columns are common.

Detector

Liquid chromatographs are equipped with a means to continuously monitor the column effluent and recognize the presence of analyte. Only small sample sizes are used with most HPLC columns, so a detector must have high sensitivity. The type of detector that has the most universal application is the **differential refractometer.** This device continuously monitors the refractive index difference between the mobile phase (pure solvent) and the mobile phase containing sample (column effluent). The sensitivity of this detector is on the order of 0.1 μg, which, compared to other detectors, is only moderately sensitive. The major advantage of the refractometer detector is its versatility; its main limitation is that there must be at least 10^{-7} refractive index units between the mobile phase and sample.

The most widely used HPLC detectors are the **photometric detectors.** These detectors measure the extent of absorption of ultraviolet or visible radiation by a sample. Since few compounds are colored, visible detectors are of limited value. Ultraviolet detectors are the most widely used in HPLC. The typical UV detector functions by focusing radiation from a low-pressure mercury lamp on a flow cell that contains column effluent. The mercury lamp provides a primary radiation at 254 nm. The use of filters or other lamps provides radiation at 220, 280, 313, 334, and 365 nm. Many compounds absorb strongly in this wavelength range, and sensitivities on the order of 1 ng are possible. Most biochemicals are detected, including proteins, nucleic acids, pigments, vitamins, some steroids, and aromatic amino acids. Aliphatic amino acids, carbohydrates, lipids, and other biochemicals that do not absorb UV can be detected by chemical derivatization with UV-absorbing functional groups. UV detectors have many positive characteristics, including high sensitivity, small sample volumes, linearity over wide concentration ranges, nondestructiveness to sample, and suitability for gradient elution.

A third type of detector that has only limited use is the **fluorescence detector.** This type of detector is extremely sensitive: its use is limited to samples containing trace quantities of biological materials. Its response is not linear over a wide range of concentrations, but it may be up to 100 times more sensitive than the UV detector.

Collection of Eluent

All of the detectors described here are nondestructive to the samples, so column effluent can be collected for further chemical and physical analysis.

Analysis of HPLC Data

Most HPLC instruments are on line with an integrator and a computer for data handling. For quantitative analysis of HPLC data, operating parameters such as rate of solvent flow must be controlled. In modern instruments, the whole system (including the pump, injector, detector, and data system) is under the control of a computer.

Figure 5.11 illustrates the separation by HPLC of several phenylhydantoin derivatives of amino acids.

Column: Microsorb Cyano, 5 μm, 4.6 mm ID × 25 cm L
Mobile phase: 15.5% THF and 17.1% acetonitrile in 6m*M* phosphate, pH 3.2
Flow: 1 mL/min
Temperature: 35°C
Detection: UV, 254 nm

Peak identification

1. Cys-A	12. Cys
2. Asn	13. Tyr
3. Gln	14. Pro
4. Ser	15. Val
5. Thr	16. Met
6. Asp	17. Ile
7. Gly	18. Phe
8. His	19. Leu
9. Glu	20. nor-Leu
10. Ala	21. Trp
11. Arg	22. Lys

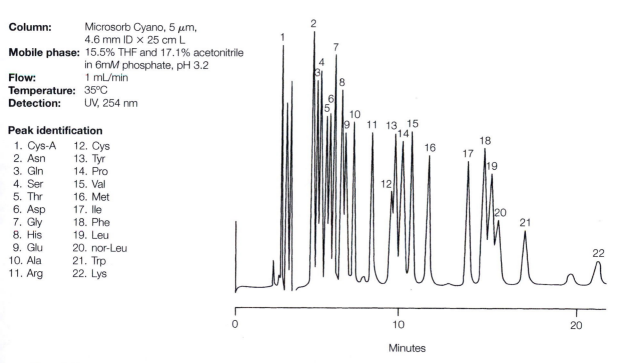

Figure 5.11

The separation of several amino acid phenylhydantoins by HPLC. *Courtesy of Rainin Instrument Co., Woburn, MA; www.rainin.com/.*

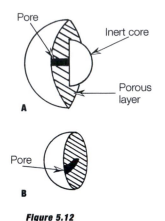

Figure 5.12

Adsorbents used in HPLC.
A Porous layer with short pores.
B Microporous particles with
longer pores.

Stationary Phases in HPLC

The adsorbents in HPLC are typically small-diameter, porous materials. Two types of stationary phases are available. **Porous layer beads** (Figure 5.12A) have an inert solid core with a thin porous outer shell of silica, alumina, or ion-exchange resin. The average diameter of the beads ranges from 20 to 45 μm. They are especially useful for analytical applications, but, because of their short pores their capacities are too low for preparative applications.

Microporous particles are available in two sizes: 20 to 40 μm diameter with longer pores and 5 to 10 μm with short pores (see Figure 5.12B). These are now more widely used than the porous layer beads because they offer greater resolution and faster separations with lower pressures. The microporous beads are prepared from alumina, silica, ion-exchanger resins, and chemically bonded phases (see next section).

The HPLC can function in several chromatographic modes. Each type of chromatography will be discussed in the following subsections, with information about stationary phases.

Liquid-Solid (Adsorption) Chromatography

HPLC in the adsorption mode can be carried out with silica or alumina porous-layer-bead columns. Small glass beads are often used for the inert core. Some of the more widely used packings are μ Porasil (Waters Associates), BioSilA (Bio-Rad Laboratories), LiChrosorb Si-100 Partisil, Vydac, ALOX 60D (several suppliers), and Supelcosil (Supelco).

In high-pressure adsorption chromatography, analytes adsorb with different affinities to binding sites in the solid stationary phase. Separation of analytes in a sample mixture occurs because polar molecules adsorb more strongly than nonpolar molecules. Therefore, the various components in a sample are eluted with different retention times from the column. This form of HPLC is usually called **normal phase** (polar stationary phase and a nonpolar mobile phase).

Liquid-Liquid (Partition) Chromatography

In the early days of HPLC (1970–78), solid supports were coated with a liquid stationary phase. Columns with these packings had short lifetimes and a gradual decrease in resolution because there was continuous loss of the liquid stationary phase with use of the column.

This problem was remedied by the discovery of methods for chemically bonding the active stationary phase to the inert support. Most chemically bonded stationary phases are produced by covalent modification of the surface silica. Three modification processes are shown in Equations 5.4–5.6.

>> Silicate esters

$$\text{—Si—OH} + \text{ROH} \longrightarrow \text{—Si—O—R}$$

Equation 5.4

>> Silica-carbon

$$-Si-Cl + \begin{array}{c} RMgBr \\ or \\ RLi \end{array} \longrightarrow -Si-R \qquad \textit{Equation 5.5}$$

>> Siloxanes

$$-Si-OH + \begin{array}{c} ClSiR_3 \\ or \\ ROSiR_3 \end{array} \longrightarrow -Si-O-SiR_3 \qquad \textit{Equation 5.6}$$

The major advantage of a bonded stationary phase is stability. Since it is chemically bonded, there is very little loss of stationary phase with column use. The siloxanes are the most widely used silica supports. Functional groups that can be attached as siloxanes are alkylnitriles ($-Si-CH_2CH_2-CN$), phenyl ($-Si-C_6H_5$), alkylamines ($-Si(CH_2)_nNH)_2-$, and alkyl side chains ($-Si-C_8H_{17}$; $-Si-C_{18}H_{37}$).

The use of nonpolar chemically bonded stationary phases with a polar mobile phase is referred to as **reverse-phase HPLC.** This technique separates sample components according to hydrophobicity. It is widely used for the separation of all types of biomolecules, including peptides, nucleotides, carbohydrates, and derivatives of amino acids. Typical solvent systems are water-methanol, water-acetonitrile, and water-tetrahydrofuran mixtures. Figure 5.13 shows the results of protein separation on a silica-based reverse-phase column.

Ion-Exchange Chromatography

Ion-exchange HPLC uses column packings with charged functional groups. Structures of typical ion exchangers are shown in Figure 5.14. They are prepared by chemically bonding the ionic groups to the support via silicon atoms or by using polystyrene-divinylbenzene resins. These stationary phases may be used for the separation of proteins, peptides, and other charged biomolecules.

Gel Exclusion Chromatography

The combination of HPLC and gel exclusion chromatography is used extensively for the separation of large biomolecules, especially proteins and nucleic acids.

The exclusion gels discussed in Section E are not appropriate for HPLC use because they are soft and, except for small-pore beads (G-25 and less), collapse under high-pressure conditions. Semirigid gels based on cross-linked styrene-divinylbenzene, polyacrylamide, and vinyl-acetate copolymer are available with various fractionation ranges useful for the separation of molecules up to 10,000,000 daltons.

Rigid packings for HPLC gel exclusion are prepared from porous glass or silica. They have several advantages over the semirigid gels, including

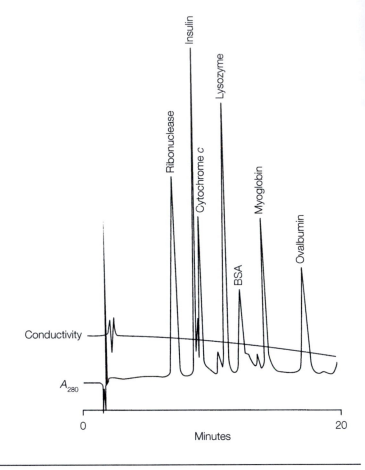

Conditions

column:	Hi-Pore RP-304 column, 250 × 4.6 mm
Sample:	Protein standard
Buffers:	A. H_2O, 0.1% B
	B. 95% Acetonitrile, 5% H_2O, 0.1% TFA
Gradient:	25%–100% B in 30 minutes
Flow rate:	1.5 mL/min
Temperature:	Ambient
Detection:	UV @ 280 nm; conductivity

Figure 5.13

Reverse-phase chromatography of a mixture of standard proteins. *Separation courtesy of Bio-Rad Laboratories, Hercules, CA; www.bio-rad.com/.*

several fractionation ranges, ease of packing, and compatibility with water and organic solvents.

Chiral Chromatography

HPLC has been extremely effective in separating and analyzing a broad range of biological molecules, including amino acids, proteins, lipids, nucleic acids, and carbohydrates, as described in the previous subsections. One class of biologically important compounds that has shown resistance to separation are the enantiomorphic forms of biomolecules that exist because of the presence of a chiral center or stereocenter. For most molecules a chiral center is a carbon atom that is surrounded by four different groups. Compounds that have a stereocenter exist in two molecular forms (**enantiomers**) that have the

Figure 5.14

Ion exchangers used in HPLC.

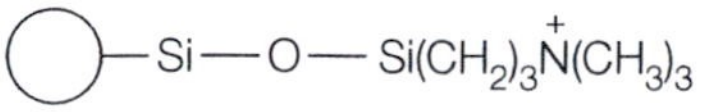

Anion exchanger

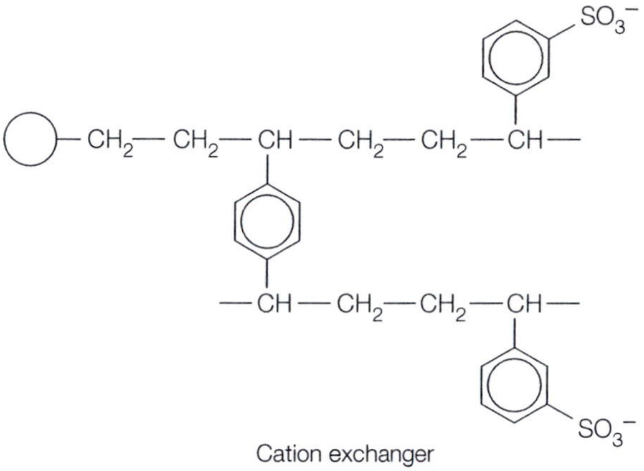

Cation exchanger

same physical and chemical properties. They differ only in the way they interact with plane-polarized light—one enantiomer rotates the light to the left (L) and one enantiomer rotates light to the right (D). We now know that enantiomers also differ in their biological actions or physiological effects. This has become increasingly important in medical treatment because many drugs are chiral and the enantiomers differ in how they interact with receptors. One enantiomeric form of a drug may interact to cause the desired effect, and one may lead to no effect or even to toxicity. For example, the drug thalidomide was widely used in Europe during the late 1950s and early 1960s as a sedative/tranquilizer, especially for pregnant women. The original drug form was the racemic mixture containing two thalidomide enantiomers. Thousands of the women taking the drug later delivered malformed babies. We now know that only one enantiomer (D, R) causes the desired sedative effect. The other enantiomer (L, S) is teratogenic.

It has become essential to develop techniques for the separation and analysis of enantiomers by chromatography. Some current drugs whose active ingredient is a single enantiomer include Lipitor (Atorvastatin), Zocor (Simvastatin), Nexium (Esomeprazole), Plavix (Clopidogrel), Advair (Fluticasone), and Zoloft (Sertraline).

A logical approach to the separation of enantiomers is to use a chiral stationary phase, because it is known that chiral compounds are able to interact selectively, favoring one enantiomer over the other. Because of the different spatial arrangements around the stereocenters of the enantiomers, they interact differently with a chiral surface. The most effective chiral stationary phases contain proteins that are, of course, composed of amino acids, each of which has a stereocenter (except gly). Many protein sorbents have been tested, and the most successful results have come from alpha1-acid glycoproteins (AGP), human serum albumin (HSA), and cellobiohydrolase (CBH). For example, the proteins AGP and HSA on HPLC columns have been used to separate the enantiomers of the antiinflammatory drug ibuprofen (see Figure 5.15).

The Mobile Phase

Selection of a column packing that is appropriate for a given analysis does not ensure a successful HPLC separation. A suitable solvent system must also be chosen. Several critical solvent properties will be considered here.

Purity Very-high-purity solvents with no particulate matter are required. Many laboratory workers do not purchase expensive prepurified solvents, but rather they purify lesser grade solvents by microfiltration through a Millipore system or distillation in glass.

Reactivity The mobile phase must not react with the analytical sample or column packing. This does not present a major limitation since many relatively unreactive hydrocarbons, alkyl halides, and alcohols are suitable.

Detector Compatibility A solvent must be carefully chosen to avoid interference with the detector. Most UV detectors monitor the column effluent at 254 nm. Any UV-absorbing solvent, such as benzene or olefins, would be unacceptable because of high background. Since refractometer detectors monitor the difference in refractive index between solvent and column effluent, a greater difference leads to greater ability to detect the solute.

Solvents for HPLC Operation

It has long been recognized that the eluting power of a solvent is related to its polarity. Chromatographic solvents have been organized into a list according to their ability to displace adsorbed molecules (eluting power, $\varepsilon°$). This list of solvents, called an **eluotropic series,** is shown in Table 5.5. It should be noted that $\varepsilon°$ increases with an increase in polarity. Using the eluotropic series makes solvent choice less a matter of trial and error.

Occasionally, a single solvent does not provide suitable resolution of analytes. Solvent **binary mixtures** can be prepared with eluent strengths intermediate between the $\varepsilon°$ values for the individual solvents.

Gradient Elution in HPLC

Figure 5.16A illustrates the separation of a multicomponent sample using a single solvent and a silica column. Note that some earlier components (1 to 5) are poorly resolved and later peaks (9 to 11) display extensive broadening. This common difficulty encountered in the analysis of multicomponent samples is referred to as the **general elution problem.** It is due to the fact that the com-

IBUPROFEN

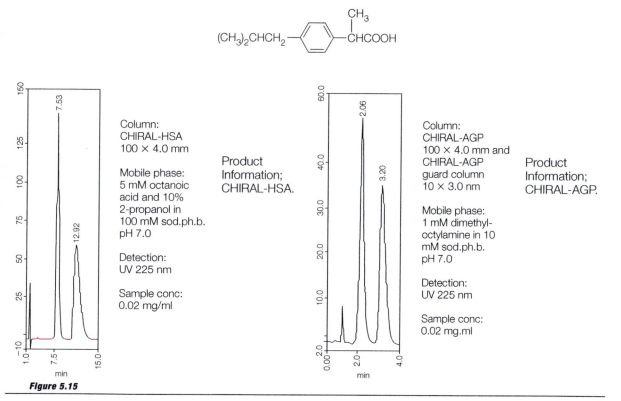

Figure 5.15

Separation of the enantiomers of the painkiller, ibuprofen on two different chiral supports. *Courtesy of ChromTech; www.chromtech.se.*

ponents have a wide range of K_D values and no single solvent system is equally effective in displacing all components from the column.

The general elution problem is solved by the use of **gradient elution.** This is achieved by varying the composition of the mobile phase during elution. In practice, gradient elution is performed by beginning with a weakly eluting solvent (low $\varepsilon°$) and gradually increasing the concentration of a more strongly eluting solvent (higher $\varepsilon°$). The weaker solvent is able to improve resolution of components having low K_D values (peaks 1 to 5, Figure 5.16B). In addition, the gradual increase in $\varepsilon°$ of the solvent mixture decreases line broadening of components 9 to 11 and provides a more effective separation.

The choice of solvents for gradient elution is still somewhat empirical; however, using the data from Table 5.5 narrows the choices. Modern HPLC instruments are equipped with **solvent programming units** that control gradient elution in a stepwise or continuous manner.

Table 5.5

Eluotropic Series of HPLC Solvents

Solvent	$\varepsilon^{\circ 1}$	$n_0{}^2$	UV Cutoff (nm)
n-Pentane	0.00	1.358	210
n-Hexane	0.01	1.375	210
Cyclohexane	0.04	1.427	210
Carbon tetrachloride	0.18	1.466	265
2-Chloropropane	0.29	1.378	225
Toluene	0.29	1.496	285
Ethyl ether	0.38	1.353	220
Chloroform	0.40	1.443	245
Tetrahydrofuran	0.45	1.408	230
Acetone	0.56	1.359	330
Ethyl acetate	0.58	1.370	260
Dimethylsulfoxide	0.62	1.478	270
Acetonitrile	0.65	1.344	210
2-Propanol	0.82	1.380	210
Ethanol	0.88	1.361	210
Methanol	0.95	1.329	210
Ethylene glycol	1.11	1.427	210

[1]Eluent strength.

[2]Refractive index.

Sample Preparation and Selection of HPLC Operating Conditions

During the initial stages of biochemical sample preparation, the sample is often quite crude; it may contain hundreds of components in addition to the desired biomolecules. Most samples must be pretreated before optimum HPLC results can be expected. The following procedures may be needed in order to convert a crude sample into a clean one: desalting, removal of anions and cations, removal of metal ions, concentration of the desired macromolecules, removal of detergent, and particulate removal. Sample preparation techniques used to achieve these results are gel exclusion chromatography, ion-exchange chromatography, microfiltration, and metal affinity chromatography. These procedures may be completed by commercially available prefilters or precolumns.

Each type of HPLC instrument has its own characteristics and operating directions. It is not feasible to describe those here. However, it is appropriate to outline the general approach taken when an HPLC analysis is desired. The following items must be considered:

1. Chemical nature and proper preparation of the sample.

2. Selection of type of chromatography (partition, adsorption, ion exchange, gel exclusion).

3. Choice of solvent system and mode of elution.

Figure 5.16

The general elution problem. **A** Normal elution. **B** Gradient elution. From L. Snyder, *J. Chromatogr. Sci.* **8,** 692 (1970).

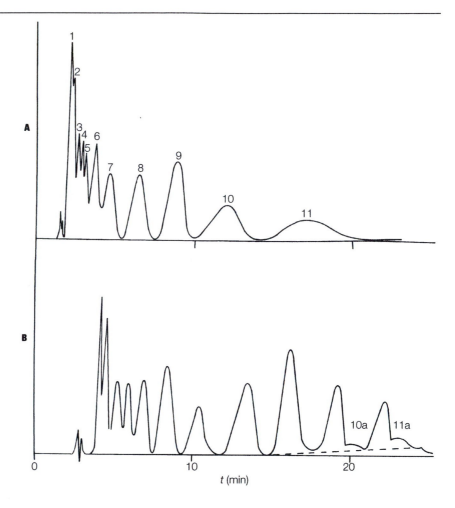

4. Selection of column packing.
5. Choice of equipment (type of detector).

FPLC—A Modification of HPLC

In 1982 Pharmacia introduced an innovative chromatographic method that provides a link between classical column chromatography and HPLC. This technique, called **fast protein liquid chromatography (FPLC),** uses experimental conditions intermediate between those of column chromatography and HPLC. The typical FPLC system requires a pump that will deliver solvent to the column in the flow rate range 1–499 mL/hr with operating pressures of 0–40 bar. (HPLC pumps deliver solvent in a flow rate range of 0.010–10 mL/min with operating pressures of 1–400 bar. Classical chromatography columns are operated at atmosphere pressure.) Also required for FPLC are a controller, detector, and fraction collector. Since lower pressures are used in FPLC than in HPLC, a wider range of column supports is possible. Chromatographic techniques incorporated in an FPLC system are gel filtration, ion exchange, affinity (see next section), hydrophobic interaction, reversed phase, and chromatofocusing.

Perfusion Chromatography

A separation method that improves resolution and decreases the time required for analysis of biomolecules has recently been introduced. This method, called **perfusion chromatography,** relies on a type of particle support called POROS, which may be used in low-pressure and high-pressure liquid chromatography applications. In conventional chromatographic separations, some biomolecules in the sample move rapidly around and past the media particles while other molecules diffuse slowly through the particles (Figure 5.17A). The result is loss of resolution because some biomolecules exit the column before others. To improve resolution, the researcher with conventional media found it necessary to reduce the flow rate to allow for diffusion processes, increasing the time required for analysis. In other words, before the development of perfusion chromatography, the researcher had to choose between high speed–low resolution and low speed–high resolution. POROS particles have two types of pores—**through pores** (6000–8000 Å in diameter), which provide channels through the particles, and connected **diffusion pores** (800–1500 Å in diameter), which line the through pores and have very short diffusion path lengths (Figure 5.17B). This combination pore system increases the porosity and the effective surface area of the particles and results in improved resolution and shorter analysis times (30 seconds to 3 minutes for POROS versus 30 minutes to several hours for conventional media).

POROS media, made by copolymerization of styrene and divinylbenzene, have high mechanical strength and are resistant to many solvents and

Figure 5.17

Transport of biomolecules through chromatographic media. **A** Conventional support particles. **B** POROS particles for perfusion chromatography. *Courtesy of PerSeptive Biosystems, Cambridge, MA; www.appliedbiosystems.com/.*

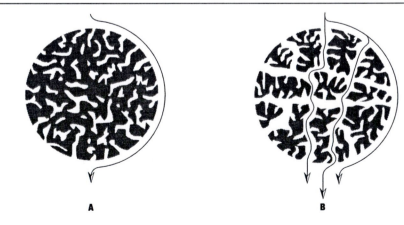

A   B

chemicals. The functional surface chemistry of the particles can be modified to provide supports for many types of chromatography, including ion exchange, hydrophobic interaction, immobilized metal affinity, reversed phase, group-selective affinity, and conventional bioaffinity. Perfusion chromatography has been applied with success to the separation of peptides, proteins, and polynucleotides on both preparative and analytical scales.

In addition to high resolution and short analysis times, perfusion chromatography has the advantage of improved recovery of biological activity because active biomolecules spend less time on the column, where denaturing conditions may exist.

G.   AFFINITY CHROMATOGRAPHY AND IMMUNOADSORPTION

The more conventional chromatographic procedures that we have studied up to this point rely on rather nonspecific physicochemical interactions between a sorbent and an analyte. The molecular characteristics of net charge, size, and polarity do not provide a basis for high selectivity in the separation and isolation of biomolecules. The desire for more specificity in chromatographic separations has led to the development of **affinity chromatography.** This technique offers the ultimate in specificity—separation on the basis of biological interactions. The biological function displayed by most macromolecules (antibodies, transport proteins, enzymes, nucleic acids, polysaccharides, receptor proteins, etc.) is a result of recognition of and interaction with specific molecules called **ligands.** This is illustrated by Equation 5.7, where M represents a macromolecule and L a smaller molecule or ligand. The two molecules interact in a specific manner to form a complex, L:M

>>   L + M $\rightleftharpoons$ L:M ----> biological response   *Equation 5.7*

In a biological system, the formation of the complex often triggers some response such as immunological action, control of a metabolic process, hormone action, catalytic breakdown of a substrate, or membrane transport. The biological response depends on proper molecular recognition and binding as shown in the reaction. The most common example of Equation 5.7 is the interaction that occurs between an enzyme molecule, E, and a substrate, S, with reversible formation of an ES complex. The biological event resulting from this interaction is the transformation of S to a metabolic product, P. Only the first step in Equation 5.7, formation of the complex, is of concern in affinity chromatography. (See Chapter 8.)

In practice, affinity chromatography requires the preparation of an insoluble sorbent, to which appropriate ligand molecules (L) are covalently affixed. Thus, ligand molecules are immobilized on the stationary support. The affinity support is packed into a column through which a mixture containing the desired macromolecule, M, is allowed to percolate. There are many types of molecules in the mixture, especially if it is a crude cell extract, but only macromolecules that recognize and bind to immobilized L are retarded in their movement through the column. After the nonbinding molecules have washed through the column, the desired macromolecules are eluted by gentle disruption of the L:M complex. Study Figure 5.18 for an illustration of affinity chromatography.

Affinity chromatography can be applied to the isolation and purification of virtually all biological macromolecules. It has been used to purify nucleic acids, enzymes, transport proteins, antibodies, hormone receptor proteins, drug-binding proteins, neurotransmitter proteins, and many others.

Successful application of affinity chromatography requires careful design of experimental conditions. The essential components, which are outlined in the following subsections, are (1) creation and preparation of a stationary matrix with immobilized ligand and (2) design of column development and eluting conditions.

Chromatographic Media

Selection of the matrix used to immobilize a ligand requires consideration of several properties. The stationary supports used in gel exclusion chromatography are found to be quite suitable for affinity chromatography because (1) they are physically and chemically stable under most experimental conditions, (2) they are relatively free of nonspecific adsorption effects, (3) they have satisfactory flow characteristics, (4) they are available with very large pore sizes, and (5) they have reactive functional groups to which an appropriate ligand may be attached.

Four types of media possess most of these desirable characteristics: agarose, polyvinyl, polyacrylamide, and controlled-porosity glass (CPG) beads. Highly porous agarose beads such as Sepharose 4B (Amersham-Pharmacia-LKB-Hoefer) and Bio-Gel A-150 m (Bio-Rad Laboratories) have virtually all of these characteristics and are the most widely used matrices. Polyacrylamide gels such as Bio-Gel P-300 (Bio-Rad) display many of the recommended features; however, the porosity is not especially high.

Figure 5.18

Purification of a macromolecule, **A,** by affinity chromatography. Ligand **B,** which has a specific affinity for **A,** is immobilized on the gel.

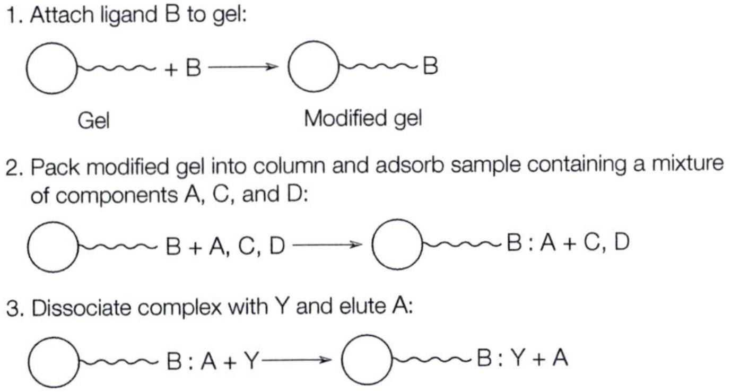

1. Attach ligand B to gel:

Gel   Modified gel

2. Pack modified gel into column and adsorb sample containing a mixture of components A, C, and D:

3. Dissociate complex with Y and elute A:

The Immobilized Ligand

The ligand, B, in Figure 5.18 can be selected only after the nature of the macromolecule to be isolated is known. When a hormone receptor protein is to be purified by affinity chromatography, the hormone itself is an ideal candidate for the ligand. For antibody isolation, an antigen or hapten may be used as ligand. If an enzyme is to be purified, a substrate analog, inhibitor, cofactor, or effector may be used as the immobilized ligand. The actual substrate molecule may be used as a ligand, but only if column conditions can be modified to avoid catalytic transformation of the bound substrate.

In addition to the foregoing requirements, the ligand must display a strong, specific, but reversible interaction with the desired macromolecule and it must have a reactive functional group for attachment to the matrix. It should be recognized that several types of ligand may be used for affinity purification of a particular macromolecule. Of course, some ligands will work better than others, and empirical binding studies can be performed to select an effective ligand.

Attachment of Ligand to Matrix

Several procedures have been developed for the covalent attachment of the ligand to the matrix. All procedures for gel modification proceed in two separate chemical steps: (1) activation of the functional groups on the matrix and (2) joining of the ligand to the functional group on the matrix. The attachment method must leave the ligand in a form and position capable of binding the desired macromolecule.

A wide variety of activated gels is now commercially available. The most widely used are described as follows.

Cyanogen Bromide-Activated Agarose

This gel is especially versatile because all ligands containing primary amino groups are easily attached to the agarose. It is available under the trade name CNBr-activated Sepharose 4B. Since the gel is extremely reactive, very gentle conditions may be used to couple the ligand. One disadvantage of CNBr activation is that small ligands are coupled very closely to the matrix surface; macromolecules, because of steric repulsion, may not be able to interact fully with the ligand. The procedure for CNBr activation and ligand coupling is outlined in Figure 5.19A.

6-Aminohexanoic Acid (CH)-Agarose and 1,6-Diaminehexane (AH)-Agarose

These activated gels overcome the steric interference problems stated above by positioning a six-carbon spacer arm between the ligand and the matrix. Ligands with free primary amino groups can be covalently attached to CH-agarose, whereas ligands with free carboxyl groups can be coupled to AH-agarose. The attachment of ligands to AH and CH gels is outlined in Figure 5.19B,C.

Figure 5.19

Attachment of specific ligands to activated gels. R = ligands.

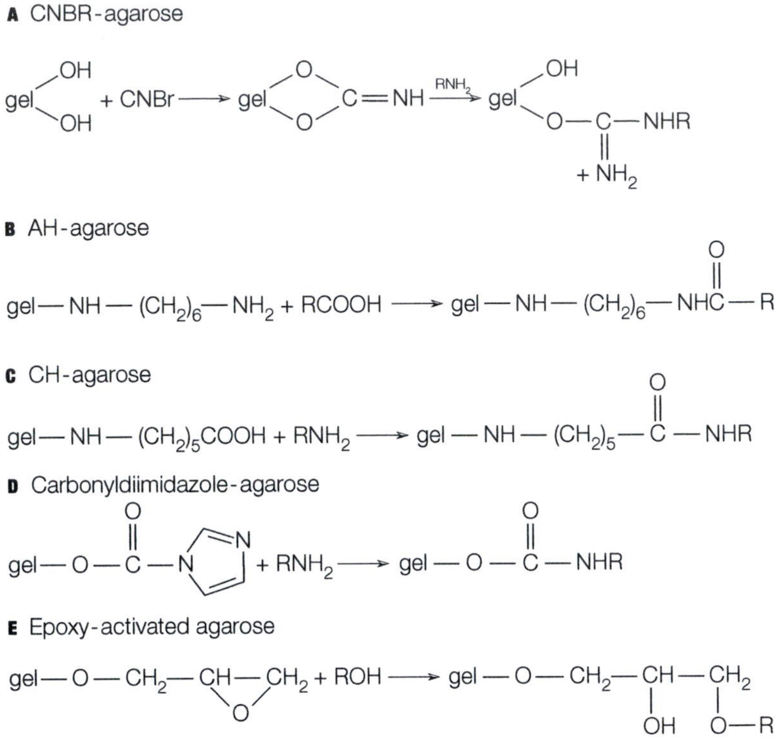

A CNBR-agarose

B AH-agarose

C CH-agarose

D Carbonyldiimidazole-agarose

E Epoxy-activated agarose

Carbonyldiimidazole (CDI)-Activated Supports

Reaction with CDI produces gels that contain uncharged N-alkylcarbamate groups (see Figures 5.19D). CDI-activated agarose, dextran, and polyvinyl acetate are sold by Pierce Biotechnology, Co. under the trade name Reacti-Gel.

Epoxy-Activated Agarose

The structure of this gel is shown in Figure 5.19E. It provides for the attachment of ligands containing hydroxyl, thiol, or amino groups. The hydroxyl groups of mono-, oligo-, and polysaccharides can readily be attached to the gel.

Group-Specific Adsorbents

The affinity materials described up to this point are modified with a ligand having specificity for a particular macromolecule. Therefore, each time a biomolecule is to be isolated by affinity chromatography, a new adsorbent must be designed and prepared. Ligands of this type are called substance specific. In contrast, **group-specific adsorbents** contain ligands that have affinity for a class of biochemically related substances. Table 5.6 shows several commercially available group-specific adsorbents and their specificities. The principles

Table 5.6

Group-Specific Adsorbents Useful in Biochemical Applications

Group-Specific Adsorbent	Group Specificity
5'-AMP-agarose	Enzymes that have NAD$^+$ cofactor; ATP-dependent kinases
Benzamidine–Sepharose	Serine proteases
Boronic acid–agarose	Compounds with *cis*-diol groups; sugars, catecholamines, ribonucleotides, glycoproteins
Cibracron blue–agarose	Enzymes with nucleotide cofactors (dehydrogenases, kinases, DNA polymerases); serum albumin
Concanavalin A–agarose	Glycoproteins and glycolipids
Heparin–Sepharose	Nucleic acid–binding proteins, restriction endonucleases, lipoproteins
Iminodiacetate–agarose	Proteins with affinity for metal ions, serum proteins, interferons
Lentil lectin–Sepharose	Detergent-soluble membrane proteins
Lysine–Sepharose	Nucleic acids
Octyl-Sepharose	Weakly hydrophobic proteins, membrane proteins
Phenyl-Sepharose	Strongly hydrophobic proteins
Poly(A)-agarose	Nucleic acids containing poly(U) sequences, mRNA-binding proteins
Poly(U)-agarose	Nucleic acids containing poly(A) sequences, poly(U)-binding proteins
Protein A–agarose	IgG-type antibodies
Thiopropyl-Sepharose	—SH containing proteins

behind binding of nucleic acids and proteins to group-specific adsorbents depend on the actual affinity adsorbent. In most cases, the immobilized ligand and macromolecule (protein or nucleic acid) interact through one or more of the following forces: hydrogen bonding, hydrophobic interactions, and/or covalent interactions. Some group-specific adsorbents deserve special attention. Phenyl- and octyl-Sepharose are gels used for **hydrophobic interaction chromatography.** These adsorbent materials separate proteins on the basis of their hydrophobic character. Because most proteins contain hydrophobic amino acid side chains, this method is widely used. Octyl-Sepharose is strongly hydrophobic; hence it binds strongly to nonpolar proteins. Phenyl-Sepharose is more weakly hydrophobic; therefore, it is more likely to reversibly bind strongly hydrophobic proteins.

The use of thiopropyl-Sepharose and boronic acid–agarose is an example of **covalent chromatography,** since relatively strong but reversible covalent bonds are formed between the affinity gel and specific macromolecules.

Metal affinity chromatography is a relatively new method that separates proteins on the basis of metal binding.

The availability of a great variety of group-specific adsorbents in prepacked columns makes possible the combination of FPLC and affinity chromatography for the separation and purification of proteins.

Immunoadsorption

One of the most effective modifications of affinity chromatography is **immunoaffinity,** also called **immunoadsorption.** The unique high specificity of antibodies for their antigens is exploited for the purification of antigens. The interaction between antigen and antibody is very selective and very strong, perhaps the most specific in affinity chromatography. In practice, the antibody is immobilized on a column support. The antibody may be obtained by immunizing a rabbit, but you must have relatively large amounts of the purified protein antigen. The better choice is to make monoclonal antibodies against the antigen. When a mixture containing several other proteins along with the protein antigen is passed through the column, only antigen binds; the other proteins wash off the column. Because antigen:antibody pairs form very strong complexes, it is not especially difficult to get the protein antigen to bind to the column. There is often difficulty, though, in eluting the bound protein without denaturing it. Effective eluting agents for disrupting the complexes and eluting the antigen are discussed in the next section.

Protein A-agarose in Table 5.6 is an example of immunoaffinity; however, this adsorbent does not recognize specific antibodies but, rather, the general family of immunoglobulin G antibodies.

Experimental Procedure for Affinity Chromatography

Although the procedure is different for each type of substance isolated, a general experimental plan is outlined here. Figure 5.20 provides a step-by-step plan in flowchart form. Many types of matrix-ligand systems are commercially available and the costs are reasonable, so it is not always necessary to spend valuable laboratory time for affinity gel preparation. Even if a specific gel is not available, time can be saved by purchasing pre-activated gels for direct attachment of the desired ligand. Once the gel is prepared, the procedure is similar to that described earlier. The major difference is the use of shorter columns. Most affinity gels have high capacities and column beds less than 10 cm in length may be used. A second difference is the mode of elution. Ligand-macromolecule complexes immobilized on the column are held together by hydrogen bonding, ionic interactions, and hydrophobic effects. Any agent that diminishes these forces causes the release and elution of the macromolecule from the column. The common methods of elution are change of buffer pH, increase of buffer ionic strength, affinity elution, and chaotropic agents. The choice of elution method depends on many factors, including the types of forces responsible for complex formation and the stability of the ligand matrix and isolated macromolecule.

Buffer pH or Ionic Strength

If ionic interactions are important for complex formation, a change in pH or ionic strength weakens the interaction by altering the extent of ionization of

Figure 5.20

Experimental procedure
for affinity chromatography.

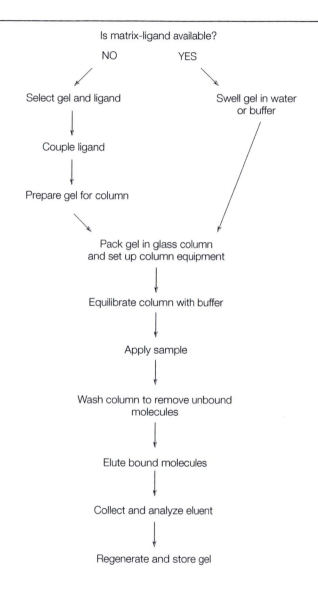

ligand and macromolecule. In practice, either a decrease in pH or a gradual increase in ionic strength (continual or stepwise gradient) is used.

Affinity Elution

In this method of elution, a selective substance added to the eluting buffer competes for binding to the ligand or for binding to the adsorbed macromolecule.

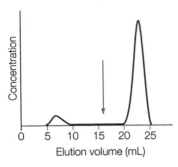

Figure 5.21

Purification of α-chymotrypsin by affinity chromatography on immobilized D-tryptophan methyl ester. From *Affinity Chromatography: Principles and Methods.* Pharmacia (Uppsala, Sweden).

Chaotropic Agents

If gentle and selective elution methods do not release the bound macromolecule, as may be the case in immunoadsorption, then mild denaturing agents can be added to the buffer. These substances deform protein and nucleic acid structure and decrease the stability of the complex formed on the affinity gel. The most useful agents are urea, guanidine $\cdot$ HCl, CNS^-, ClO_4^-, and CCl_3COO^-. These substances should be used with care, because they may cause irreversible structural changes in the isolated macromolecule.

The application of affinity chromatography is limited only by the imagination of the investigator. Every year literally hundreds of research papers appear with new and creative applications of affinity chromatography. Figure 5.21 illustrates the purification of α-chymotrypsin by affinity chromatography on immobilized D-tryptophan methyl ester. α-chymotrypsin can recognize and bind, but not chemically transform, D-tryptophan methyl ester. The enzyme catalyzes the hydrolysis of L-tryptophan methyl ester. The impure α-chymotrypsin mixture was applied to the gel, D-tryptophan methyl ester coupled to CH-Sepharose 4B, and the column washed with Tris buffer. At the point shown by the arrow, the eluent was changed to 0.1 M acetic acid. The decrease in pH caused release of α-chymotrypsin from the column.

➤ | **Study Exercise 5.7**   You have discovered that leaves from the quaking aspen tree contain an enzyme that catalyzes the transfer of a phosphoryl group from ATP to various carbohydrates. You wish to purify the enzyme, a kinase, by affinity chromatography. Select two stationary phases from Table 5.6 that might be effective in this separation.

Study Problems

▶ *1.* Amino acid analyzers are instruments that automatically separate amino acids by cation-exchange chromatography. Predict the order of elution (first to last) for each of the following sets of amino acids at pH = 4.
(a) Gly, Asp, His
(b) Arg, Glu, Ala
(c) Phe, His, Glu

▶ *2.* Predict the relative order of paper chromatography R_f values for the amino acids in the following mixture: Ser, Lys, Leu, Val, and Ala. Assume that the developing solvent is *n*-butanol, water, and acetic acid.

▶ *3.* In what order would the following proteins be eluted from a DEAE-cellulose ion exchanger by an increasing salt gradient. The pH_I is listed for each protein.
Egg albumin, 4.6   Cytochrome *c*, 10.6
Pepsinogen, 1.0   Myoglobin, 6.8
Serum albumin, 4.9   Hemoglobin, 6.8

▶ *4.* Draw the elution curve (A_{280} vs. fraction number) obtained by passing a mixture of the following proteins through a column of Sephadex G-100. The molecular mass is given for each protein.

Myoglobin, 16,900   Myosin, 524,000
Catalase, 222,000   Serum albumin, 68,500
Cytochrome c, 13,370   Chymotrypsinogen, 23,240

▶ *5.* Describe the various detection methods that can be used in HPLC. What types of biomolecules are detected by each method?

▶ *6.* Name three enzymes that you predict will bind to the affinity support, 5'-AMP-agarose.

7. Briefly describe how you would experimentally measure the exclusion limit for a Sephadex gel whose bottle has lost its label.

▶ *8.* Describe how you would use affinity elution to remove the enzyme alcohol dehydrogenase bound to a Cibracron blue–agarose column.

9. Explain the elution order of amino acids in Figure 5.11.

▶ *10.* In affinity chromatography and immunoadsorption, analytes are usually tightly bound to the sorbent. Which of the following types of bonding are important in forming these complexes?
Hydrogen bonding
Covalent bonds between carbon atoms
Ionic bonds
Hydrophobic interactions
Van der Waals forces

11. You have read in your botany textbook that bean leaves contain an enzyme that catalyzes the hydrolysis of the methyl esters of aromatic amino acids (for example, the methyl ester of phenylalanine). You wish to design a plan for isolating and purifying the enzyme. Assume that your last step of purification is to be affinity chromatography. Describe an affinity sorbent that might be effective in purifying the protein.

Further Reading

H. Aboul-Enein and I. Ali, *Chiral Separations by Liquid Chromatography: Theory and Applications* (2003), Marcel Dekker (New York).

S. Ahuja, *Chiral Separations by Chromatography* (2000), American Chemical Society (Washington, DC).

P. Bailon, G. Ehrlich, W. Fung, and W. Berthold, *Affinity Chromatography: Methods and Protocols* (2000), Humana Press (Totowa, NJ).

T. Beesley and R. Scott, *Chiral Chromatography* (1999), John Wiley & Sons (New York).

J. Berg, J. Tymoczko, and L. Stryer, *Biochemistry,* 5th ed. (2002), Freeman (New York), pp. 80–82. Introduction to chromatography.

R. Boyer, *Concepts in Biochemistry,* 2nd ed. (2001), John Wiley & Sons, (New York), pp. 88–89. An introduction to chromatography.

M. Chakravarthy, L. Snyder, T. Vanyo, J. Holbrook, and H. Jakubowski, *J. Chem. Educ.* **73,** 268–272 (1996). "Protein Structure and Chromatographic Behavior."

B. Chankvetadze, Editor, *Chiral Separations* (2001), Elsevier (Amsterdam).

J. Coligan, Editor, *Short Protocols in Protein Science* (2003), John Wiley & Sons (Hoboken, NJ).

R. Cunico, K. Gooding, and T. Wehr, *Basic HPLC and CE of Biomolecules* (1998), Bay Bioanalytical Laboratory (Hercules, CA).

R. Curtright, R. Emry, and J. Markwell, *J. Chem. Educ.* **76,** 249 (1999). "Student Understanding of Chromatography: A Hands-On Approach."

B. Fried and J. Sherma, *Thin Layer Chromatography,* 4th ed. (1999), Marcel Dekker (New York).

I. Fritz and D. Gjerde, *Ion Chromatography,* 3rd ed. (2001), John Wiley & Sons (New York).

R. Garrett and C. Grisham, *Biochemistry,* 3rd ed. (2005), Brooks/Cole (Belmont, CA), pp. 96–100, 115, 148–152. An introduction to chromatography.

D. Gjerde, C. Hanna, and D. Hornby, *DNA Chromatography* (2002), John Wiley & Sons (New York).

F. Gorga, *J. Chem. Educ.* **77,** 264 (2000). "A Problem-Solving Approach to Chromatography in the Biochemistry Lab."

E. Grushka and P. Brown, *Advances in Chromatography* (2003), Marcel Dekker (New York).

M. Gupta, Editor, *Methods for Affinity Based Separations for Enzymes and Proteins* (2002), Birkhauser (Cambridge, MA).

D. Hage, *Handbook of Affinity Chromatography* (2003), Marcel Dekker (New York).

R. Hatti-Kaul and B. Mattiasson, Editors, *Isolation and Purification of Proteins* (2003), Marcel Dekker (New York).

S. Kromidas, *Practical Problem Solving in HPLC* (2000), John Wiley & Sons (New York).

S. Lindsay and J. Barnes, *High Performance Liquid Chromatography,* 2nd ed. (2002), John Wiley & Sons (New York).

P. Matejtschuk, Editor, *Affinity Separations: A Practical Approach,* (1997), IRL Press (Oxford).

D. Nelson and M. Cox, *Lehninger Principles of Biochemistry,* 4th ed. (2005), Freeman (New York), pp. 89–92, 180–181, 364–365. Introduction to chromatography.

T. Phillips and D. Dickens, *Affinity and Immunoaffinity Purification Techniques* (2000), Eaton (Westborough, MA).

A. Poole, *The Essence of Chromatography* (2003), Elsevier (Amsterdam).

M. Pugh and E. Schultz, *Biochem. Mol. Biol. Educ.* **30,** 179–183 (2002). "Assessment of the Purification of a Protein by Ion Exchange and Gel Permeation Chromatography."

S. Roe, Editor, *Protein Purification Applications,* 2nd ed. (2001), Oxford University Press (Oxford).

K. Stewart and R. Ebel, *Chemical Measurements in Biological Systems* (2000), John Wiley & Sons (New York), pp. 121–146. Chromatography.

G. Subramanian, Editor, *Chiral Separation Techniques,* 2nd ed. (2001), John Wiley & Sons (New York).

M. Vijayalakshmi, Editor, *Biochromatography: Theory and Practice* (2002), Taylor & Francis (Hamden, CT).

D. Voet and J. Voet, *Biochemistry,* 3rd ed. (2004), John Wiley & Sons (Hoboken, NJ), pp. 133–144. Use of chromatography in protein purification.

D. Voet, J. Voet, and C. Pratt, *Fundamentals of Biochemistry,* 2nd ed. (2006), John Wiley & Sons (Hoboken NJ), pp. 102–105. Chromatography in protein purification.

J. Watson et al., *Molecular Biology of the Gene,* 5th ed. (2004), Benjamin/Cummings (San Francisco), Chapter 20 (Methods).

K. Williams, *J. Chem. Educ.* **79,** 922 (2002). "Colored Bands: History of Chromatography."

K. Wilson and J. Walker, Editors, *Principles and Techniques of Practical Biochemistry,* 5th ed. (2000), Cambridge University Press (Cambridge), pp. 619–687. Chromatographic techniques.

http://www.bio.mtu.edu/campbell/482w91a.htm
    Graphical presentation of the steps in affinity chromatography.

http://kerouac.pharm.uky.edu/ASRG/HPLC/hplcmytry.html
    A users' guide to HPLC.

http://aesop.rutgers.edu/~dbm/affinity.html
    Affinity and immunoadsorbent chromatography.

http://www.chromtech.se/
    Click on "Chiral Applications" to see a directory of enantiomeric molecules separated by chiral columns.

http://www5.amershambiosciences.com
    On the home page, click on "lab chromatography" under Protein Purification and Production for a complete discussion of gel filtration, including "How to Pack a Column."

CHARACTERIZATION OF PROTEINS AND NUCLEIC ACIDS BY ELECTROPHORESIS

Electrophoresis is an analytical tool that allows biochemists to examine the differential movement of charged molecules in an electric field. A. Tiselius, a Swede who invented the technique in the 1930's, performed experiments in free solution that were severely limited by the effects of diffusion and convection currents. Modern electrophoretic techniques use a polymerized gel-like matrix, which is more stable as a support medium. The sample to be analyzed is applied to the medium as a spot or thin band, hence the term **zonal** electrophoresis is often used. The migration of molecules is influenced by: (1) the size, shape, charge, and chemical composition of the molecules to be separated; (2) the rigid, mazelike matrix of the gel support; and (3) the applied electric field. Electrophoresis, which is a relatively rapid, inexpensive, and convenient technique, is capable of analyzing and purifying many different types of biomolecules, but especially proteins and nucleic acids. The newest version of the analytical technique, **capillary electrophoresis (CE)**, provides extremely high resolution and is useful for analysis of both large and small molecules. CE has been found to be especially useful in the analysis of pharmaceuticals.

Proteomics, the discipline that attempts systematic, large-scale studies on the structure and function of gene products in an organism or cell, is expanding rapidly because of the availability of electrophoresis, especially two-dimensional techniques, to analyze proteins and peptides.

Even though electrophoresis has been studied for over 70 years, it has been a challenge to provide an accurate, theoretical description of the electrophoretic movement of molecules in a gel support. However, the lack of theoretical understanding has not hampered growth in the use of the technique in separating and characterizing a wide variety of biomolecules.

A. THEORY OF ELECTROPHORESIS

Introduction

The movement of a charged molecule in a medium subjected to an electric field is represented by Equation 6.1.

$$ v = \frac{Eq}{f} $$

Equation 6.1

where

> E = the electric field in volts/cm
>
> q = the net charge on the molecule
>
> f = frictional coefficient, which depends on the mass and shape of the molecule
>
> v = the velocity of the molecule

The charged particle moves at a velocity that depends directly on the electrical field (E) and charge (q), but inversely on a counteracting force generated by the viscous drag (f). The applied voltage represented by E in Equation 6.1 is usually held constant during electrophoresis, although some experiments are run under conditions of constant current (where the voltage changes with resistance) or constant power (the product of voltage and current). Under constant-voltage conditions, Equation 6.1 shows that the movement of a charged molecule depends only on the ratio q/f. For molecules of similar conformation (for example, a collection of linear DNA fragments or spherical proteins), f varies with size but not shape; therefore, the only remaining variables in Equation 6.1 are the charge (q) and mass dependence of f, meaning that under such conditions molecules migrate in an electric field at a rate proportional to their charge-to-mass ratio.

Theory and Practice

The movement of a charged particle in an electric field is often defined in terms of **mobility, μ,** the velocity per unit of electric field (Equation 6.2).

$$ \mu = \frac{v}{E} $$

Equation 6.2

This equation can be modified using Equation 6.1.

$$ \mu = \frac{Eq}{Ef} = \frac{q}{f} $$

Equation 6.3

In theory, if the net charge, q, on a molecule is known, it should be possible to measure f and obtain information about the hydrodynamic size and shape of that molecule by investigating its mobility in an electric field. Attempts to define f by electrophoresis have not been successful, primarily because Equation 6.3 does not adequately describe the electrophoretic process. Important factors that are not accounted for in the equation are interaction of migrating molecules with the support medium and shielding of the molecules by buffer ions. This means that electrophoresis is not useful for describing specific details about the shape of a molecule. Instead, it has been applied to the analysis of purity and size of macromolecules. Each molecule in a mixture is expected to have a unique charge and size, and its mobility in an electric field will therefore be unique. This expectation forms the basis for analysis and separation by all electrophoretic methods. The technique is especially useful for the analysis of amino acids, peptides, proteins, nucleotides, nucleic acids, and other charged molecules, including pharmaceuticals.

B.   METHODS OF ELECTROPHORESIS

All modes of electrophoresis are based on the principles just outlined. The major difference among the various methods is the type of support medium. Cellulose and cellulose acetate are used as a support medium for low-molecular-weight biochemicals like amino acids and carbohydrates. Polyacrylamide and agarose gels are widely used as support media for larger molecules. In capillary electrophoresis, several different types of support media, including the natural, untreated surfaces inside a silica narrow bore capillary tube, are used. Geometries (vertical and horizontal), buffers, and electrophoretic conditions provide many different experimental arrangements for the variety of methods described here.

Polyacrylamide Gel Electrophoresis (PAGE)

Gels formed by polymerization of acrylamide have several positive features in electrophoresis: (1) high resolving power for small and moderately sized proteins and nucleic acids (up to approximately 1×10^6 daltons), (2) acceptance of relatively large sample sizes, (3) minimal interactions of the migrating molecules with the matrix, and (4) physical stability of the matrix. Recall from the earlier discussion of gel filtration (Chapter 5) that gels can be prepared with different pore sizes by changing the concentration of cross-linking agents. Electrophoresis through polyacrylamide gels leads to enhanced resolution of sample components because the separation is based on both molecular sieving and electrophoretic mobility. The order of molecular movement in gel filtration and PAGE is very different, however. In gel filtration (Chapter 5), large molecules migrate through the matrix faster than small molecules. The opposite is the case for gel electrophoresis, where there is no void volume in the matrix, only a continuous network of pores throughout the gel. The electrophoresis gel is comparable to a single bead in gel filtration. Therefore, large molecules do not move easily through the medium, and the rate of movement is small molecules followed by large molecules.

Preparation of Gels

Polyacrylamide gels are prepared by the free radical polymerization of acrylamide and the cross-linking agent N,N'-methylene-bis-acrylamide (Figure 6.1). Chemical polymerization is controlled by an initiator-catalyst system, ammonium persulfate–N,N,N',N'-tetramethylethylenediamine (TEMED). Photochemical polymerization may be initiated by riboflavin in the presence of ultraviolet (UV) radiation. A standard gel for protein separation is 7.5% polyacrylamide. It can be used over the molecular size range of 10,000 to 1,000,000 daltons; however, the best resolution is obtained in the range of 30,000 to 300,000 daltons. The resolving power and molecular size range of a gel depend on the concentrations of acrylamide and bis-acrylamide (see Table 6.1 for effective ranges of protein separation). Lower concentrations give gels with larger pores, allowing analysis of higher-molecular-weight biomolecules. In contrast, higher concentrations of acrylamide give gels with smaller pores, allowing analysis of lower-molecular-weight biomolecules (see Table 6.2 for effective ranges of DNA separation).

Figure 6.1

Chemical reactions illustrating the copolymerization of acrylamide and N,N'-methylene-bis-acrylamide. See text for details.

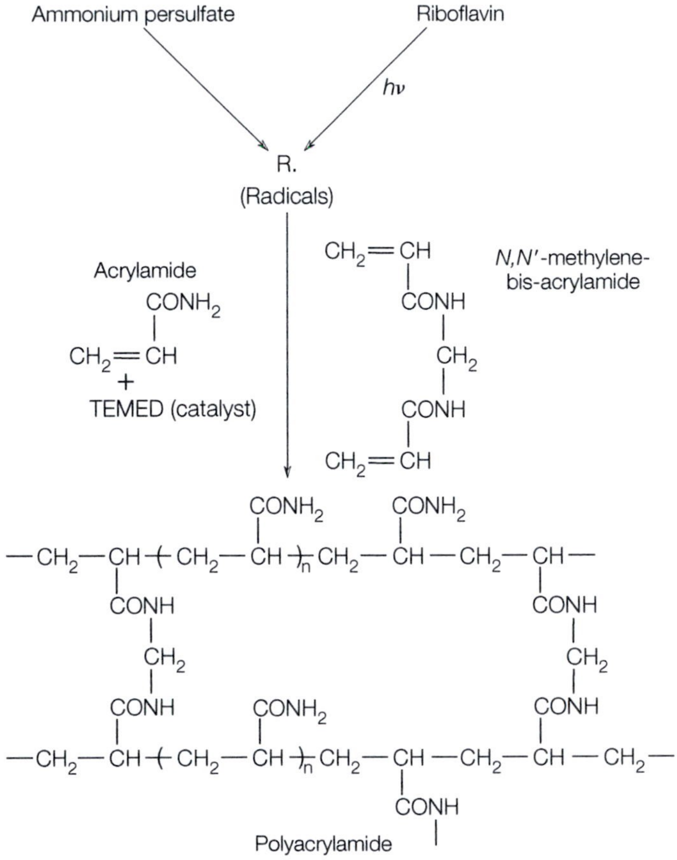

| Table 6.1 |

Separation of Proteins by PAGE. The recommended concentrations of acrylamide (% w/v) are given that will best separate proteins of different molecular weight ranges.

Protein Molecular Weight Range	Recommended % Acrylamide[1] (w/v)
<10,000	>15%
10,000–80,000	10–15%
20,000–150,000	5–10%
>100,000	3–5%

[1]Ratio of acrylamide to bis-acrylamide, 37.5:1 (g:g).

Polyacrylamide electrophoresis can be done using either of two arrangements, column or slab. Figure 6.2 shows the typical arrangement for a column gel. Glass tubes (10 cm × 6 mm i.d.) are filled with a mixture of acrylamide, N,N'-methylene-bis-acrylamide, buffer, and free radical initiator-catalyst. Polymerization occurs in 30 to 40 minutes. The gel column is inserted between two separate buffer reservoirs. The upper reservoir usually contains the cathode (−) and the lower the anode (+). Gel electrophoresis is usually carried out at basic pH, where most biological polymers are anionic; hence, they move down toward the anode. The sample to be analyzed is layered on top of the gel and voltage is applied to the system. DNA and many proteins are colorless so a reagent dye must be added to monitor the rate of electrophoresis. A "tracking dye" is applied, which moves more rapidly through the gel than the sample components. When the dye band has moved to the opposite end of the column, the voltage is turned off and the gel is removed from the column and stained with a dye. Chambers for column gel electrophoresis are commercially available or can be constructed from inexpensive materials.

| Table 6.2 |

Effective Range of Separation of DNA by PAGE

Acrylamide[1] (% w/v)	Range of Separation (bp)	Bromophenol Blue[2]	Xylene Cyanol[2]
3.5	1000–2000	100	450
5.0	80–500	65	250
8.0	60–400	50	150
12.0	40–200	20	75
20.0	5–100	10	50

[1]Ratio of acrylamide to bis-acrylamide, 20:1.

[2]The numbers (in bp) represent the size of DNA fragment with the same mobility as the dye.

Figure 6.2

A column gel for polyacrylamide gel electrophoresis.

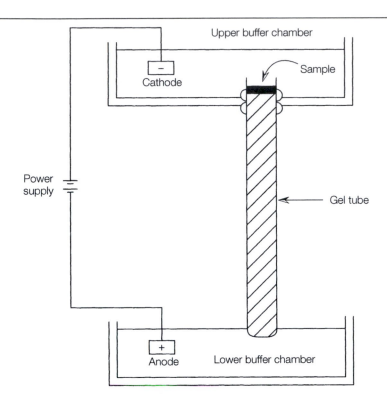

Slab Gel Electrophoresis

Slab gels are now more widely used than column gels. A slab gel on which several samples may be analyzed is more convenient to make and use than several individual column gels. Slab gels also offer the advantage that all samples are analyzed in a matrix environment that is identical in composition. A typical vertical slab gel apparatus is shown in Figure 6.3. The polyacrylamide slab is prepared between two glass plates that are separated by spacers (Figure 6.4). The spacers allow a uniform slab thickness of 0.5 to 2.0 mm, which is appropriate for analytical and some preparative procedures. Slab gels are usually 8 × 10 cm or 10 × 10 cm, but for nucleotide sequencing, slab gels as large as 20 × 40 cm are often required.

A plastic "comb" inserted into the top of the slab gel during polymerization forms indentations in the gel that serve as sample wells. Up to 20 sample wells may be formed. After polymerization, the comb is carefully removed and the wells are rinsed thoroughly with buffer to remove salts and any unpolymerized acrylamide. The gel plate is clamped into place between two buffer reservoirs, a sample is loaded into each well, and voltage is applied. For visualization, the slab is removed and stained with an appropriate dye.

Figure 6.3

A vertical electrophoresis apparatus for slab gel analysis of proteins. *Permission to use the materials has been granted by Bio-Rad Laboratories, Inc., www.bio-rad.com*

PROTEAN II xi system components
1. Tank and lid
2. Central cooling core
3. Latch (black)
4. Casting stand
5. Sandwich clamps
6. Alignment card
7. Combs

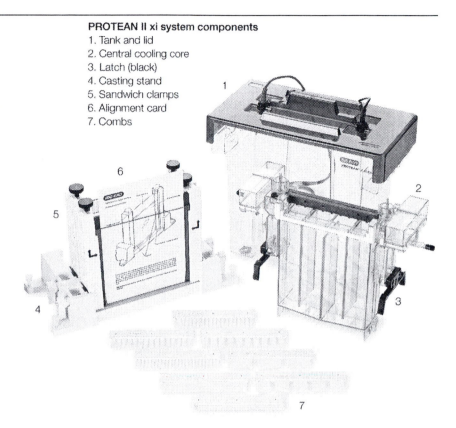

Perhaps the most difficult and inconvenient aspect of polyacrylamide gel electrophoresis is the preparation of gels. The monomer, acrylamide, is a neurotoxin and a carcinogenic agent; hence, special handling is required. Other necessary reagents including catalysts and initiators also require special handling and are unstable. In addition, it is difficult to make gels that have reproducible thicknesses and compositions. Many researchers are now turning to the use of precast polyacrylamide gels. Several manufacturers now offer gels precast in glass or plastic cassettes. Gels for all experimental operations are available including single percentage (between 3 and 27%) or gradient gel concentrations and a variety of sample well configurations and buffer chemistries. More details on precast gels will be given in Section C, Practical Aspects of Electrophoresis.

Several modifications of PAGE have greatly increased its versatility and usefulness as an analytical tool.

Discontinuous Gel Electrophoresis

The experimental arrangement for "disc" gel electrophoresis is shown in Figure 6.5. Three significant characteristics of this method are that (1) there are two gel layers, a lower or **resolving gel** and an upper or **stacking gel;** (2) the

Figure 6.4

Arrangement of two glass plates with spacers to form a slab gel. The comb is used to prepare wells for placement of samples.

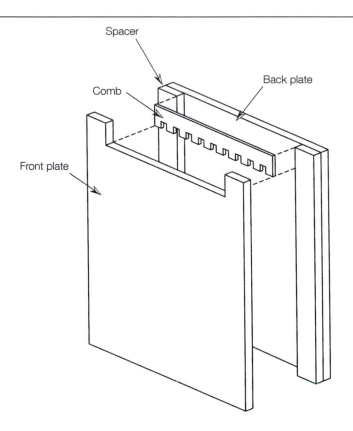

buffers used to prepare the two gel layers are of different ionic strengths and pH; and (3) the stacking gel has a lower acrylamide concentration, so its pore sizes are larger. These three changes in the experimental conditions cause the formation of highly concentrated bands of sample in the stacking gel and greater resolution of the sample components in the lower gel. Sample concentration in the upper gel occurs in the following manner. The sample is usually dissolved in glycine-chloride buffer, pH 8 to 9, before loading on the gel. Glycine exists primarily in two forms at this pH, a zwitterion and an anion (Equation 6.4).

$$\overset{+}{H_3}NCH_2COO^- \rightleftharpoons H_2NCH_2COO^- + H^+$$

Equation 6.4

The average charge on glycine anions at pH 8.5 is about -0.2. When the voltage is turned on, buffer ions (glycinate and chloride) and protein or nucleic acid sample move into the stacking gel, which has a pH of 6.9. Upon entry into the upper gel, the equilibrium of Equation 6.4 shifts toward the left, increasing the concentration of glycine zwitterion, which has no net charge and hence no electrophoretic mobility. In order to maintain a constant current in the electrophoresis system, a flow of anions must be maintained. Since most proteins and nucleic acid samples are still

Figure 6.5

The process of disc gel electrophoresis. **A** Before electrophoresis. **B** Movement of chloride, glycinate, and protein through the stacking gel. **C** Separation of protein samples by the resolving gel.

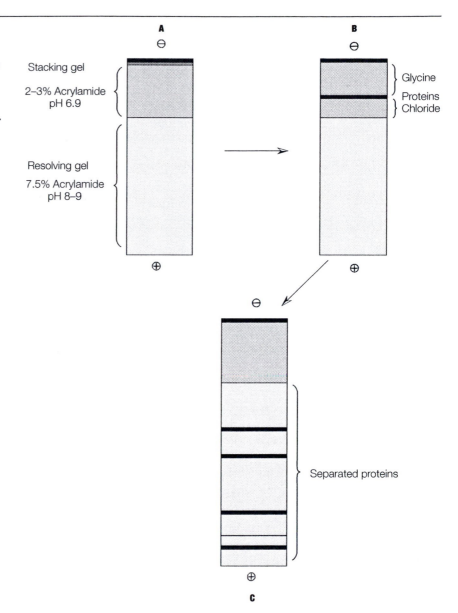

anionic at pH 6.9, they replace glycinate as mobile ions. Therefore, the relative ion mobilities in the stacking gel are chloride > protein or nucleic acid sample > glycinate. The sample will tend to accumulate and form a thin, concentrated band sandwiched between the chloride and glycinate as they move through the upper gel. Since the acrylamide concentration in the stacking gel is low (2 to 3%), there is little impediment to the mobility of the large sample molecules.

Now, when the ionic front reaches the lower gel with pH 8 to 9 buffer, the glycinate concentration increases and anionic glycine and chloride carry most of the current. The protein or nucleic acid sample molecules, now in a narrow band, encounter both an increase in pH and a decrease in pore size. The increase in pH would, of course, tend to increase electrophoretic mobility, but the smaller pores decrease mobility. The relative rate of movement of anions in the lower gel is chloride > glycinate > protein or nucleic acid sample. The separation of sample components in the resolving gel occurs as described in an earlier section on gel electrophoresis. Each component has a unique charge/mass ratio and a discrete size and shape, which directly influence its mobility. Disc gel electrophoresis yields excellent resolution and is the method of choice for analysis of proteins and nucleic acid fragments. Protein or nucleic acid bands containing as little as 1 or 2 μg can be detected by staining the gels after electrophoresis.

➤ | **Study Exercise 6.1**   You are using the technique of nondenaturing PAGE to separate a mixture of two proteins, bovine serum albumin and bovine hemoglobin. Assume that you are using a buffer of pH 8.0 and a gel of 7.5% acrylamide, and that silver staining was used for detection. Show the results of your experiment by drawing a gel with dark bands for each protein.

Protein	pH$_I$	MW
Bovine serum albumin	4.9	65,000
Bovine hemoglobin	6.8	65,000

Sodium Dodecyl Sulfate–Polyacrylamide Gel Electrophoresis (SDS-Page)

If protein samples are treated so that they have a uniform charge, electrophoretic mobility then depends primarily on size (see Equation 6.3). The molecular weights of proteins may be estimated if they are subjected to electrophoresis in the presence of a detergent, sodium dodecyl sulfate (SDS), and a disulfide bond reducing agent, mercaptoethanol. This method is often called **denaturing electrophoresis.**

The electrophoretic techniques previously discussed are called **nondenaturing** or **"native" PAGE** and are used when an investigator requires that the protein analyzed still retains its biological activity. This would be the case when the protein is an enzyme, antibody, or contains a receptor binding site. The separation of proteins under these conditions where they maintain their native conformation is influenced by both charge and size.

When protein molecules are treated with SDS, the detergent disrupts the secondary, tertiary, and quaternary structure to produce linear polypeptide chains coated with negatively charged SDS molecules. The presence of mercaptoethanol assists in protein denaturation by reducing all disulfide bonds. The detergent binds to hydrophobic regions of the denatured protein chain in a constant ratio of about 1.4 g of SDS per gram of protein. The bound detergent molecules carrying negative charges mask the native charge of the protein. In essence, polypeptide chains of a constant charge/mass ratio and uniform shape are produced. The electrophoretic mobility of the SDS-protein complexes is influenced primarily by molecular size: the larger molecules are retarded by the molecular sieving effect of the gel, and the smaller molecules have greater mobility. Empirical measurements have shown a linear relationship between the log molecular weight and the electrophoretic mobility (Figure 6.6).

In practice, a protein of unknown molecular weight and subunit structure is treated with 1% SDS and 0.1 M mercaptoethanol in electrophoresis buffer. A standard mixture of proteins with known molecular weights must also be subjected to electrophoresis under the same conditions. Two broad sets of standards are commercially available, one for low-molecular-weight proteins (molecular weight range 14,000 to 100,000) and one for high-molecular-weight proteins (45,000 to 200,000). Figure 6.7 shows a stained gel after electrophoresis of a standard protein mixture. After electrophoresis and dye staining, mobilities are measured and molecular weights determined graphically.

SDS-PAGE is valuable for estimating the molecular weight of protein subunits. This modification of gel electrophoresis finds its greatest use in characterizing the sizes and different types of subunits in oligomeric proteins. SDS-PAGE is limited to a molecular weight range of 10,000 to 200,000. Gels

Figure 6.6

Graph illustrating the linear relationship between electrophoretic mobility of a protein and the log of its molecular weight. Thirty-seven different polypeptide chains with a molecular weight of 11,000 to 70,000 are shown. Gels were run in the presence of SDS. *From K. Weber and M. Osborn, J. Biol. Chem.* **244,** *4406 (1969). By permission of the copyright owner, the American Society for Biochemistry and Molecular Biology, Inc.*

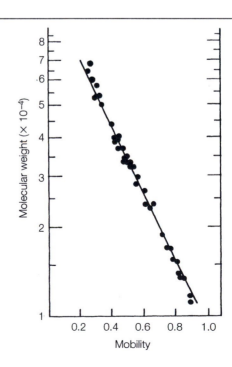

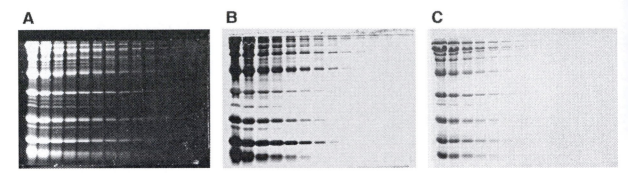

Figure 6.7

A comparison of the sensitivities achieved with three different protein stains. Identical SDS-polyacrylamide gels were stained with **A** SYPRO Red protein gel stain; **B** Silver stain; **C** Coomassie brilliant blue dye. *Courtesy of Molecular Probes; www.probes.com, a part of Invitrogen Corporation; www.invitrogen.com*

of less than 2.5% acrylamide must be used for determining molecular weights above 200,000, but these gels do not set well and are very fragile because of minimal cross-linking. A modification using gels of agarose-acrylamide mixtures allows the measurement of molecular weights above 200,000.

➤ **Study Exercise 6.2** You have just completed an experiment using SDS-PAGE to study the subunit structure of ferritin (iron storage protein). Ferritin is composed of two types of subunits, H and L, in about equal quantities. You have included in your electrophoresis several standard proteins of known molecular weight. Draw the final stained gel in the form of Figure 6.7.

Protein	MW
Ferritin	500,000
Subunit H	22,000
Subunit L	19,000
α-lactalbumin	14,200
Trypsinogen	24,000
Egg albumin	45,000
Bovine serum albumin	65,000

Nucleic Acid Sequencing Gels

Sequence analysis of nucleic acids is based on the generation of sets of DNA or RNA fragments with common ends and the separation of these oligonucleotide fragments by polyacrylamide electrophoresis. Two methods have been developed for sequencing nucleic acids: (1) the **partial chemical**

degradation method of Maxam and Gilbert, which uses four specific chemical reactions to modify bases and cleave phosphodiester bonds, and (2) the **chain termination method** developed by Sanger, which requires a single-stranded DNA template and chain extension processes, followed by chain termination caused by the presence of dideoxynucleoside triphosphates. Both sequencing methods result in nested sets of DNA or RNA fragments that have one common end and chains varying in length. The smallest possible size difference of nucleic acid fragments is one nucleotide. Separation of the nucleic acid fragments by polyacrylamide electrophoresis allows one to "read" the sequence of nucleotides from the gel (Chapter 9).

Using Sequencing Gels

The experimental arrangement is the same as that previously described for PAGE; however, the gel is prepared with many sample wells to accommodate a large number of samples. Sequence gels of 6, 8, 12, and 20% polyacrylamide are routinely used. Gels of 20% may be used to sequence the first 50 to 100 nucleotides of a nucleic acid, and lower percentage gels allow sequencing out to 250 nucleotides. Sequencing gels are large (up to 40 cm) and power supplies must provide more power than for conventional methods. Precast sequencing gels are now commercially supplied by Stratagene. They have a gel concentration of 5.5%, have 32 sample wells, and will sequence up to 500 nucleotides. Denaturants such as urea and formamide are required to prevent renaturing of the nucleic acid fragments during electrophoresis. For detection, nucleic acid chains for sequencing must be end labeled with ^{32}P, ^{35}S or a fluorescent tag. ^{32}P- and ^{35}S-labeled nucleic acids on gels are detected by autoradiography (see later). Nucleic acids end labeled with fluorescent molecules are detected by fluorimeter scanning of the gels. Many researchers working on the large and expensive human genome project[1] have generated huge amounts of DNA sequence data. Much of this information is stored in computer data banks for use by researchers around the world.

Agarose Gel Electrophoresis

The electrophoretic techniques discussed up to this point are useful for analyzing proteins and small fragments of nucleic acids up to 350,000 daltons (500 bp) in molecular size; however, the small pore sizes in the gel are not appropriate for analysis of large nucleic acid fragments or intact DNA molecules. The standard method used to characterize RNA and DNA in the range 200 to 50,000 base pairs (50 kilobases) is electrophoresis with agarose as the support medium.

    Agarose, a product extracted from seaweed, is a linear polymer of galactopyranose derivatives. Gels are prepared by dissolving agarose in

---

[1] *The human genome project was a federal government-sponsored program to sequence all DNA in human chromosomes. The project was completed in 2001.*

Figure 6.8

An apparatus for horizontal slab gel electrophoresis of nucleic acids. *Courtesy of Hoefer, Inc. www.hoeferinc.com*

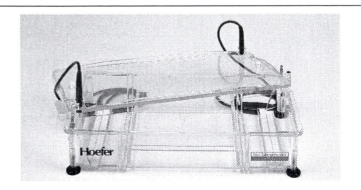

warm electrophoresis buffer. After cooling the gel mixture to 50°C, the agarose solution is poured between glass plates as described for polyacrylamide. Gels with less than 0.5% agarose are rather fragile and must be used in a horizontal arrangement (Figure 6.8). The sample to be separated is placed in a sample well made with a comb, and voltage is applied until separation is complete. Precast agarose gels of all shapes, sizes, and percent composition are commercially available.

Nucleic acids can be visualized on the slab gel after separation by soaking in a solution of ethidium bromide, a dye that displays enhanced fluorescence when intercalated between stacked nucleic acid bases. Ethidium bromide may be added directly to the agarose solution before gel formation. This method allows monitoring of nucleic acids during electrophoresis. Irradiation of ethidium bromide–treated gels by UV light results in orange-red bands where nucleic acids are present. Nucleic acids may also be stained with the new, fluorescent SYBR dyes, which are less toxic than the mutagenic ethidium bromide.

The mobility of nucleic acids in agarose gels is influenced by the agarose concentration and the molecular size and molecular conformation of the nucleic acid. Agarose concentrations of 0.3 to 2.0% are most effective for nucleic acid separation (Table 6.3). Adding ethidium bromide to the gel can retard DNA mobility. Like proteins, nucleic acids migrate at a rate that is

Table 6.3

Effective Range of Separation of DNA by Agarose

Agarose (% w/v)	Effective Range (kb)
0.3	5–50
0.5	2–25
0.7	0.8–10
1.2	0.4–5
1.5	0.2–3
2.0	0.1–2

inversely proportional to the logarithm of their molecular weights; hence, molecular weights can be estimated from electrophoresis results using standard nucleic acids or DNA fragments of known molecular weight. The DNA conformations most frequently encountered are superhelical circular (form I), nicked circular (form II), and linear (form III). The small, compact, supercoiled form I molecules usually have the greatest mobility, followed by the rodlike, linear form III molecules. The extended, circular form II molecules migrate more slowly. The relative electrophoretic mobility of the three forms of DNA, however, depends on experimental conditions such as agarose concentration and ionic strength.

The versatility of agarose gels is obvious when one reviews their many applications in nucleic acid analysis. The rapid advances in our understanding of nucleic acid structure and function in recent years are due primarily to the development of agarose gel electrophoresis as an analytical tool. Two of the many applications of agarose gel electrophoresis will be described here.

Analysis of DNA Fragments after Digestion by Restriction Endonucleases

Restriction endonucleases are enzymes that recognize a specific base sequence in double-stranded DNA and catalyze cleavage (hydrolysis of phosphodiester bonds) in or near that specific region (see Chapter 10, Section B). Many viral, bacterial, or animal DNA molecules are substrates for the enzymes. When each type of DNA is treated with a restriction endonuclease, a specific number of DNA fragments is produced. The base sequence recognized by the enzyme occurs only a few times in any particular DNA molecule; therefore, the smaller the DNA molecule, the fewer specific cleavage sites there are. Viral or phage DNA, for example, is cleaved into about 50 fragments depending on the enzyme used, whereas larger bacterial or animal DNA may be cleaved into hundreds or thousands of fragments. Smaller DNA molecules, upon cleavage with a particular enzyme, will produce a limited set of fragments. It is unlikely that this set of fragments will be the same for any two different DNA molecules, so the fragmentation pattern can be considered a "fingerprint" of the DNA substrate. The **restriction pattern** is produced by electrophoresis of the cleavage reaction mixture through agarose gels, followed by staining with ethidium bromide or SYBR dyes (Figure 6.9). The separation of the fragments is based on molecular size, with large fragments remaining near the origin and smaller fragments migrating farther down the gel. In addition to characterization of DNA structure, endonuclease digestion coupled with agarose gel electrophoresis is a valuable tool for plasmid mapping and DNA recombination experiments.

Characterization of Superhelical Structure of DNA

The structure of plasmid, viral, and bacterial DNA is often closed circular with negative superhelical turns. It is possible under various experimental conditions to induce reversible changes in the conformation of DNA.

Figure 6.9

Restriction patterns produced by agarose electrophoresis of DNA fragments after endonuclease action. DNA molecular weight ladders have been electrophoresed on a 1% agarose gel and then stained with SYBR Green I nucleic acid gel stain. Lanes **1** and **8** contain *Hind*III-cut λ DNA; lanes **2** and **7**, *Hae*III-cut ΦX174 RF DNA; lanes **3** and **6**, 1 kilobase pair DNA ladder; lane **4**, 100 base pair DNA ladder; lane **5**, *Eco*R-I-cut pUC19 DNA mixed with *Pst*I-cut ΦX174 RF DNA.

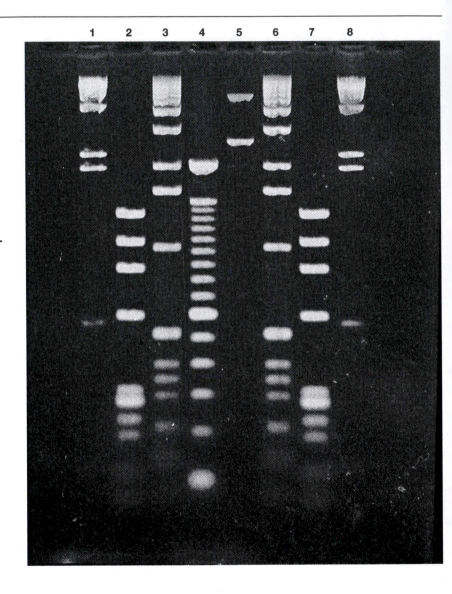

The intercalating dye ethidium bromide causes an unwinding of supercoiled DNA that affects its electrophoretic mobility. Electrophoresis of DNA on agarose in the presence of increasing concentrations of ethidium bromide provides an unambiguous method for distinguishing between closed circular and other DNA conformations.

Closed circular, negatively supercoiled DNA (form I) usually has the greatest electrophoretic mobility of all DNA forms because supercoiled DNA molecules tend to be compact. If ethidium bromide is added to form I

DNA, the dye intercalates between the stacked DNA bases, causing unwinding of some of the negative supercoils. As the concentration of ethidium bromide is increased, more and more of the negative supercoils are removed until no more are present in the DNA. The conformational change of the DNA supercoil can be monitored by electrophoresis because the mobility decreases with each unwinding step. With increasing concentration of ethidium bromide, the negative supercoils are progressively unwound and the electrophoretic mobility decreases to a minimum. This minimum represents the free dye concentration necessary to remove all negative supercoils. (The free dye concentration at this minimum has been shown to be related to the superhelix density, which is a measure of the extent of supercoiling in a DNA molecule.) The circular DNA at this point is equivalent to the "relaxed" form. If more ethidium bromide is added to the relaxed DNA, positive superhelical turns are induced in the structure and the electrophoretic mobility increases. Forms II and III DNA, under the same conditions of increasing ethidium bromide concentration, show a gradual decrease in electrophoretic mobility throughout the entire concentration range.

Agarose gel electrophoresis is able to resolve topoisomers of native, covalently closed, circular DNA that differ only in their degree of supercoiling. This technique has proved useful in the analysis and characterization of enzymes that catalyze changes in the conformation or topology of native DNA. These enzymes, called **topoisomerases,** have been isolated from bacterial and mammalian cells. They change DNA conformations by catalyzing nicking and closing of phosphodiester bonds in circular duplex DNA. Agarose gel electrophoresis is an ideal method for identifying and assaying topoisomerases because the intermediate DNA molecules can be resolved on the basis of the extent of supercoiling. Topoisomerases may be assayed by incubating native DNA with an enzyme preparation, removing aliquots after various periods of time, and subjecting them to electrophoresis on an agarose gel with standard supercoiled and relaxed DNA.

Pulsed Field Gel Electrophoresis (PFGE)

Conventional agarose gel electrophoresis is limited in use for the separation of nucleic acid fragments smaller than 50,000 bp (50 kb). In practice, that limit is closer to 20,000 to 30,000 bp if high resolution is desired. Since chromosomal DNA from most organisms contains thousands and even millions of base pairs, the DNA must be cleaved by restriction enzymes before analysis by standard electrophoresis. In the early 1980s it was discovered by Schwartz and Cantor at Columbia University that large molecules of DNA (yeast chromosomes, 200–3000 kb) could be separated by **pulsed field gel electrophoresis (PFGE).** There is one major distinction between standard gel electrophoresis and PFGE. In PFGE, the electric field is not constant as in the standard method but is changed repeatedly (pulsed) in direction and strength during the separation (Figure 6.10). The physical mechanism for separation of the large DNA molecules as they

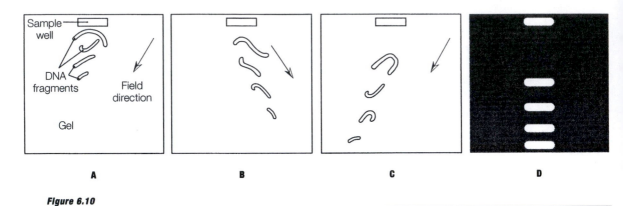

Figure 6.10

Pulsed field gel electrophoresis. Generalized PFGE separation of four DNA fragments of different sizes in one lane. **A** DNA molecules of various shapes and configurations move toward the positive field. **B** The new field orientation pulls the DNA in a different direction, realigning the molecules. **C** The field returns to the original configuration. **D** The bands show the final position of a large collection of the molecules. *Reprinted with permission from the Journal of NIH Research 1, 115 (1999). Illustration by Terese Winslow.*

move through the gel under these conditions is not yet well understood. An early explanation was that the electrical pulses abruptly perturbed the conformation of the DNA molecules. They would be oriented by the influence of the electric field coming from one direction and then reoriented as a new electric field at a different angle to the first was turned on. According to this explanation, it takes longer for larger molecules to reorient, so smaller fragments respond faster to the new pulse and move faster. More recent experiments on dyed DNA moving in gels have shown that conformational changes of the DNA are not abrupt but more gradual in response to the electrical pulse, and DNA molecules tend to "slither" through the gel matrix. In addition, it has been discovered that the gel becomes more fluid during electrical pulsing.

Applications of PFGE

Even though our theoretical understanding of PFGE is lacking, practical applications and experimental advances are expanding rapidly. The availability of PFGE has sparked changes in DNA research. New methods for isolating intact DNA molecules have been developed. Because of mechanical breakage, the average size of DNA isolated from cells in the presence of lysozyme, detergent, and EDTA is about 400–500 kb (see Chapter 9, Section B). Intact chromosomal DNA can be isolated by embedding cells in an agarose matrix and disrupting the cells with detergents and enzymes. Slices or "plugs" of the agarose with intact DNA are then placed on the gel for PFGE analysis. Newly discovered restriction endonucleases that cut DNA only rarely can now be used to subdivide chromosome-sized DNA. Two important endonucleases with eight-base recognition sites are *Not* I and *Sfi* I.

There are also many instrumental advances that allow changes in the experimental design of PFGE. Some of the variables that can be changed for each experiment are voltage, pulse length, number of electrodes, relative angle of electrodes, gel box design, temperature, agarose concentration, buffer pH, and time of electrophoresis.

Like all laboratory techniques, PFGE has its disadvantages and problems. Long periods of electrophoresis are often required for good resolution, and migration of fragments is extremely dependent on experimental conditions. Therefore, it is difficult to compare gels even when they are run under similar conditions. In spite of these shortcomings, PFGE will continue to advance as a significant tool for the characterization of very large molecules. The technique, which was widely used in the human genome project, will greatly increase our understanding of chromosome structure and function.

Isoelectric Focusing of Proteins

Another important and effective use of electrophoresis for the analysis of proteins is **isoelectric focusing (IEF),** which examines electrophoretic mobility as a function of pH. The net charge on a protein is pH dependent. Proteins below their **isoelectric pH** (pH_I, or sometimes P_I; the pH at which they have zero net charge) are positively charged and migrate in a medium of fixed pH toward the negatively charged cathode. At a pH above its isoelectric point, a protein is deprotonated and negatively charged and migrates toward the anode. If the pH of the electrophoretic medium is identical to the pH_I of a protein, the protein has a net charge of zero and does not migrate toward either electrode. Theoretically, it should be possible to separate protein molecules and to estimate the pH_I of a protein by investigating the electrophoretic mobility in a series of separate experiments in which the pH of the medium is changed. The pH at which there is no protein migration should coincide with the pH_I of the protein. Because such a repetitive series of electrophoresis runs is a rather tedious and time-consuming way to determine the pH_I, IEF has evolved as an alternative method for performing a single electrophoresis run in a medium of gradually changing pH (i.e., a pH gradient).

Separating Proteins by IEF

Figure 6.11 illustrates the construction and operation of an IEF pH gradient. An acid, usually phosphoric, is placed at the cathode; a base, such as triethanolamine, is placed at the anode. Between the electrodes is a medium in which the pH gradually increases from 2 to 10. The pH gradient can be formed before electrophoresis is conducted or formed during the course of electrophoresis. The pH gradient can be either broad (pH 2–10) for separating several proteins of widely ranging pH_I values or narrow (pH 7–8) for precise determination of the pH_I of a single protein. P in Figure 6.11 represents

Figure 6.11

Illustration of isoelectric
focusing. See text for details.

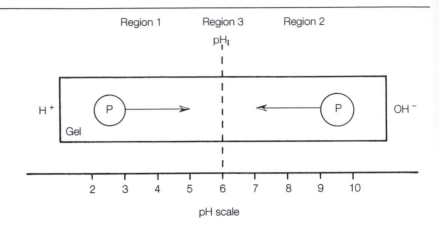

different molecules of the same protein in two different regions of the pH
gradient. Assuming that the pH in region 1 is less than the pH_I of the protein
and the pH in region 2 is greater than the pH_I of the protein, molecules of P
in region 1 will be positively charged and will migrate in an applied electric
field toward the cathode. As P migrates, it will encounter an increasing pH,
which will influence its net charge. As it migrates *up* the pH gradient, P will
become increasingly deprotonated and its net charge will decrease toward
zero. When P reaches a region where its net charge is zero (region 3), it will
stop migrating. The pH in this region of the electrophoretic medium will
coincide with the pH_I of the protein and can be measured with a surface
microelectrode, or the position of the protein can be compared to that of a
calibration set of proteins of known pH_I values. P molecules in region 2 will
be negatively charged and will migrate toward the anode. In this case, the net
charge on P molecules will gradually decrease to zero as P moves *down* the
pH gradient, and P molecules originally in region 2 will approach region 3
and come to rest. The P molecules move in opposite directions, but the
final outcome of IEF is that P molecules located anywhere in the gradient will
migrate toward the region corresponding to their isoelectric point and will
eventually come to rest in a sharp band; that is, they will "focus" at a point
corresponding to their pH_I.

Since different protein molecules in mixtures have different pH_I values,
it is possible to use IEF to separate proteins. In addition, the pH_I of each
protein in the mixture can be determined by measuring the pH of the region
where the protein is focused.

Practical Aspects of IEF

The pH gradient is prepared in a horizontal glass tube or slab. Special
precautions must be taken so that the pH gradient remains stable and is not
disrupted by diffusion or convective mixing during the electrophoresis

experiment. The most common stabilizing technique is to form the gradient in a polyacrylamide, agarose, or dextran gel. The pH gradient is formed in the gel by electrophoresis of synthetic polyelectrolytes, called **ampholytes,** which migrate to the region of their pH_I values just as proteins do and establish a pH gradient that is stable for the duration of the IEF run. Ampholytes are low-molecular-weight polymers that have a wide range of isoelectric points because of their numerous amino and carboxyl or sulfonic acid groups. The polymer mixtures are available in specific pH ranges (pH 5–7, 6–8, 3.5–10, etc.). It is critical to select the appropriate pH range for the ampholyte so that the proteins to be studied have pH_I values in that range. The best resolution is, of course, achieved with an ampholyte mixture over a small pH range (about two units) encompassing the pH_I of the sample proteins. If the pH_I values for the proteins under study are unknown, an ampholyte of wide pH range (pH 3–10) should be used first and then a narrower pH range selected for use.

The gel medium is prepared as previously described except that the appropriate ampholyte is mixed prior to polymerization. The gel mixture is poured into the desired form (column tubes, horizontal slabs, etc.) and allowed to set. Immediately after casting of the gel, the pH is constant throughout the medium, but application of voltage will induce migration of ampholyte molecules to form the pH gradient. The standard gel for proteins with molecular sizes up to 100,000 daltons is 7.5% polyacrylamide; however, if larger proteins are of interest, gels with larger pore sizes must be prepared. Such gels can be prepared with a lower concentration of acrylamide (about 2%) and 0.5 to 1% agarose to add strength. Precast gels for isoelectric focusing are also commercially available.

The protein sample can be loaded on the gel in either of two ways. A concentrated, salt-free sample can be layered on top of the gel as previously described for ordinary gel electrophoresis. Alternatively, the protein can be added directly to the gel preparation, resulting in an even distribution of protein throughout the medium. The protein molecules move more slowly than the low-molecular-weight ampholyte molecules, so the pH gradient is established before significant migration of the proteins occurs. Very small protein samples can be separated by IEF. For analytical purposes, 10 to 50 μg is a typical sample size. Larger sample sizes (up to 20 mg) can be used for preparative purposes.

Two-Dimensional Electrophoresis (2-DE) of Proteins

The separation of proteins by IEF is based on charge, whereas SDS-PAGE separates molecules based on molecular size. A combination of the two methods leads to enhanced resolution of complex protein mixtures. Such an experiment was first reported by O'Farrell (1975), and the combined method has since become a routine and powerful separatory technique. Figure 6.12 shows the results of O'Farrell's analysis of total *Escherichia coli* protein. The sample was first separated in one dimension by IEF. The

Figure 6.12

Two-dimensional electrophoresis of total *E. coli* proteins. *Photo courtesy of Dr. P. O'Farrell.*

IEF ⟶   SDS

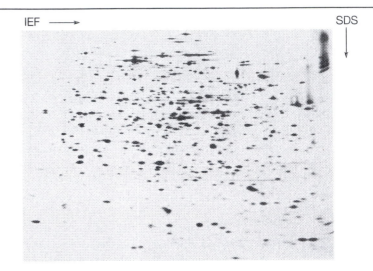

sample gel was then transferred to an SDS-PAGE slab and electrophoresis was continued in the second dimension. At least 1000 discrete protein spots are visible.

The technique of 2-DE is a powerful tool for researchers in the new field of **proteomics,** whose goal is the systematic study of the structure, interactions, and biological function of the proteins expressed by the genome (proteome) of an organism. We see that the 1000 or so proteins in the *E. coli* cell are well resolved by 2-DE (Figure 6.12), but what about the numerous proteins expressed in higher organisms like humans? It is currently estimated from data generated by the Human Genome Project that humans have about 20,000–25,000 genes. This number does not relate directly to the number of proteins because many of the genes are alternately spliced and because of numerous post-translational modification processes. It is now estimated that there may be as many as one million different protein products in the human. This is a daunting number even for the high resolution displayed in 2-DE. It is not realistic to imagine that 2-DE techniques will ever be able to analyze this many proteins, not only because of the complexity, but also because the technique is tedious and time consuming, and the gels are difficult to analyze. However, when 2-DE data are combined and analyzed by modern computer software, valuable on-line databases on protein expression may be constructed. In addition, when proteins separated by 2-DE are further analyzed by mass spectrometry (see Chapter 7, Section D), partial protein sequence data can be obtained.

The 2-DE technique is also of value in developmental biochemistry, where the increase or decrease in intensity of a spot representing a specific protein may be monitored as a function of cell growth.

Capillary Electrophoresis (CE)

Capillary electrophoresis is a new technique that combines the high resolving power of electrophoresis with the speed, versatility, and automation of high-performance liquid chromatography (HPLC). It offers the ability to analyze very small samples (5–10 nL) utilizing up to 1 million theoretical plates to achieve high resolution and sensitivity to the attomole level (10^{-18} mole). It will become a widely used technique in the analysis of amino acids, peptides, proteins, nucleic acids, and pharmaceuticals.

CE in Practice

A general experimental design is diagrammed in Figure 6.13. The equipment consists of a power supply, two buffer reservoirs, a buffer-filled capillary tube, and an on-line detector. Platinum electrodes connected to the power supply are immersed in each buffer reservoir. A high voltage is applied along the capillary and a small plug of sample solution is injected into one end of the capillary. Components in the solution migrate along the length of the capillary under the influence of the electric field. Molecules are detected as they exit from the opposite end of the capillary. The detection method used depends on the type of molecules separated, but the most common are UV-VIS fixed-wavelength detectors and diode-array detectors (see Chapter 7). The capillaries used are flexible, fused, silica tubes of $50-100$ μm i.d. and 25–100 cm length that may or may not be filled with chromatographic matrix.

Figure 6.13

Experimental setup for capillary electrophoresis. *Courtesy of Bio-Rad Laboratories, Life Science Research Group, Hercules, CA.*

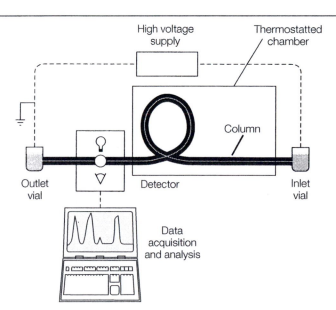

A major advantage of capillary electrophoresis is that many analytical experimental designs are possible, just as in the case of HPLC. In HPLC, a wide range of molecules can be separated by changing the column support (see Chapter 5, Section F). In CE, the capillary tube may be coated or filled with a variety of materials. For separation of small, charged molecules, bare silica or polyimide-coated capillaries are often used. If separation by molecular sieving is desired, the tube is filled with polyacrylamide or SDS-polyacrylamide. If the capillary is filled with electrolyte and an ampholyte pH gradient, isoelectric focusing experiments on proteins may be done. We can expect to see numerous applications of CE in all aspects of biochemistry, and molecular biology. New applications will include DNA sequencing, analysis of single cells, and separations of neutral molecules. (For example, see nucleotide separation in Figure 6.14.)

Figure 6.14

Separation of a series of oligonucleotides using capillary electrophoresis. The compounds are separated on a capillary coated on the inside with polyacryloylaminoethoxyethanol. The structures of the standard oligonucleotides are shown in the lower diagram. Adapted from Bio-Rad Laboratories, Inc. www.bio-rad.com

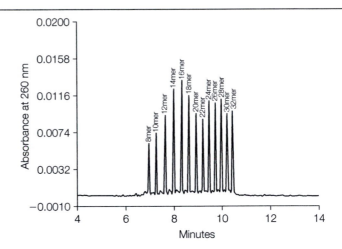

8 mer	GACTGACT
10 mer	GACTGACTGT
12 mer	GACTGACTGACT
14 mer	GACTGACTGACTGT
16 mer	GACTGACTGACTGACT
18 mer	GACTGACTGACTGACTGT
20 mer	GACTGACTGACTGACTGACT
22 mer	GACTGACTGACTGACTGACTGT
24 mer	GACTGACTGACTGACTGACTGACT
26 mer	GACTGACTGACTGACTGACTGACTGT
28 mer	GACTGACTGACTGACTGACTGACTGACT
30 mer	GACTGACTGACTGACTGACTGACTGACTGT
32 mer	GACTGACTGACTGACTGACTGACTGACTGACT

Immunoelectrophoresis (IE)

In immunoelectrophoresis two sequential procedures are applied to the analysis of complex protein mixtures: (1) separation of the protein mixture by agarose gel electrophoresis, followed by (2) interaction with specific antibodies to examine the antigenic properties of the separated proteins.

The technique of IE was first reported by Grabar and Williams in 1953 for the separation and immunoanalysis of serum proteins, but it can be applied to the analysis of any purified protein or complex mixture of proteins. In practice (Figure 6.15), a protein mixture is separated by standard electrophoresis in an agarose gel prepared on a small glass plate. This is followed by exposing the separated proteins to a specific antibody preparation. The antibody is added to a trough cut into the gel, as shown in Figure 6.15, and is allowed to diffuse through the gel toward the separated proteins. If the antibody has a specific affinity for one of the proteins, a visible precipitin arc forms. This is an insoluble complex formed at the boundary of antibody and antigen protein. The technique is most useful for the analysis of protein purity, composition, and antigenic properties. The basic IE technique described here allows only qualitative examination of antigenic proteins. If quantitative results in the form of protein antigen concentration are required, the advanced modifications, rocket immunoelectrophoresis and two-dimensional (crossed) immunoelectrophoresis, may be used.

Figure 6.15

Immunoelectrophoresis. **A** Antigen is placed in sample wells. **B** Electrophoresis. **C** Antiserum containing antibody is placed in trough. **D** Insoluble antigen–antibody complexes from precipitin arcs.

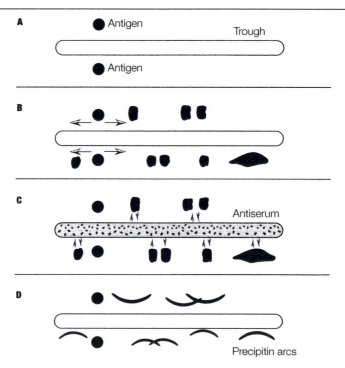

C. PRACTICAL ASPECTS OF ELECTROPHORESIS

Instrumentation

The basic components required for electrophoresis are a power supply and an electrophoresis chamber (gel box). A power supply that provides a constant current is suitable for most conventional electrophoresis experiments. Power supplies that generate both constant voltage (up to 4 kV) and constant current (up to 200 mA) are commercially available. In order to implement the many modifications of electrophoresis, such versatile power supplies are essential in a research laboratory. If isoelectric focusing experiments are planned, a power supply that furnishes a constant voltage is necessary. DNA sequencing experiments require power supplies capable of generating 2–4 kV. Because of the high power requirements of these experiments, the glass plates sandwiching the gel must be covered with conductive aluminum plates to dissipate heat and prevent gel melting and glass plate breakage. CE also requires the use of high voltage, which generates heat as well. However, the heat is quickly dissipated through the thin walls of the capillary tubing. Pulsed field electrophoresis experiments have special requirements, including a high-voltage power supply, electrical switching devices for control of the field, equipment for temperature control, and a specially designed gel box with an array of electrodes. All types of electrophoresis chambers for both horizontal and vertical placements of gels are available from commercial suppliers.

Reagents

High-quality, electrophoresis grade chemicals must be used, since impurities may influence both the gel polymerization process and electrophoretic mobility. The reagents used for gel formation should be stored in a refrigerator.

CAUTION

Acrylamide, N,N,N',N'-tetramethylethylenediamine, N,N'-methylene-bis-acrylamide, and ammonium persulfate are toxic and must be used with care. Acrylamide is a neurotoxin, a carcinogenic agent, and a potent skin irritant, so gloves and a mask must be worn while handling it in the unpolymerized form.

Buffers appropriate for electrophoresis gels include Tris-glycine, Tris-acetate, Tris-phosphate, and Tris-borate at concentrations of about 0.05 M (Chapter 3 Section A).

Several advances in gel electrophoresis have recently been made that have streamlined and improved many of the established techniques. Some of those improvements will be outlined here.

1. **Precast Gels**   As previously mentioned, precast gels of polyacrylamide and agarose are now commercially available. A wide variety of gel sizes, types, configurations, and compositions may be purchased. Costs are reasonable, beginning at about $10 for an 8×10 cm single percentage polyacrylamide gel. Precast gels not only offer convenience but also increase safety, save time, and provide more reproducible electrophoretic runs.

2. **Bufferless Precast Gels**   One of the major inconveniences of running gels is the necessity of having liquid buffer reservoirs to saturate gels during electrophoresis. There is always the chance of leaking reservoirs and spilling solutions perhaps containing acrylamide or ethidium bromide. Precast gels (agarose and acrylamide) are now available in dry, plastic cassettes. They do not require liquid buffers because the gels contain ion-exchange matrices, which sustain the electric field.

3. **Reusable Precast Gels**   Precast agarose gels are now available that may be recycled. After a run, the DNA samples on the gel are removed by reversing the direction of the electric field. The gels are then reloaded with new samples and reused.

Staining and Detecting Electrophoresis Bands

Tracking Dyes

During the electrophoretic process, it is important to know when to stop applying the voltage or current. If the process is run for too long, the desired components may pass entirely through the medium and into the buffer; if too short a period is used, the components may not be completely resolved. It is common practice to add a "tracking dye," usually bromphenol blue and/or xylene cyanol, to the sample mixture. These dyes, which are small and anionic, move rapidly through the gel ahead of most proteins or nucleic acids. After electrophoresis, bands have a tendency to widen by diffusion. Because this broadening may decrease resolution, gels should be analyzed as soon as possible after the power supply has been turned off. In the case of proteins, gels can be treated with an agent that "fixes" the proteins in their final positions, a process that is often combined with staining.

Staining Proteins with Coomassie Blue

Reagent dyes that are suitable for visualization of biomolecules after electrophoresis were mentioned earlier. The most commonly used general stain for proteins is **Coomassie Brilliant Blue.** This dye can be used as a 0.25% aqueous solution. It is followed by destaining (removing excess background dye) by repeated washing of the gel with 7% acetic acid. Alternatively, gels may be stained by soaking in 0.25% dye in H_2O, methanol, acetic acid (5:5:1) followed by repeated washings with the same solvent.

The most time-consuming procedure in the Coomassie Blue visualization process is destaining, which often requires days of washing. Rapid destaining of gels may be brought about electrophoretically. The gel, after soaking in the stain, is subjected to electrophoresis again, using a buffer of higher concentration to remove excess stain.

The search for more rapid and sensitive methods of protein detection after electrophoresis has led to the development of fluorescent staining techniques. Two commonly used fluorescent reagents are fluorescamine and anilinonaphthalene sulfonate. New dyes based on silver salts (silver diamine or silver–tungstosilicic acid complex) have been developed for protein staining. They are 10 to 100 times more sensitive than Coomassie Blue (Fig. 6.7).

There is often a need for a visualization procedure that is specific for a certain biomolecule, for example, an enzyme. If the enzyme remains in an active form while in the gel, any substrate that produces a colored product could be used to locate the enzyme on the gel. Although it is less desirable for detection, the electrophoresis support medium may be cut into small segments and each part extracted with buffer and analyzed for the presence of the desired component.

SYPRO Dyes

The new, fluorescent SYPRO dyes, produced by Molecular Probes, Inc., promise to challenge the use of silver staining for protein gel staining. SYPRO Orange stain, SYPRO Ruby stain, and SYPRO Red stain may be used for 1-D and 2-D gels, and their sensitivities are equal to silver staining. However, the SYPRO dyes are much easier and faster to use (they require no fixing or destaining) than either Coomassie Blue or silver. Gels treated with the SYPRO dyes may be viewed using a standard UV light or with a laser scanner.

Viewing Nucleic Acids on Gels

Nucleic acids may be visualized in agarose and polyacrylamide gels using the fluorescent dye ethidium bromide. The gel is soaked in a solution of the dye and washed to remove excess dye. Illumination of the rinsed slab with UV light reveals red-orange stains where nucleic acids are located.

Although ethidium bromide stains both single- and double-stranded nucleic acids, the fluorescence is much greater for double-stranded molecules. The electrophoresis may be performed with the dye incorporated in the gel and buffer, although this does influence mobility. This has the advantage that the gel can be illuminated with UV light during electrophoresis to view the extent of separation. Precast gels are made in plastic cassettes that are UV transparent. The mobility of double-stranded DNA may be reduced 10 to 15% in the presence of ethidium bromide. Destaining of the gel is not necessary because ethidium bromide–DNA complexes have a much greater fluorescent yield than free ethidiurn bromide, so relatively small amounts of DNA can be detected in the presence of free ethidium bromide. The detection limit for DNA is 10 ng.

> **CAUTION**
>
> Ethidium bromide must be used with great care as it is a potent mutagen. Gloves should be worn at all times while using dye solutions or handling gels.

The new, fluorescent SYBR dyes, produced by Molecular Probes, Inc., offer several advantages over ethidium bromide. They are far less toxic than the mutagenic ethidium bromide and some are as much as five times more sensitive. The most widely used SYBR dye for staining DNA gels is SYBR Green I, which is a cyanine dye and has a sensitivity similar to EtBr. The detailed mechanism for SYBR dye binding to DNA is unknown, but it is most likely not base intercalation as is the case for ethidium bromide. Other new dyes now being tested include Blueview, Carolina Blu, and 4',6-diamidino-2-phenylindole dihydrochloride hydrate (DAPI).

Because single-stranded nucleic acids do not stain deeply with ethidium bromide, and other fluorescent dyes, alternate techniques must be used for detection, especially when only small amounts of biological material are available for analysis. One of the most sensitive techniques is to use radiolabeled molecules. For nucleic acids, this usually means labeling the 5' or 3' end with ^{32}P, a strong β emitter. Bands of labeled nucleic acids on an electrophoresis gel can easily be located by **autoradiography.** For this technique, the electrophoresed slab gel is transferred to heavy chromatography paper. After covering the gel and paper with plastic wrap, they are placed on X-ray film and wrapped in a cassette to avoid external light exposure. This procedure must be done in a darkroom. The gel-film combination is stored at −70°C for exposure. The low temperature maintains the gel in a rigid form and prevents diffusion of gel bands. The exposure time depends on the amount of radioactivity but can range from a few minutes to several days. Figure 6.16 shows the autoradiogram of a typical DNA sequencing gel. The autoradiogram also provides a permanent record of the gel for storage and future analysis. The actual gel is very fragile and difficult to store. Also, because of the short half-life of ^{32}P (14 days), an autoradiogram of the gel cannot be obtained in the distant future. Proteins labeled with ^{32}P or ^{125}I can be dealt with in a manner similar to nucleic acids. If an autoradiogram of a gel can be prepared, a permanent record of the experimental data is available.

Because of concerns about the safety of radioisotope use, researchers are developing fluorescent and chemiluminescent methods for detection of small amounts of biomolecules on gels. One attractive approach is to label biomolecules before analysis with the coenzyme biotin. Biotin forms a strong complex with enzyme-linked streptavidin. Some dynamic property of the enzyme is then measured to locate the biotin-labeled biomolecule on

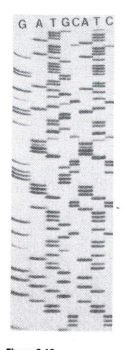

G A T G C A T C

Figure 6.16

Autoradiogram of a DNA sequencing gel. *From Zyskind and Bernstein, DNA Laboratory Manual (1999), Academic Press (San Diego, CA).*

the gel. These new methods approach the sensitivity of methods involving radiolabeled molecules, and rapid advances are being made.

Protein and Nucleic Acid Blotting

Only a minute amount of protein or nucleic acid is present in bands on electropherograms. In spite of this, there is often a need to extract the desired biomolecule from the gel for further investigation. This sometimes involves the tedious and cumbersome process of crushing slices of the gel in a buffer to release the trapped proteins or nucleic acids. Techniques are now available for removing nucleic acids and proteins from gels and characterizing them using probes to detect certain structural features or functions. After electrophoresis, the desired biomolecules are transferred or "blotted" out of the gel onto a nitrocellulose filter or nylon membrane. The desired biomolecule is now accessible on the filter for further analysis. The first blotting technique was reported by E. Southern in 1975. Using labeled complementary DNA probes, he searched for certain nucleotide sequences among DNA molecules blotted from the gel. This technique of detecting DNA-DNA hybridization is called **Southern blotting.** The general blotting technique has now been extended to the transfer and detection of specific RNA with labeled complementary DNA probes **(Northern blotting)** and the transfer and detection of proteins that react with specific antibodies **(Western blotting)**. In practice, the electropherogram is alkali treated, neutralized, and placed in contact with the filter or nylon membrane. A buffer is used to facilitate the transfer. Figure 6.17 shows the setup for a blotting experiment. The location of the desired nucleic acid or protein is then detected by incubation of the membrane with a radiolabeled probe and autoradiography, by use of a biotinylated probe or by linkage to an enzyme-catalyzed reaction that generates a color.

Blotting techniques have many applications, including mapping the genes responsible for inherited diseases by using restriction fragment length polymorphisms (RFLPs), screening collections of cloned DNA fragments (DNA libraries), "DNA fingerprinting" for analysis of biological material remaining at the scene of a crime, and identification of specific proteins. More details on blotting is found in Chapter 10, Section C.

The Western Blot

Western blotting has become an important, modern technique for analysis and characterization of proteins. The procedure consists of the electrophoretic transfer (blotting) of proteins from a polyacrylamide gel to a synthetic membrane. The transferred blots are then probed using various methods involving antibodies that bind to specific proteins.

PAGE is indeed a very effective analytical tool to achieve fractionation of protein mixtures, to analyze purity, and to estimate molecular weight, but it provides no experimental data to prove the identity of any of the protein bands. Stained spots simply indicate the presence and location of each and every protein on the gel. It is often possible to identify proteins by treating gel

Figure 6.17

Diagram of a Southern
blotting experiment.

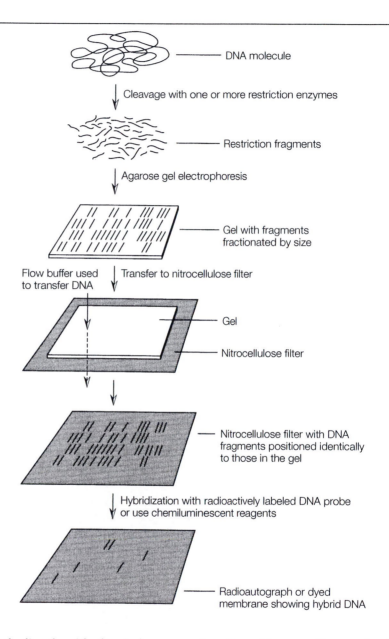

DNA molecule

Cleavage with one or more restriction enzymes

Restriction fragments

Agarose gel electrophoresis

Gel with fragments
fractionated by size

Flow buffer used
to transfer DNA

Transfer to nitrocellulose filter

Gel

Nitrocellulose filter

Nitrocellulose filter with DNA
fragments positioned identically
to those in the gel

Hybridization with radioactively labeled DNA probe
or use chemiluminescent reagents

Radioautograph or dyed
membrane showing hybrid DNA

bands directly with chemical reagents that react with a specific protein; for example, enzymes on the gel could be treated with substrates that form a colored product. However, proteins are deeply embedded in the polyacrylamide gel matrix and are not readily accessible to most analytical reagents. This hinders specific analysis of the protein bands in order to identify individual proteins. Proteins separated by PAGE may be transferred (or blotted) from the gel to a thin support matrix, usually a nitrocellulose membrane, which strongly binds and immobilizes proteins. The protein blots on the membrane surface are more accessible to chemical or biochemical reagents for further analysis. When the transfer process is coupled with protein identification using highly specific and sensitive immunological detection techniques, the procedure is called **Western blotting.** Western blotting or immunoblotting assays of proteins have many advantages, including the need for only small reagent volumes, short processing times, relatively inexpensive equipment, and ease of performance (see immunoadsorption, Chapter 5, Section G).

The Western Blotting Procedure

To begin the Western blot, a protein mixture for analysis and further characterization is fractionated by PAGE. Since denaturing SDS-PAGE results in better resolution than native PAGE, the SDS version is usually preferred; however, the detection method used at the conclusion of the blotting experiment must be able to recognize denatured protein subunits. The next step involves selection of the membrane matrix for transfer. Three types of support matrices are available for use: nitrocellulose, nylon, and polyvinyldifluoride (PVDF). Nitrocellulose membranes, currently the most widely used supports, have a satisfactory general protein binding capacity ($100 \ \mu g/cm^2$), but they display weak binding of proteins of molecular weights smaller than 14,000 and they are subject to tearing. Binding of proteins to nitrocellulose membranes is noncovalent, most likely involving hydrophobic interactions. Nylon membranes are stronger than nitrocellulose and some have a binding capacity up to $450 \ \mu g/cm^2$. However, since they are cationic, they only weakly bind basic proteins. During detection procedures, nylon membranes often display high background colors, so it is difficult to visualize proteins of interest. PVDF membranes bind proteins strongly ($125 \ \mu g/cm^2$) and, because of their hydrophobic nature, give light background color after analysis. For overall general use in protein transfer and immunoblotting, nitrocellulose membranes are the most common choice.

The actual blotting process may be accomplished by one of two methods: **passive (or capillary) transfer** and **electroblotting.** In passive transfer, the membrane is placed in direct contact with the polyacrylamide gel and organized in a sandwich-like arrangement consisting of (from bottom to top) filter paper soaked with transfer buffer, gel, membrane, and more filter paper (see Figure 6.17). The sandwich is compressed by a heavy weight. Buffer passes by capillary action from the bottom filter paper through the gel, transferring the protein molecules to the membrane, where the macromolecules are immobilized. Passive transfer is very time consuming, sometimes requiring 1–2 days for complete protein transfer. Faster and more efficient transfer is afforded by the use of an electroblotter. Here a sandwich of filter paper, gel, membrane and more filter paper is prepared in a cassette, which is placed between platinum electrodes. An electric current is passed through the gel, causing the proteins to electrophorese out of the gel and onto the membrane. The electroblotting process usually is complete in 1–4 hours.

Detection of Blotted Proteins

The Western blot procedure is concluded by probing the blotted protein bands and detecting a specific protein or group of proteins among the blots. In other words, visualization of specific protein blots is now possible. The most specific identification techniques are based on immunology (antigen–antibody interactions; see Chapter 6, Section G). A general procedure for immunoblotting is outlined in Figure 6.18. Before the protein detection process can begin, it is necessary to block protein binding sites on the membrane that are not occupied by blotted proteins. This is essential because antibodies used to detect blotted proteins are also proteins and will bind to the membrane and interfere with detection procedures. Protein binding sites still remaining on blotted membranes may be blocked by treatment with solutions of casein (major protein in milk), gelatin, or bovine serum albumin.

The blotted membrane, with all protein binding sites occupied, can now be treated with analytical reagents for detection of specific proteins. Typically, the blotted membrane is incubated with an antibody specific for the protein of interest. This is called the **primary antibody,** which is a protein of the immunoglobulin G (IgG) class (see Figure 6.18). The primary antibody binds to the desired protein, forming an antigen–antibody complex. The interaction between the protein and its antibody does not usually result in a visible signal. The blot is then incubated with a **secondary antibody,** which is directed against the general class of primary antibody. For example, if the primary antibody was produced in rabbit serum, then the second antibody would be anti-rabbit IgG, usually from a goat or horse. The

---

Figure 6.18

Specific detection of nitrocellulose membrane-bound proteins using a conjugated enzyme. **1** Proteins are transferred from electrophoresis gel to nitrocellulose membrane. Blocker proteins bind to unoccupied sites on the membrane. **2** The membrane is incubated with a primary antibody directed against the protein of interest. **3** A secondary antibody is directed against the primary antibody. **4** The second antibody is conjugated with an enzyme to provide a detection mechanism. Substrate solution is added to the blot. The conjugated enzyme (HRP or AP) catalyzes the conversion of substrate (S) to product (P) to form a colored precipitate at the site of the protein–antibody complex.

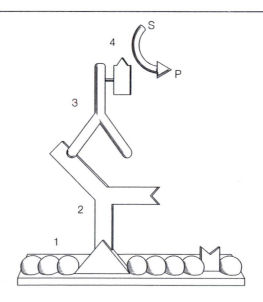

Figure 6.19

Results from a Western blot. A SDS-PAGE gels, 12%, were run and transferred to nitrocellulose. Lane **1**, MW standards; lane **2**, biotinylated standards; lane **3**, human transferrin; lane **4**, *E. coli* lysate; lane **5**, total human serum; lane **6**, biotinylated standards. Gel A was stained with a protein dye. Blot **B** was assayed using rabbit anti-human transferrin as the first antibody. The second antibody solution contained anti-rabbit HRP conjugates. Only the transferrin bands and the prestained biotinylated standards were detected by the antibodies and the avidin-HRP treatment.

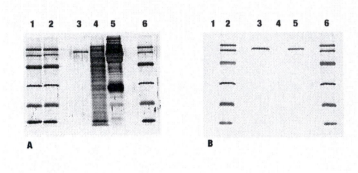

second antibody is labeled (conjugated) so that the interaction of the second antibody with the primary antibody produces some visual signal. For most detection procedures, the secondary antibody may be tagged with an enzyme, usually horseradish peroxidase (HRP) or alkaline phosphatase (AP). When the treated blot is incubated in a substrate solution, the conjugated enzyme catalyzes the conversion of the substrate into a visible product that precipitates at the blot site. The presence of a colored band indicates the position of the protein of interest (see Figure 6.19). This general procedure is also the basis for the widely used enzyme-linked immunosorptive assay (ELISA). The reactions catalyzed by conjugated enzymes are shown below:

$$\text{4-chloro-1-naphthol}_{reduced} + H_2O_2 \xrightleftharpoons{\text{HRP}}$$

$$\text{4-chloro-1-naphthol}_{oxidized} + H_2O + \tfrac{1}{2}O_2$$

Equation 6.5

$$\text{5-bromo-4-chloro-3-indolylphosphate} + \text{nitroblue tetrazolium}_{oxidized} \xrightleftharpoons{\text{AP}}$$

$$\text{5-bromo-4-chloro-3-indole} + P_i + \text{nitroblue tetrazolium}_{reduced}$$

Equation 6.6

A modification of these coloring systems has recently been developed that leads to more sensitive detection. Chemiluminescent substrates have been designed that are converted by the enzymes to products that generate a light signal that can be captured on photographic film. This increases the level of sensitivity about 1000-fold over standard color detection methods.

Even though primary and secondary antibodies are widely used in Western blotting detection systems, they do have some disadvantages. For proteins to be detected, specific antibodies must be available. It is often very time-consuming and expensive for a research laboratory to generate the proper antibodies if they are not available commercially. Even if antibodies are commercially available, they are very expensive.

The Western blot has gained widespread use in biochemical and clinical investigations. It is one of the best methods for identifying the presence of specific proteins in complex biological mixtures. The Western blot procedure has been modified to develop a diagnostic assay that detects the presence in serum of antibodies to the AIDS virus. The presence of AIDS antibodies in a patient is an indication of viral infection.

Analysis of Electrophoresis Results

By separating biochemicals on the basis of charge, size, and conformation, electrophoresis can provide valuable information, such as purity, identity, and molecular weight. Purity is indicated by the number of stained bands in the electropherogram. One band usually means that only one detectable component is present; that is, the sample is homogeneous or "electrophoretically pure." Two or more bands usually indicate that the sample contains two or more components, contaminants, or impurities and is therefore heterogeneous or impure. There are, of course, exceptions to this description. Other proteins or nucleic acids may be present in what appears to be a homogeneous sample, but they may be below the limit of detection of the staining method. Occasionally a homogeneous sample may result in two or more bands because of degradation during the electrophoresis process. Information on purity may be obtained by all of the electrophoresis methods discussed.

The identity of unknown biomolecules can be confirmed by electrophoresis on the same gel, the unknown alongside known standards. This is similar to the identification of unknowns by HPLC as discussed in Chapter 5, Section F.

As previously discussed in this chapter, the molecular size of protein or nucleic acid samples may be determined by electrophoresis. This requires the preparation of standard curves of log molecular weight versus μ (mobility) using standard proteins or nucleic acids.

Study Problems

> *1.* What physical characteristics of a biomolecule influence its rate of movement in an electrophoresis matrix?

> *2.* Draw a slab gel to show the results of nondenaturing electrophoresis of the following mixture of proteins. The molecular weight is given for each.
> Lysozyme (13,930)   Egg white albumin (45,000)
> Chymotrypsin (21,600)   Serum albumin (65,400)

3. Each of the proteins listed below is treated with sodium dodecyl sulfate and separated by electrophoresis on a polyacrylamide slab gel. Draw pictures of the final results.

(a) Myoglobin

(b) Hemoglobin (two α subunits, molecular weight = 15,500; two β subunits, molecular weight = 16,000)

4. Explain the purpose of each of the chemical reagents that are used for PAGE.

(a) acrylamide

(b) N, N'-methylene-bis-acrylamide

(c) TEMED

(d) sodium dodecyl sulfate

(e) Coomassie Blue dye

(f) bromophenol blue

▶ 5. What is the main advantage of slab gels over column gels for PAGE?

▶ 6. Is it possible to use polyacrylamide as a matrix for electrophoresis of nucleic acids? What are the limitations, if any?

7. Explain the purposes of protein and nucleic acid "blotting."

8. Can polyacrylamide gels be used for the analysis of plasmid DNA with greater than 3000 base pairs? Why or why not?

9. Describe the toxic characteristics of acrylamide and outline precautions necessary for its use.

10. The dye ethidium bromide is often used to detect the presence of nucleic acids on electrophoresis supports. Explain how it functions as an indicator.

▶ 11. The name "Western blot" is derived from other blotting procedures called the Southern blot (developed by Earl Southern) and the Northern blot. What are the differences among these three types of blotting techniques and what is the purpose of each?

▶ 12. An economical way to "block" a Western blot membrane is to incubate it in a 10% solution of nonfat milk powder. How does this solution function as a blocking reagent?

▶ 13. If a protein you wish to analyze by Western blotting is acidic (anionic under blotting conditions), what type of membrane would be best to ensure the tightest binding?

▶ 14. Design a diagnostic test based on the Western blot that would give an indication of infection by the AIDS virus. Assume that a blood serum sample is available from the patient.

▶ 15. A protein mixture can be fractionated either by native PAGE or by denaturing SDS-PAGE, before Western blotting. What factors would determine your choice of electrophoresis method?

▶ 16. Bromophenol blue dye is often used as a marker to tell you when to stop the electrophoresis process. What assumptions must you make about the relative movement of the dye versus sample proteins during electrophoresis?

17. Define each of the following items in terms of their use in Western blotting.

(a) SDS-PAGE

(b) Nylon membrane

(c) Electroblotting

(d) Primary antibody

(e) Secondary antibody

(f) Protein molecular weight standard mixture

(g) Conjugated enzyme

18. In your biochemistry research project, you have isolated a new protein from spinach leaves. You wish to do a Western blot experiment with the protein to help determine its chemical structure and/or biological function. An amino acid analysis of the protein showed a great abundance of Phe, Leu, and Val. What would be your choice of membrane for the blotting experiment?

Further Reading

J. Berg, J. Tymoczko, and L. Stryer, *Biochemistry,* 5th ed. (2002), Freeman (New York), pp. 83–87. Analysis of proteins by electrophoresis.

K. Bourzac, L. LaVine, and M. Rice, *J. Chem. Educ.* **80**, 1292–1297 (2003). "Analysis of DAPI and SYBR Green I as Alternatives to Ethidium Bromide for Nucleic Acid Staining in Agarose Gel Electrophoresis."

R. Boyer, *Concepts in Biochemistry,* 2nd ed. (2002), John Wiley & Sons, (New York), pp. 88–89. Introduction to electrophoresis.

R. Boyer, *Modern Experimental Biochemistry,* 3rd ed. (2000), Benjamin/ Cummings (San Francisco), pp. 321–331. Identification of serum glycoproteins by SDS-PAGE and Western blotting.

P. Cummings, *Biochem. Educ.* **25**, 39–40 (1997). "Simulated Western Blot for the Science Curriculum."

R. Cunico, K. Gooding, and T. Wehr, *Basic HPLC and CE of Biomolecules* (1998), Bay Bioanalytical Laboratory (Hercules, CA).

E. Eberhardt *et al., Biochem. Mol. Biol. Educ.* **31**, 402–409 (2003). "Preparing Undergraduates to Participate in the Post-Genome Era: A Capstone Laboratory Experience in Proteomics."

S. Farrell and L. Farrell, *J. Chem. Educ.* **73**, 740–742 (1995). "A Fast and Inexpensive Western Blot Experiment for the Undergraduate Laboratory."

R. Garrett and C. Grisham, *Biochemistry,* 3rd ed. (2005), Brooks Cole (Belmont, CA), pp. 124–125, 149–150, 388. An introduction to electrophoresis.

D. Nelson and M. Cox, *Lehninger Principles of Biochemistry,* 4th ed. (2005), Freeman (New York), pp. 92–94, 296–298, 314. Introduction to electrophoresis.

N. Price, R. Dwek, M. Wormald, and R. Ratcliffe, *Principles and Problems in Physical Chemistry for Biochemists,* 3rd ed. (2002), Oxford University Press (New York).

K. Stewart and R. Ebel, *Chemical Measurements in Biological Systems,* (2000), John Wiley & Sons (New York), pp. 147–168. Theory and uses of electrophoresis.

I. Tinoco, Jr., K. Sauer, J. Wang, and J. Puglisi, *Physical Chemistry: Principles and Applications in Biological Sciences,* 4th ed. (2002), Prentice Hall (Upper Saddle River, NJ). Chapter 6 covers electrophoresis.

D. Voet and J. Voet, *Biochemistry,* 3rd ed. (2004), John Wiley & Sons (Hoboken, NJ), pp. 144–151, 156–157. Introduction to electrophoresis.

D. Voet, J. Voet, and C. Pratt, *Fundamentals of Biochemistry*, 2nd ed. (2006), John Wiley & Sons (Hoboken, NJ), pp. 105-107. An introduction to electrophoresis.

J. Walker, Editor, *The Protein Protocols Handbook*, 2nd ed. (2002), Humana Press (Totowa, NJ). Contains several chapters on electrophoresis of proteins and peptides.

J. Yan et al., *Electrophoresis* **21**, 3657–3665 (2000). "Postelectrophoretic Staining of Proteins Separated by Two-Dimensional Gel Electrophoresis Using SYPRO Dyes."

http://www.uct.ac.za/microbiology/sdspage.html
> Discussion of SDS-PAGE.

http://biotech.biology.arizona.edu/labs/Electrophoresis_dyes_stude.html
> The University of Arizona Biotech Project.

http://www.ceandcec.com
> CE and CEC Web site.

http://www.bio.davidson.edu/Biology/courses/Molbio/tips/trblDNAgel.html
> Troubleshooting DNA agarose gel electrophoresis.

http://javalab.chem.virginia.edu/lab3d OR lab3d.chem.virginia.edu
> The University of Virginia Lab 3D Project, a virtual SDS-PAGE lab.

http://www.probes.com
> This Web site for Molecular Probes Inc. has extensive information on gel staining for proteins and nucleic acids.

http://www.tropix.com/westbak.htm
> Overview of Western blotting.

http://www.med.unc.edu/wrkunits/3ctrpgm/pmbb/mbt/WESTERN.htm
> For introduction and procedures read Unit 1: Western Blotting.

http://www.biology.arizona.edu/immunology/activities/western_blot/w_main.html
> Reviews of procedures in Western blotting.

http://www.uct.ac.za/microbiology/western.html
> Review of the technique of Western blotting.

http://www.uct.ac.za/microbiology/sdspage.html
> Review of SDS-PAGE.

http://www.bio-rad.com
> Information on all areas of electrophoresis and blotting.

http://proteomics.cancer.dk
> Danish Center for Human Genome Research: Human 2-D PAGE Databases.

SPECTROSCOPIC ANALYSIS OF BIOMOLECULES

Some of the earliest experimental measurements on biomolecules involved studies of their interactions with electromagnetic radiation of all wavelengths, including X-ray, ultraviolet–visible, and infrared. It was experimentally observed that when light impinges on solutions or crystals of molecules, at least two distinct processes occur: **light scattering** and **light absorption.** Both processes have led to the development of fundamental techniques for characterizing and analyzing biomolecules. We now use the term **spectroscopy** to label the discipline that studies the interaction of electromagnetic radiation with matter.

Absorption of ultraviolet–visible light by molecules is an especially valuable procedure for measuring concentration and for molecular structure elucidation. The absorption process is dependent upon two factors: (1) the properties of the radiation (wavelength, energy, etc.), and (2) the structural characteristics of the absorbing molecules (atoms, functional groups, etc.). The interaction of electromagnetic radiation with molecules is a quantum process and described mathematically by quantum mechanics; that is, the radiation is subdivided into discrete energy packets called **photons.** In addition, molecules have quantized excitation levels and can accept packets of only certain quantities of energy, thus allowing only certain electronic transitions.

With some molecules, the process of absorption is followed by emission of light of a longer wavelength. This process, called **fluorescence,** depends on molecular structure and environmental factors and assists in the characterization and analysis of biologically significant molecules and dynamic processes occurring between molecules.

Nuclear magnetic resonance spectroscopy and **mass spectrometry** techniques are also now being applied to the study of biological macromolecules and processes. NMR is especially versatile because, in addition to proton spectra, monitoring the presence of ^{13}C, ^{19}F, ^{15}N, and ^{31}P nuclei in biomolecules is possible. Multidimensional NMR is being applied to the study of protein secondary and tertiary structure. Accurate measurements of protein molecular weight and sequence analysis may now be carried out by mass spectrometry.

A. ULTRAVIOLET-VISIBLE ABSORPTION SPECTROPHOTOMETRY

Wavelength and Energy

The electromagnetic spectrum, as shown in Figure 7.1, is composed of a continuum of waves with different properties defined by wavelength and energy. Several regions of the electromagnetic spectrum are of importance in biochemical studies including X-ray (X-ray crystallography, up to 7 nm), the ultraviolet (UV, 180–340 nm), the visible (VIS, 340–800 nm), the infrared (IR, 1000–100,000 nm), and radio waves (NMR, 10^6–10^{10} nm). In this section, we will concentrate on the UV and VIS regions. Light in these regions has energy sufficient to excite the valence electrons of molecules and thus move the electrons from one energy level (or state) to a higher energy level (excited state).

Figure 7.2 shows that the propagation of light is due to an electrical field component E and a magnetic field component H that are perpendicular to each other. The **wavelength** of light, defined by Equation 7.1, is the distance between adjacent wave peaks as shown in Figure 7.2.

   $\lambda = c/v$ *Equation 7.1*

where

λ = wavelength

c = speed of light, 3×10^8 m/s

v = frequency, the number of waves passing a certain point per unit time

---

Figure 7.1

The electromagnetic spectrum. The wavelengths of light associated with important spectroscopy techniques are shown in the figure.

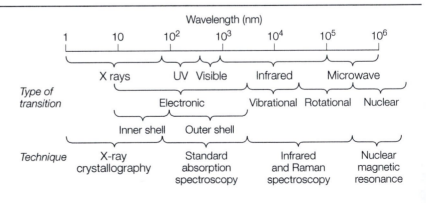

Figure 7.2

An electromagnetic wave, showing the E and H components.

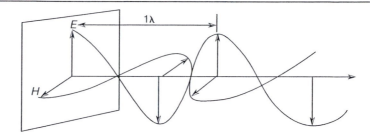

Light also behaves as though it were composed of energetic particles. The amount of energy E associated with these particles (or **photons**) is given by Equation 7.2:

>> $E = h\upsilon$ *Equation 7.2*

where

h = Planck's constant, 6.63×10^{-34} Joules · s

Equation 7.1 and 7.2 may be combined to yield Equation 7.3:

>> $E = hc/\lambda$ *Equation 7.3*

Note the inverse relationship between wavelength and energy. In Figure 7.1, X-rays have the shortest wavelength, but the most energy, and microwaves have long wavelengths, but the least energy.

➤   **Study Exercise 7.1**   What is the amount of energy (E) associated with light of wavelength 300 nm in the ultraviolet range?

Solution: Using Equation 7.3:

$$E = hc/\lambda$$
$$E = \frac{(6.63 \times 10^{-34}\,\text{J} \cdot \text{s})(3 \times 10^{8}\,\text{m/s})}{300 \times 10^{-9}\,\text{m}}$$
$$= 6.63 \times 10^{-19}\,\text{J/photon} = 398\,\text{kJ/mole}$$

This is the approximate amount of energy associated with carbon-based covalent bonds.

Light Absorption

When a photon of specified energy interacts with a molecule, one of two processes may occur. The photon may be **scattered,** or it may transfer its

Figure 7.3

Energy-level diagram showing the ground state G and the first excited state S_1.

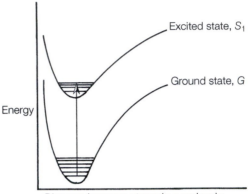

energy to the molecule, producing an **excited state** of the molecule. The former process, called **Rayleigh scattering,** occurs when a photon collides with a molecule and is diffracted or scattered with unchanged frequency. Light scattering is the physical basis of several experimental methods used to characterize macromolecules. Before the development of electrophoresis, light scattering techniques were used to measure the molecular weights of macromolecules. The widely used techniques of X-ray diffraction (crystal and solution), electron microscopy, laser light scattering, and neutron scattering all rely in some way on the light scattering process.

The other process mentioned above, the transfer of energy from a photon to a molecule, is **absorption.** For a photon to be absorbed, its energy must match the energy difference between two energy levels of the molecule. Molecules possess a set of quantized energy levels, as shown in Figure 7.3. Although several states are possible, only two electronic states are shown, a **ground state,** *G,* and the **first excited state,** S_1. These two states differ in the distribution of valence electrons. When electrons are promoted from a ground state orbital in *G* to an orbital of higher energy in S_1, an **electronic transition** is said to occur. The energy associated with ultraviolet and visible light is sufficient to promote molecules from one electronic state to another, that is, to move electrons from one quantized level to another.

Within each electronic energy level is a set of **vibrational levels.** These represent changes in the stretching and bending of covalent bonds. The importance of these energy levels will not be discussed here, but transitions between the vibrational levels are the basis of infrared spectroscopy.

The electronic transition for a molecule from *G* to S_1, represented by the vertical arrow in Figure 7.3, has a high probability of occurring if the energy of the photon corresponds to the energy necessary to promote an electron from energy level E_1 to energy level E_2:

$$E_2 - E_1 = \Delta E = \frac{hc}{\lambda}$$

Equation 7.4

A transition may occur from any vibrational level in G to some other vibrational level in S_1, for example, $v = 3$; however, not all transitions have equal probability. The probability of absorption is described by quantum mechanics and will not be discussed here.

Electronic Transitions in Biomolecules

Absorption of UV and VIS light occurs when the energy associated with the electromagnetic radiation matches the energy difference between molecular states in a molecule. But what are the actual electronic transitions that occur in a molecule? If this question can be answered, then it should be possible to predict what molecules will or will not absorb UV-VIS light. The reverse is also true—by observing the wavelengths of light absorbed by a molecule, it should be possible to speculate on molecular structure (what atoms and functional groups are present). Consider the electronic properties of a carbonyl group, $\diagdown$C$=$O, a functional group present in many types of biomolecules including amino acids, fatty acids, carbohydrates (open chain), nucleotides, proteins, and nucleic acids. A molecular orbital diagram showing the valence electrons in the higher-energy orbitals of the carbonyl group is displayed in Figure 7.4. There are three types of orbitals shown:

• n = nonbonding orbitals with lone pairs of electrons in atoms like O, N, and S.

• π = bonding orbitals formed by overlap of p atomic orbitals

• π^* = antibonding orbitals

Two kinds of electronic transitions are possible:

• $\pi \rightarrow \pi^*$, which are "allowed" by quantum mechanics and have a high probability of occurring, and

• $n \rightarrow \pi^*$, which are "forbidden" and have a low probability of occurring.

These two types of electronic transitions are shown in Figure 7.4.

Figure 7.4

Molecular orbital diagram for the carbonyl group. Only valence electrons in higher-energy orbitals are shown. Two types of electronic transitions are illustrated: $\pi \rightarrow \pi^*$ and $n \rightarrow \pi^*$.

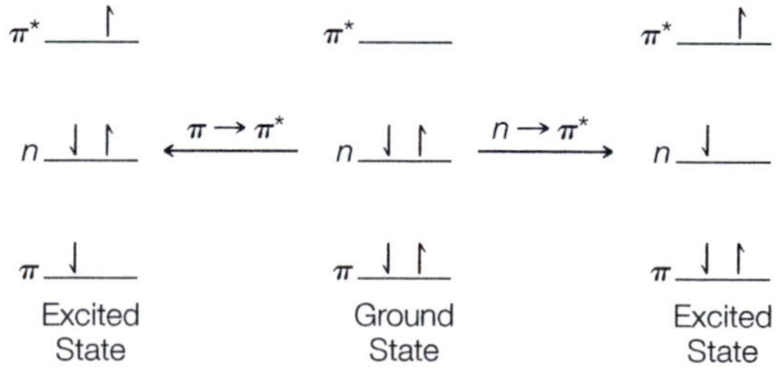

Proteins

The following functional groups in proteins are responsible for absorption of UV-VIS light:

1. Peptide bonds:

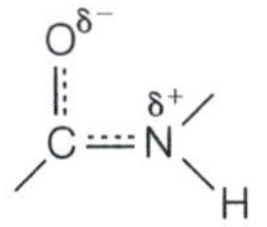

-   $\pi \rightarrow \pi^*$, intense absorption at 190 nm
-   $n \rightarrow \pi^*$, very weak absorption at 210–220 nm

2. Amino acid residues containing aromatic phenyl groups: Phe, Tyr, Trp

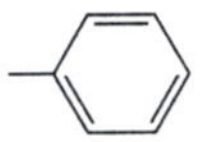

-   $\pi \rightarrow \pi^*$

    Phe: weak absorption at 250 nm

    Tyr: weak absorption at 274 nm

    Trp: strong absorption at 280 nm

    The absorbance at 280 nm is often used to measure protein concentration (Chapter 3, Section B, The Spectrophotometric Assay).

3. Prosthetic groups:

    Nucleotide cofactors: NAD, FAD, FMN, etc.

    Heme

    Chlorophyll

    Metal ions: Cu(II), Fe(II), Fe(III), Ni(II), etc.

    Highly conjugated compounds: retinal, etc.

Many of the cofactors and prosthetic groups above absorb in both the UV and VIS regions. Absorption in the VIS region causes many of the biomolecules in Group 3 to be colored.

Nucleic Acids

Absorption is due to the presence of pyrimidine and purine bases which have several different types of electronic transitions (see structures, Chapter 9, Section A):

1. Extensive aromatic character, $(\pi \rightarrow \pi^*)$
2. Carbonyl groups in bases, $(\pi \rightarrow \pi^*; n \rightarrow \pi^*)$
3. Several nitrogen and oxygen atoms, $(n \rightarrow \pi^*)$.

Nucleic acids display intense absorption in the 200–300-nm range with a maximum at 260 nm. This is the basis for the spectrophotometric assay of DNA and RNA solutions (Chapter 3, Section C).

➤ | **Study Exercise 7.2**   Predict the types of electronic transitions in each of the following biomolecules and suggest whether UV-VIS spectroscopy might be useful in their analysis and characterization. You may need to review their structures in your biochemistry textbook.
(a) Alanine   (f) Adenine
(b) Lysine   (g) β-carotene
(c) Cysteine   (h) Chlorophyll
(d) Glucose   (i) Bilirubin
(e) Lauric acid   (j) Ferritin, a protein containing Fe(III)

The Absorption Spectrum

A UV-VIS spectrum is obtained by measuring the light absorbed by a sample as a function of wavelength. Since only discrete packets of energy (specific wavelengths) are absorbed by molecules in the sample, the spectrum theoretically should consist of sharp discrete lines. However, the many vibrational levels of each electronic energy level increase the number of possible transitions. This results in several spectral lines, which together make up the familiar spectrum of broad peaks as shown in Figure 7.5.

An absorption spectrum can aid in the identification of a molecule because the wavelength of absorption depends on the functional groups or arrangement

Figure 7.5

The visible absorbance spectrum of hemoglobin.

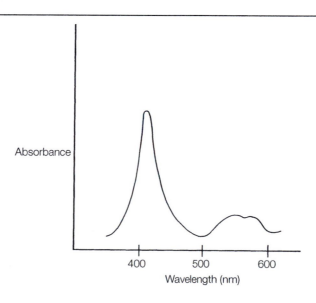

of atoms in the sample. The spectrum of oxyhemoglobin in Figure 7.5 is due to the presence of the iron porphyrin moiety and is useful for the characterization of heme derivatives or hemoproteins. Note that the spectrum consists of several peaks at wavelengths where absorption reaches a maximum (415, 542, and 577 nm). These points, called λ_{max}, are of great significance in the identification of unknown molecules and will be discussed later in the chapter.

The Beer-Lambert Law

Quantitative measurements in spectrophotometry are evaluated using the Beer-Lambert law:

>> $A = Elc$ **Equation 7.5**

where

A = absorbance or $-\log(I/I_0)$
I_0 = intensity of light irradiating the sample
I = intensity of light transmitted through the sample
E = absorption coefficient or absorptivity
l = path length of light through the sample, or thickness of the cell
c = concentration of absorbing material in the sample

Since the absorbance, A, is derived from a ratio ($-\log I/I_0$), it is unitless. The term E, which is a proportionality constant, defines the efficiency or extent of absorption. If this is defined for a particular chromophore at a specific wavelength, the term **absorption coefficient** or **absorptivity** is used. However, students should be aware that in the older biochemical literature, the term extinction coefficient is often used. The units of E depend on the units of l (usually cm) and c (usually molar) in Equation 7.5. For biomolecules, E is often used in the form **molar absorption coefficient, ε,** which is defined as the absorbance of a 1 M solution of pure absorbing material in a 1-cm cell under specified conditions of wavelength and solvent. The units of ε are M^{-1} cm^{-1}. To illustrate the use of Equation 7.5, consider the following calculation.

➤ **Study Exercise 7.3**   The absorbance, A, of a 5×10^{-4} M solution of the amino acid tyrosine, at a wavelength of 280 nm, is 0.75. The path length of the cuvette is 1 cm. What is the molar absorption coefficient, ε?

Solution:

$A = \varepsilon lc = 0.75$
$l = 1$ cm
$c = 5 \times 10^{-4}$ M

$$\varepsilon = \frac{0.75}{(1 \text{ cm})(5 \times 10^{-4} \text{ mole/liter})}$$

$$= 1500 \frac{\text{liter}}{\text{mole} \times \text{cm}} = 1500 \; M^{-1} \; \text{cm}^{-1}$$

Notice that the units of ε are defined by the concentration units of the tyrosine solution (M) and the dimension units of the cuvette (cm). Although E is most often expressed as a molar absorption coefficient, you may encounter other units such as $E_\lambda^{1\%}$, which is the absorbance of a 1% (w/v) solution of pure absorbing material in a 1-cm cuvette at a specified wavelength, λ.

> **Study Exercise 7.4** In the older biochemical literature, absorption coefficients are often reported as **extinction coefficients ($E_\lambda^{1\%}$)**. The extinction coefficient of the enzyme, yeast alcohol dehydrogenase is:
>
> $$E_\lambda^{1\%} = 12.6$$
>
> Calculate the molar absorption coefficient for yeast alcohol dehydrogenase. The molecular weight for the enzyme is 141,000.

Instrumentation

The **spectrophotometer** is used to measure absorbance experimentally. This instrument produces light of a preselected wavelength, directs it through the sample (usually dissolved in a solvent and placed in a cuvette), and measures the intensity of light transmitted by the sample. The major components are shown in Figure 7.6. These consist of a light source, a monochromator (including various filters, slits, and mirrors), a sample chamber, a detector, and a meter or recorder. All of these components are usually under the control of a computer.

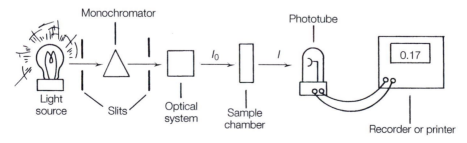

Figure 7.6

A diagram of a conventional, single-beam spectrophotometer.

Light Source

For absorption measurements in the ultraviolet region, a high-pressure hydrogen or deuterium lamp is used. These lamps produce radiation in the 200 to 340 nm range. The light source for the visible region is the tungsten-halogen lamp, with a wavelength range of 340 to 800 nm. Instruments with both lamps have greater flexibility and can be used for the study of most biologically significant molecules.

Monochromator

Both lamps discussed above produce continuous emissions of all wavelengths within their range. Therefore, a spectrophotometer must have an optical system to select monochromatic light (light of a specific wavelength). Modern instruments use a prism or, more often, a diffraction grating to produce the desired wavelengths. It should be noted that light emitted from the monochromator is not entirely of a single wavelength, but is enhanced in that wavelength. That is, most of the light is of a single wavelength, but shorter and longer wavelengths are present.

Before the monochromatic light impinges on the sample, it passes through a series of slits, lenses, filters, and mirrors. This optical system concentrates the light, increases the spectral purity, and focuses it toward the sample. The operator of a spectrophotometer has little control over the optical manipulation of the light beam, except for adjustment of slit width. Light passing from the monochromator to the sample encounters a "gate" or slit. The slit width, which is controlled by a computer, determines both the intensity of light impinging on the sample and the spectral purity of that light. Decreasing the slit width increases the spectral purity of the light, but the amount of light directed toward the sample decreases. The efficiency or sensitivity of the detector then becomes a limiting factor.

In instruments equipped with photodiode array detectors (see Figure 7.7 and the following section on detectors), polychromatic light from a source passes through the sample and is focused on the entrance slit of a polychromator (a holographic grating that disperses the light into its wavelengths). The wavelength-resolved light beam is then focused onto the photodiode array detector. The relative positions of the sample and grating are reversed compared to conventional spectrometry; hence the new configuration is often called reversed optics. (Compare Figures 7.6 and 7.7.)

Sample Chamber

The processed monochromatic light is then directed into a sample chamber, which can accommodate a wide variety of sample holders. Most UV-VIS measurements on biomolecules are taken on solutions of the molecules. The sample is placed in a tube or cuvette made of glass, quartz, or other transparent

Figure 7.7

Optical diagram for a diode-array spectrophotometer.

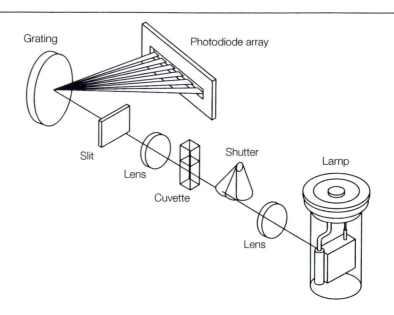

material. Figure 7.8 shows the transmission properties of several transparent materials used in cuvette construction.

Glass cuvettes are inexpensive but, because they absorb UV light, they can be used only above 340 nm. Quartz or fused silica cuvettes may be used throughout the UV and visible regions (~200–800 nm). Disposable plastic cuvettes are now commercially available in polymethacrylate (280–800 nm) and polystyrene (350–800 nm). The care and use of cuvettes were discussed in Chapter 1, Section C.

Figure 7.8

The transmission properties of several materials used in cuvettes. **A** = silica (quartz), **B** = NIR silica, **C** = polymethacrylate, **D** = polystyrene, **E** = glass.

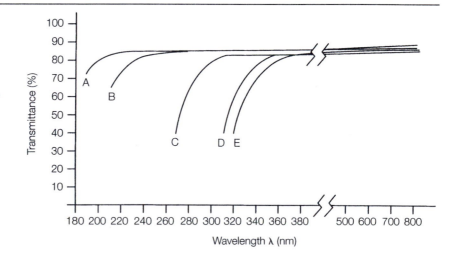

Sample chambers for spectrometers come in two varieties—those holding only one cuvette at a time (single-beam) and those holding two cuvettes, one for a reference, usually solvent, and one for sample (double-beam). In the past, single-beam instruments were usually less expensive but more cumbersome to use because reference and sample cuvettes required constant exchange. However, modern single-beam instruments with computer control and analysis can be programmed to correct automatically for the reference spectrum, which may be stored in a memory file. In double-beam optics, the light beam is split into two paths by directing it through a monochromator. The two beams, which are of identical wavelength and intensity, pass through the sample cell (analyte plus solvent) and reference chamber (solvent only). This allows for the correction of sample absorbance by continuously subtracting the reference spectrum.

Detector

The intensity of the light that passes through the sample under study depends on the amount of light absorbed by the sample. Intensity is measured by a light-sensitive detector, usually a photomultiplier tube (PMT). The PMT detects a small amount of light energy, amplifies this by a cascade of electrons accelerated by dynodes, and converts it into an electrical signal that can be fed into a meter or recorder.

A new technology has been introduced during the past few years that greatly increases the speed of spectrophotometric measurements. New detectors called **photodiode arrays** are being used in modern spectrometers. Photodiodes are composed of silicon crystals that are sensitive to light in the wavelength range 170–1100 nm. Upon photon absorption by the diode, a current is generated in the photodiode that is proportional to the number of photons. Linear arrays of photodiodes are self-scanning and have response times on the order of 100 milliseconds; hence, an entire UV-VIS spectrum can be obtained with an extremely brief exposure of the sample to polychromatic light. New spectrometers designed by Hewlett-Packard, Perkin-Elmer, and others use this technology and can produce a full spectrum from 190 to 820 nm in one-tenth of a second.

Printers and Recorders

Less expensive instruments give a direct readout of absorbance and/or transmittance in analog or digital form. These instruments are suitable for single-wavelength measurements; however, if a scan of absorbance vs. wavelength (Figure 7.5) is desired, some type of device to display the spectrum must be available.

Modern, research-grade spectrometers are available that offer the latest in technology. All of the components discussed above are integrated into a single package and are completely under the control of a computer. By simply pushing a button, one can obtain the UV-VIS spectrum of a sample displayed on a computer screen in less than 1 second. In addition, these modern instruments with computers can be programmed to carry out several functions, such as subtraction of solvent spectrum, spectral overlay, storage, difference spectra, derivative spectra, and calculation of concentrations and rate constants.

Applications of UV–VIS Spectroscopy

Now that you are familiar with the theory and instrumentation of absorption spectrophotometry, you will more easily understand the actual operation and typical applications of a spectrophotometer. Since virtually all UV-VIS measurements are made on samples dissolved in solvents, only those applications will be described here. Although many different types of operations can be carried out on a spectrophotometer, all applications fall in one of two categories:

1. measurement of absorbance at a fixed wavelength
2. measurement of absorbance as a function of wavelength.

Measurements at a fixed wavelength are most often used to obtain quantitative information such as the concentration of a solute in solution or the absorption coefficient of a chromophore. Absorbance measurements as a function of wavelength provide qualitative information that assists in solving the identity and structure of a pure substance by detecting characteristic groupings of atoms in a molecule.

For fixed-wavelength measurements with a single-beam instrument, a cuvette containing solvent only is placed in the sample beam and the instrument is adjusted to read "zero" absorbance. A matched cuvette containing sample plus solvent is then placed in the sample chamber and the absorbance is read directly from the display. The adjustment to zero absorbance with only solvent in the sample chamber allows the operator to obtain a direct reading of absorbance for the sample.

Fixed-wavelength measurements using a double-beam spectrophotometer are made by first zeroing the instrument with no cuvette in either the sample or reference holder. Alternatively, the spectrophotometer can be balanced by placing matched cuvettes containing water or solvent in both sample chambers. Then, a cuvette containing pure solvent is placed in the reference position and a matched cuvette containing solvent plus sample is set in the sample position. The absorbance reading given by the instrument is that of the sample; that is, the absorbance due to solvent is subtracted by the instrument.

An **absorbance spectrum** of a compound is obtained by scanning a range of wavelengths and plotting the absorbance at each wavelength. Most double-beam spectrophotometers automatically scan the desired wavelength range and record the absorbance as a function of wavelength. If solvent is placed in the reference chamber and solvent plus sample in the sample position, the instrument will continuously and automatically subtract the solvent absorbance from the total absorbance (solvent plus sample) at each wavelength; hence, the recorder output is really a difference spectrum (absorbance of sample plus solvent, minus absorbance of solvent).

Both types of measurements (fixed wavelength and absorbance spectrum) are common in biochemistry, and you should be able to interpret results from each. The following four examples are typical of the kinds of problems readily solved by spectrophotometry.

Measurement of the Concentration of an Analyte in Solution

According to the Beer-Lambert law, the absorbance of a material in solution is directly dependent on the concentration of that material. Two methods are commonly used to measure concentration. If the absorption coefficient is known for the absorbing species, the concentration can be calculated after experimental measurement of the absorbance of the solution.

➤ | **Study Exercise 7.5**   A solution of the nucleotide base uracil, in a 1-cm cuvette, has an absorbance at λ_{max}(260 nm) of 0.65. Pure solvent in a matched quartz cuvette has an absorbance of 0.07. What is the molar concentration of the uracil solution? Assume the molar absorption coefficient, ε, is $8.2 \times 10^3\ M^{-1}\ cm^{-1}$.

Solution:

$$A = \varepsilon lc$$
$$A = (\text{absorbance of solvent} + \text{sample}) - (\text{absorbance of solvent})$$
$$A = 0.65 - 0.07 = 0.58$$
$$\varepsilon = 8.2 \times 10^3\ M^{-1}\ cm^{-1}$$
$$l = 1\ cm$$
$$c = \frac{A}{\varepsilon l} = \frac{0.58}{(8.2 \times 10^3\ M^{-1}\ cm^{-1})(1\ cm)}$$
$$c = 7.1 \times 10^{-5}\ M$$

If the absorption coefficient for an absorbing species is known, the concentration of that species in solution can be calculated as outlined. However, there are limitations to this application. Most spectrophotometers are useful for measuring absorbances up to 1, although more sophisticated instruments can measure absorbances as high as 2. (If the absorbance is above 1, dilute sample with the same solvent). Most absorbance readings below 0.1 are not accurate. Also, some substances do not obey the Beer-Lambert law; that is, absorbance may not increase in a linear fashion with concentration. Reasons for deviation from the Beer-Lambert law are many; however, the majority are instrumental, chemical, or physical. Spectrophotometers often display a nonlinear response at high absorption levels because of stray light. Physical reasons for nonlinearity include hydrogen bonding of the absorbing species with the solvent and intermolecular interactions at high concentrations. Chemical reasons may include reaction of the solvent with the absorbing species and the presence of impurities. Linearity is readily tested by preparing a series of concentrations of the absorbing species and measuring the absorbance of each. A plot of A vs. concentration should be linear if the Beer-Lambert law is valid. If the absorption coefficient for a species is unknown, its concentration in solution can be measured if the absorbance of a standard solution of the compound is known.

➤ **Study Exercise 7.6**   The absorbance of a 1% (w/v) solution of the enzyme tyrosinase, in a 1-cm cell at 280 nm, is 24.9. What is the concentration of a tyrosinase solution that has an A_{280} of 0.25?

Since the absorption coefficient, $E\%$, is the same for both solutions, the concentration can be calculated by a direct ratio:

Solution:

$$\frac{A_{std}}{C_{std}} = \frac{A_x}{C_x}$$

A_{std} = absorbance of the 1% standard solution = 24.9

C_{std} = concentration of the standard solution = 1% (1 g/dL)

A_x = absorbance of the unknown solution = 0.25

C_x = concentration of the unknown solution in %

$$\frac{24.9}{1\%} = \frac{0.25}{C_x}$$

$$C_x = 0.01\% = 0.01 \text{ g/dL} = 0.1 \text{ mg/mL}$$

Alternatively, the concentration of a species in solution can be determined by preparing a standard curve of absorbance vs. concentration.

➤ **Study Exercise 7.7**   The Bradford protein assay is one of the most used spectrophotometric assays in biochemistry. (For a discussion of the Bradford assay, see Chapter 3, Section B). Solutions of varying amounts of a standard protein are mixed with reagents that cause the development of a color. The amount of color produced depends on the amount of protein present. The absorbance at 595 nm of each reaction mixture is plotted against the known protein concentration. A protein sample of unknown concentration is treated with the Bradford reagents and the color is allowed to develop.

The following absorbance measurements are typical for the standard curve of a protein:

Protein (μg per assay)	A_{595}
15	0.07
25	0.15
50	0.28
100	0.55
150	0.90
0.1 mL unknown protein solution	0.10
0.2 mL unknown protein solution	0.22

A standard curve for the data is plotted in Figure 3.6 on page 73. Note the linearity, indicating that the Beer-Lambert law is obeyed over this concentration range of standard protein. Two different volumes of unknown protein were tested. This was to ensure that one volume would be in the concentration range of the standard curve. Since the accuracy of the assay is dependent on identical times for color development, the unknowns must be assayed at the same time as the standards. From graphical analysis, an A_{595} of 0.22 corresponds to 40 μg of protein per 0.20 mL. This indicates that the original protein solution concentration was approximately 200 μg/mL or 0.20 mg/mL. Using computer graphics, students can now quickly visualize experimental data and determine the need for further analysis. Most modern computer programs use the method of least squares to calculate automatically the slope, intercepts, and correlation coefficients.

Identification of Unknown Biomolecules by Spectrophotometry

The UV-VIS spectrum of a biomolecule reveals much about its molecular structure. Therefore, a spectral analysis is one of the first experimental measurements made on an unknown biomolecule. Natural molecules often contain chromophoric (color-producing) functional groups that have characteristic spectra. Figure 7.9 displays spectra of the well-characterized biomolecules DNA, FMN, $FMNH_2$, NAD, NADH, a nucleotide, and pyrimidine base.

The procedure for obtaining a UV-VIS spectrum begins with the preparation of a solution of the species under study. A standard solution should be prepared in an appropriate solvent. An aliquot of the solution is transferred to a cuvette and placed in the sample chamber of a spectrophotometer. A cuvette containing solvent is placed in the reference holder. The spectrum is scanned over the desired wavelength range and an absorption coefficient is calculated for each major λ_{max}.

Kinetics of Biochemical Reactions

Spectrophotometry is one of the best methods available for measuring the rates of biochemical reactions. Consider a general reaction as shown in Equation 7.6.

$$A + B \rightleftharpoons C + D$$

Equation 7.6

If reactants A or B absorb in the UV-VIS region of the spectrum at some wavelength λ_1 the rate of the reaction can be measured by monitoring the decrease of absorbance at λ_1 due to loss of A or B. Alternatively, if products C or D absorb at a specific wavelength λ_2, the kinetics of the reaction can be evaluated by monitoring the absorbance increase at λ_2. According to the Beer-Lambert law, the absorbance change of a reactant or product is proportional to the concentration change of that species occurring during the reaction. This method is widely used to assay enzyme-catalyzed processes. Since the rates of chemical reactions vary with temperature, the sample cuvette containing the reaction mixture must be held in a thermostated chamber.

Figure 7.9

UV-VIS absorbance spectra
of significant biomolecules.
A DNA, **B** FMN and FMNH$_2$,
C NAD$^+$ and NADH, **D** d-GMP,
E thymine.

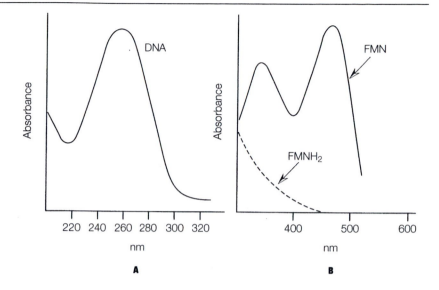

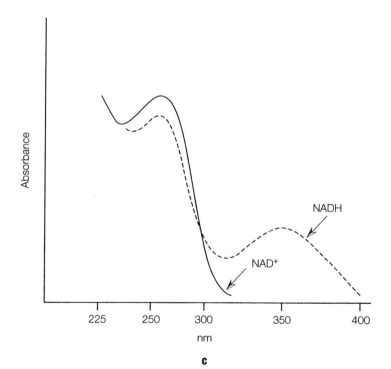

Figure 7.9

continued

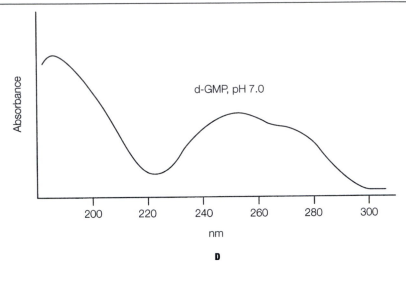

D

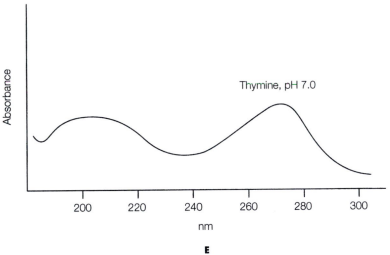

E

Characterization of Macromolecule-Ligand Interactions by Difference Spectroscopy

Many biological processes depend on a specific interaction between molecules. The interaction often involves a macromolecule (protein or nucleic acid) and a smaller molecule, a **ligand.** Specific examples include enzyme-substrate interactions and receptor protein–hormone interactions (see Chapter 8). One of the most effective and convenient methods for detecting and characterizing such interactions is **difference spectroscopy.** The interaction of small molecules with the transport protein hemoglobin is a classic example of the utility of difference spectroscopy. If a small molecule or ligand, such as inositol hexaphosphate, binds to hemoglobin, there is a change in the heme spectral properties. The spectral change is small and would be difficult to detect if the experimenter recorded the spectrum in the usual fashion. Normally, one would obtain a spectrum of a heme protein by placing a solution of the protein in a cuvette in the sample compartment of a spectrophotometer and the neat solvent in the reference compartment. Any absorption due to solvent is subtracted because the solvent is present in both light beams, so the spectrum is that due to the heme protein. Then the ligand to be tested would be added to the heme protein, and the spectrum would be obtained for this mixture vs. solvent. If the free or bound ligand molecule did not absorb light in the wavelength range studied, there would be no need to have ligand in the reference cell. The two spectra (heme protein in solvent vs. solvent and heme protein and ligand in solvent vs. solvent) can then be compared and differences noted.

A difference spectrum is faster than the preceding method because only one spectral recording is necessary. Two cuvettes are prepared in the following manner. The reference cuvette contains heme protein and solvent, whereas the sample cuvette contains heme protein, solvent, and ligand. There must be equal concentrations of the heme protein in the two cuvettes. (Why?) The two cuvettes are placed in a double-beam spectrophotometer and the spectrum is recorded. If the spectrum of the heme protein is not influenced by the ligand, the result would be a zero difference spectrum, that is, a straight line (Figure 7.10A). Both samples have identical spectral properties, indicating that there is probably little or no interaction between heme protein and ligand. However, such data should be treated with caution

Figure 7.10

Difference spectroscopy.
A Hemoglobin vs. hemoglobin.
B Hemoglobin vs. hemoglobin + inositol hexaphosphate.

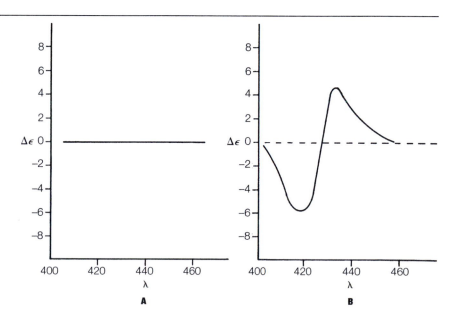

because it is possible that the heme group is not affected by ligand binding. A nonzero difference spectrum indicates that the ligand interacts with the heme protein and induces a change in the environment of the heme group (Figure 7.10B).

A difference spectrum can be analyzed and used in several ways. This is a useful technique for demonstrating qualitatively whether an interaction occurs between a macromolecule and a ligand. Quantitative analysis of difference spectra requires measurement of λ_{max} and ΔA at λ_{max}. This method can be used to quantify the strength of ligand-protein interaction. Note that every time you record a spectrum in a double-beam spectrometer, you are obtaining a difference spectrum between sample and reference. Although a heme protein was used in this example, this does not imply that only heme interactions can be characterized by difference spectroscopy. Any protein that contains a chromophoric group, whether it be an aromatic amino acid, cofactor, prosthetic group, or metal ion, can be studied by difference spectroscopy.

Limitations and Precautions in Spectrophotometry

The use of a spectrophotometer is relatively straightforward and can be mastered in a short period of time. There are, however, difficulties that must be considered. A common problem encountered with biochemical measurements is turbidity or cloudiness of biological samples. This can lead to great error in absorbance measurements because much of the light entering the cuvette is not absorbed but is scattered. This causes artificially high absorbance readings. Occasionally, absorbance readings on turbid solutions are desirable (as in measuring the rate of bacterial growth in a culture), but in most cases turbid solutions must be avoided or clarified by filtration or centrifugation.

A difficulty encountered in measuring the concentration of an unknown absorbing species in solution is deviation from the Beer-Lambert law. For reasons stated earlier in this chapter, some absorbing species do not demonstrate an increase in absorbance that is proportional to an increase in concentration. (In reality, most compounds follow the Beer-Lambert relationship over a relatively small concentration range.) When measuring solution concentration, adherence to the Beer-Lambert law must always be tested in the concentration range under study.

B. FLUORESCENCE SPECTROPHOTOMETRY

Principles

In our discussion of absorption spectroscopy, we noted that the interaction of photons with molecules resulted in the promotion of valence electrons from ground state orbitals to higher energy level orbitals. The molecules were said to be in an excited state.

Figure 7.11

Energy-level diagram outlining the fluorescence process. See text for details.

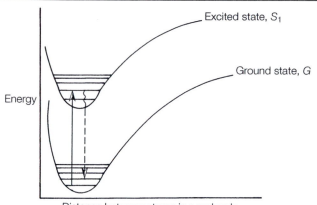

Molecules in the excited state do not remain there long, but spontaneously relax to the more stable ground state. With most molecules, the relaxation process is brought about by collisional energy transfer to solvent or other molecules in the solution. Some excited molecules, however, return to the ground state by emitting the excess energy as light. This process, called **fluorescence,** is illustrated in Figure 7.11. The solid vertical arrow in the figure indicates the photon absorption process in which the molecule is excited from G to some vibrational level in S. The excited molecule loses vibrational energy by collision with solvent and ground state molecules. This relaxation process, which is very rapid, leaves the molecule in the lowest vibrational level of S, as indicated by the wavy arrow. The molecule may release its energy in the form of light (fluorescence, dashed arrow) to return to some vibrational level of G.

Two important characteristics of the emitted light should be noted: (1) The emitted light is of longer wavelength (lower energy) than the excitation light. This is because part of the energy initially associated with the S state is lost as heat energy, and the energy lost by emission may be sufficient only to return the excited molecule to a higher vibrational level in G. (2) The emitted light is composed of many wavelengths, which results in a fluorescence spectrum as shown in Figure 7.12. This is due to the fact that fluorescence from any particular excited molecule may return the molecule to one of many vibrational levels in the ground state. Just as in the case of an absorption spectrum, a wavelength of maximum fluorescence is observed, and the spectrum is composed of a wavelength distribution centered at this emission maximum.

Quantum Yield

In the foregoing discussion, it was pointed out that a molecule in the excited state can return to lower energy levels by collisional transfer or by light emission. Since these two processes are competitive, the **fluorescence intensity** of a fluorescing system depends on the relative importance of each process.

Figure 7.12

Absorption (——) and fluorescence (–––) spectra of the amino acid tryptophan.

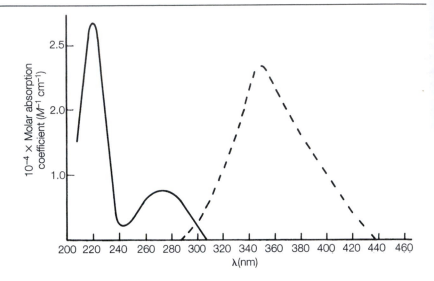

The fluorescence intensity is often defined in terms of **quantum yield,** represented by Q. This describes the efficiency or probability of the fluorescence process. By definition, Q is the ratio of the number of photons emitted to the number of photons absorbed (Equation 7.7).

>> $$Q = \frac{\text{number of photons emitted}}{\text{number of photons absorbed}}$$

Equation 7.7

Measurement of quantum yield is often the goal in fluorescence spectroscopy experiments. Q is of interest because it may reveal important characteristics of the fluorescing system. Two types of factors affect the intensity of fluorescence, internal and external (environmental) influences. Internal factors, such as the number of vibrational levels available for transition and the rigidity of the molecules, are associated with properties of the fluorescent molecules themselves. Internal factors will not be discussed in detail here because they are of more interest in theoretical studies. The external factors that affect Q are of great interest to biochemists because information can be obtained about macromolecule conformation and molecular interactions between small molecules (ligands) and larger biomolecules (proteins, nucleic acids). Of special value is the study of experimental conditions that result in **quenching** or **enhancement** of the quantum yield. Quenching in biochemical systems can be caused by chemical reactions of the fluorescent species with added molecules, transfer of energy to other molecules by collision (actual contact between molecules), and transfer of energy over a distance (no contact, resonance energy transfer). The reverse of quenching, enhancement of fluorescent intensity, is also observed in some situations. Several fluorescent dye molecules are quenched in aqueous solution, but their fluorescence is greatly enhanced in a nonpolar or rigidly bound environment (the interior of a protein, for example). This is a convenient method for characterizing ligand binding. Both fluorescence quenching and fluorescence enhancement studies can yield important information about biomolecular structure and function.

Instrumentation

The basic instrument for measuring fluorescence is the **spectrofluorometer.** It contains a light source, two monochromators, a sample holder, and a detector. A typical experimental arrangement for fluorescence measurement is shown in Figure 7.13. The setup is similar to that for absorption measurements, with two significant exceptions. First, there are two monochromators, one for selection of the excitation wavelength, another for wavelength analysis of the emitted light. Second, the detector is at an angle (usually 90°) to the excitation beam. This is to eliminate interference by the light that is transmitted through the sample. Upon excitation of the sample molecules, the fluorescence is emitted in all directions and is detected by a photocell at right angles to the excitation light beam.

The lamp source used in most instruments is a xenon arc lamp that emits radiation in the ultraviolet, visible, and near-infrared regions (200 to 1400 nm). The light is directed by an optical system to the excitation monochromator, which allows either preselection of a wavelength or scanning of a certain wavelength range. The exciting light then passes into the sample chamber, which contains a fluorescence cuvette with dissolved sample.

Figure 7.13

A diagram of a typical spectrofluorometer.

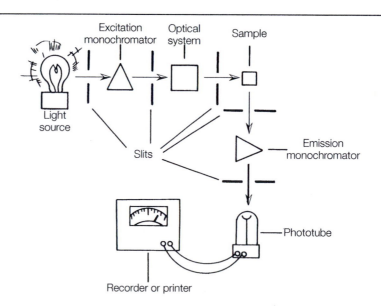

Because of the geometry of the optical system, a typical fused absorption cuvette with two opaque sides cannot be used; instead, special fluorescence cuvettes with four translucent quartz or glass sides must be used. When the excitation light beam impinges on the sample cell, molecules in the solution are excited and some will emit light.

Light emitted at right angles to the incoming beam is analyzed by the emission monochromator. In most cases, the wavelength analysis of emitted light is carried out by measuring the intensity of fluorescence at a preselected wavelength (usually the wavelength of emission maximum). The analyzer monochromator directs emitted light of only the preselected wavelength toward the detector. A photomultiplier tube serves as a detector to measure the intensity of the light. The output current from the photomultiplier is fed to some measuring device that indicates the extent of fluorescence. The final readout is not in terms of Q but in units of the photomultiplier tube current (microamperes) or in relative units of percent of full scale. Therefore, the scale must be standardized with a known. Some newer instruments provide, as output, the ratio of emitted light to incident light intensity.

Applications of Fluorescence Spectroscopy

Two types of measurements are most common in fluorescence experiments, measurements of **relative fluorescence intensities** and measurements of the **quantum yield**. Most experiments require only relative fluorescence intensity measurements, and they proceed as follows. The fluorometer is set to "zero" or "full scale" fluorescence intensity (microamps or %) with the desired biochemical system under standard conditions. Some perturbation is then made in the system (pH change, addition of a chemical agent in varying concentrations, change of ionic strength, etc.) and the fluorescence intensity is determined relative to the standard conditions. This is a straightforward type of experiment because it consists of replacing one solution with another in the fluorometer and reading the detector output for each. For these experiments, the excitation wavelength and the emission wavelength are preselected and set for each monochromator.

The measurement of quantum yield is a more complicated process. Before these measurements can be made, the instrument must be calibrated. A thermopile or chemical actinometer may be used to measure the absolute intensity of incident light on the sample. Alternatively, quantum yields may be measured relative to some accepted standard. Two commonly used fluorescence standards are quinine sulfate in 0.5 M H_2SO_4 ($Q = 0.70$) and fluorescein in 0.1 M NaOH ($Q = 0.93$). The quantum yield of the unknown, Q_x, is then calculated by Equation 7.8.

$$\gg \qquad \frac{Q_x}{Q_{std}} = \frac{F_x}{F_{std}} \qquad \qquad \textit{Equation 7.8}$$

where

Q_x = quantum yield of unknown

Q_{std} = quantum yield of standard

F_x = experimental fluorescence intensity of unknown

F_{std} = experimental fluorescence intensity of standard

Some biomolecules are **intrinsic fluors;** that is, they are fluorescent themselves. The amino acids with aromatic groups (phenylalanine, tyrosine, and tryptophan) are fluorescent; hence, proteins containing these amino acids have intrinsic fluorescence. The purine and pyrimidine bases in nucleic acids (adenine, guanine, cytosine, uracil) and some coenzymes (NAD, FAD) are also intrinsic fluors. Intrinsic fluorescence is most often used to study protein conformational changes (protein folding) and to probe the location of active sites and coenzymes in enzymes.

Valuable information can also be obtained by the use of **extrinsic fluors.** These are fluorescent molecules that are added to the biochemical system under study. Many fluorescent dyes have enhanced fluorescence when they are in a nonpolar solution or bound in a rigid hydrophobic environment. Some of these dyes bind to specific sites on proteins or nucleic acid molecules, and the resulting fluorescence intensity depends on the environmental conditions at the binding site. Extrinsic fluorescence is of value in characterizing the binding of natural ligands to biochemically significant macromolecules. This is because many of the extrinsic fluors bind in the same sites as natural ligands. Extrinsic fluorescence has been used to study the binding of fatty acids to serum albumin, to characterize the binding sites for cofactors and substrates in enzyme molecules, to characterize the heme binding site in various hemoproteins, and to study the intercalation of small molecules into the DNA double helix.

Figure 7.14 shows the structures of extrinsic fluors that have been of value in studying biochemical systems. ANS, dansyl chloride, and fluorescein are used for protein studies, whereas ethidium, proflavine, and various acridines are useful for nucleic acid characterization. Ethidium bromide has the unique characteristic of enhanced fluorescence when bound to double-stranded DNA but not to single-stranded DNA. Aminomethyl coumarin (AMC) is of value as a fluorogenic leaving group in measuring peptidase activity.

Difficulties in Fluorescence Measurements

Fluorescence measurements have much greater sensitivity than absorption measurements. Therefore, the experimenter must take special precautions in making fluorescence measurements because any contaminant or impurity in the system can lead to inaccurate results. The following factors must be considered when preparing for a fluorescence experiment.

Figure 7.14

Molecules useful as extrinsic fluors in biochemical studies.

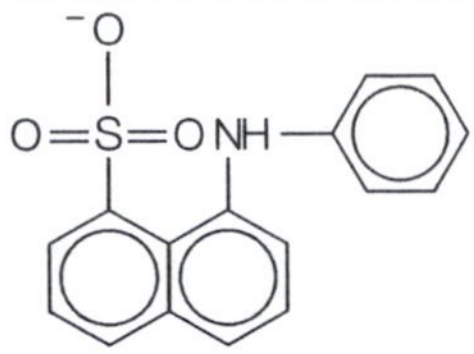

1-Anilino-8-naphthalene sulfonate (ANS)

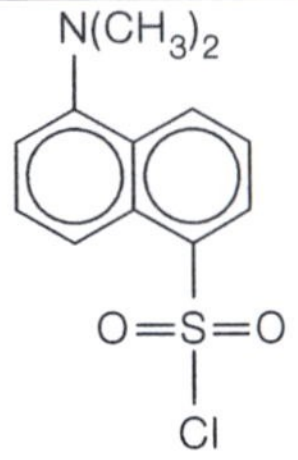

Dansyl chloride

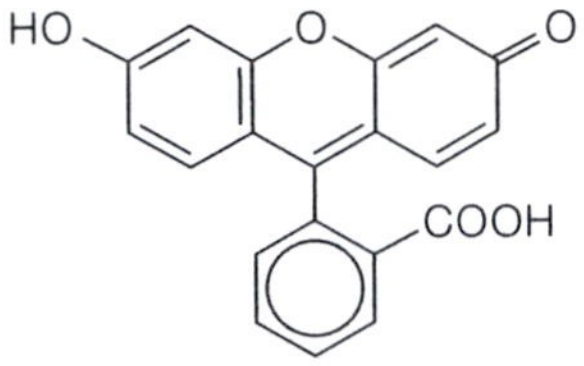

Fluorescein

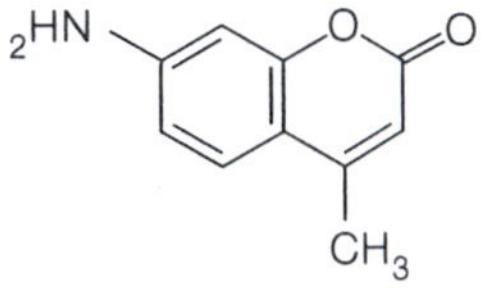

Aminomethyl coumarin (AMC)

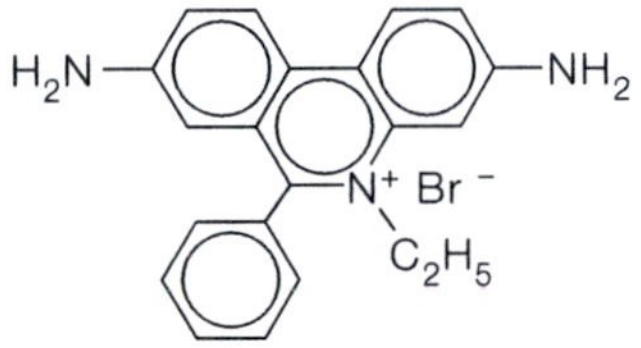

Ethidium bromide

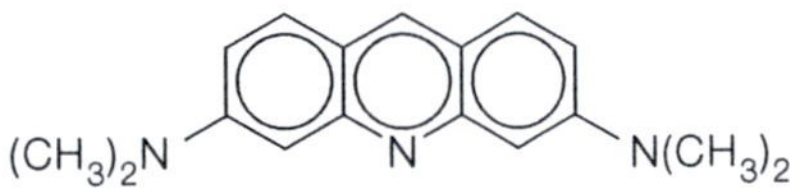

Acridine orange

Preparation of Reagents and Solutions

Since fluorescence measurements are very sensitive, dilute solutions of biomolecules and other reagents are appropriate. Special precautions must be taken to maintain the integrity of these solutions. All solvents and reagents must be checked for the presence of fluorescent impurities, which can lead to large errors in measurement. "Blank" readings should be taken on all solvents and solutions, and any background fluorescence must be subtracted from the fluorescence of the complete system under study. Solutions should be stored in the dark, in clean glass-stoppered containers, in order to avoid photochemical breakdown of the reagents and contamination by corks and rubber stoppers. Some biomolecules, especially proteins, tend to adsorb to glass surfaces, which can lead to loss of fluorescent material or to contamination of fluorescence cuvettes. All glassware must be scrupulously cleaned. Turbid solutions must be clarified by centrifugation or filtration.

Control of Temperature

Fluorescence measurements, unlike absorption, are temperature dependent. All solutions, especially if relative fluorescent measurements are taken, must be thermostated at the same temperature.

C. NUCLEAR MAGNETIC RESONANCE SPECTROSCOPY

Nuclear magnetic resonance (NMR) spectroscopy is a method of absorption spectroscopy that has some characteristics similar to ultraviolet and visible spectroscopy but also some that are unique. In NMR, a molecular sample, usually dissolved in a liquid solvent, is placed in a magnetic field and the absorption of radio-frequency waves by certain nuclei (protons and others) is measured. NMR spectroscopy was originally developed as an analytical tool to determine molecular structure by monitoring the environment of individual protons. The relative positions and intensity of absorption signals provide detailed information from which chemical structures may be elucidated. Early proton NMR studies focused primarily on small organic molecules, some of which had biological origin and significance. Since most biomolecules have a very large number of protons, the observed spectra are extremely complex and difficult to interpret. With the advent of modern techniques—superconducting magnets, Fourier transform analysis, multidimensional spectra, and powerful computer control—NMR has become an important method for the study of biological macromolecules. The fundamental principles, instrumentation, and biochemical applications for NMR will be outlined here.

NMR Theory

All nuclei possess a positive charge. For some nuclei, this charge confers the property of spin, which causes the nuclei to act like tiny magnets. The angular momentum of the spin is described by the quantum spin number I. If I is an integral number ($I = 0, 1, 2$, etc.), then there is no net spin and no NMR signal for that nucleus. However, if I is half-integral ($I = {}^1\!/_2, {}^3\!/_2, {}^5\!/_2$, etc.), the nuclei have spins, and when placed in a magnetic field, the spins orient themselves with (parallel) or opposed (antiparallel) to the external magnetic field. The nuclei aligned with the magnetic field have lower energy (are more stable) than those opposed. Energy in the radio-frequency range is sufficient to flip the nuclei from the parallel to the antiparallel alignment. The NMR instrument is designed to measure the energy difference between the nuclear spin states. Absorption of energy may be detected by scanning the radio-frequency range and measuring the absorption that causes spin state transition (resonance). Modern NMR instruments instead maintain a constant radio-frequency and electrically induce small changes in the strength of the magnetic field until resonance is attained. The point of resonance for a nucleus is dependent upon the electronic environment of that nucleus, so an NMR spectrum provides information that helps elucidate biochemical structures.

NMR in Biochemistry

NMR has found a wide variety of applications in biochemistry. Proton NMR has a long and rich history in organic chemistry and biochemistry. The structures of many small but significant biomolecules were elucidated by proton

Table 7.1

Nuclei Important in Biochemical NMR

Isotope[1]	Natural Abundance (%)[2]	Relevant Biomolecules
^1H	100	Most biomolecules
^{13}C	1	Most biomolecules
^{15}N	0.4	Amino acids, proteins, nucleotides
^{19}F	100	Substitute for H
^{31}P	100	Nucleotides, nucleic acids, and other phosphorylated compounds

[1]All have a spin of $^1/_2$.

[2]The percentage of this isotope in naturally occurring molecules containing this element.

NMR. Protons on different atoms and in different molecular environments absorb energy of different levels (measured by radio-frequency units or magnetic field units). Two experimentally measured characteristics, **chemical shifts** and **spin-spin coupling,** provide important structural information. The chemical shift (δ) of an absorbing nucleus, measured in parts per million (ppm), is the spectral position of resonance relative to a standard signal, usually tetramethylsilane. The NMR signal for a proton is "split" by interactions with neighboring protons. This characteristic, called spin-spin coupling, helps to determine positions and numbers of equivalent and nonequivalent protons.

NMR experiments are not limited to the study of protons. Resonance signals from other atomic nuclei including ^{13}C, ^{15}N, ^{19}F, and ^{31}P can be detected and measured (see Table 7.1). ^{13}C NMR techniques have been especially valuable for the study of carbohydrate, amino acid, and fatty acid structures. Just as with protons, each distinct carbon atom in a molecule yields a signal that is split by neighboring interacting nuclei (see Figure 7.15). Because of the low natural abundance of the ^{13}C isotope, it is necessary either to enrich the ^{13}C content by chemical synthesis or to use powerful computers and magnets (500–600 MHz). The ^{31}P spectra of phosphorylated biomolecules display a peak for each type of phosphorus (Figure 7.16). NMR instruments now have the sensitivity necessary to measure the *in vivo* concentrations and reactions of biomolecules in cells. For example, the catabolism of glucose by glycolysis in erythrocytes has been monitored by ^{13}C NMR, and the involvement of ATP in phosphoryl-group transfer processes has been studied by ^{31}P NMR (see Figure 7.16).

NMR and Protein Structures

New computer techniques in data analysis and improvements in instrumentation have now made it possible to carry out detailed structural and conformational studies of all biopolymers, but especially proteins. NMR, which may be done on noncrystalline materials in solution, provides a technique complementary to X-ray diffraction, which requires crystals for analysis.

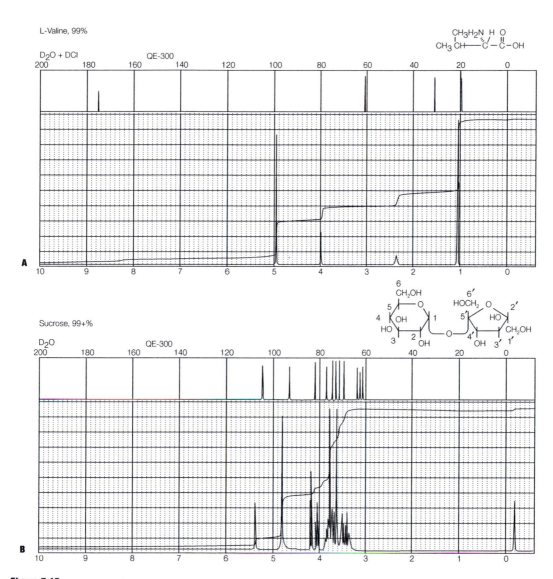

Figure 7.15

¹H and ¹³C FT-NMR spectra of biomolecules. The lower spectra are for ¹H showing a ppm scale of 0–10. TMS standard is at 0 ppm. The upper spectra are for ¹³C showing a ppm scale of 0–200. **A** L-valine in D_2O + DCl. Can you assign each peak to the correct protons and carbon atoms in the valine structure? Hint: The carboxyl carbon of valine has a peak at about 175 ppm. **B** Sucrose in D_2O. Carbon numbers in the chemical structure correspond to the following peaks in order from 0 to 120 ppm: C-6; C-1′; C-6′; C-4; C-2; C-5; C-3; C-4′; C-3′; C-5′; C-1; C-2′. *Reprinted with permission of Aldrich Chemical Co., Inc.*

CHAPTER 7

Figure 7.16

^{31}P NMR spectra of human forearm muscle showing the effect of exercise. **A** Before exercise; **B** and **C** during 19 minutes of exercise; **D** 5–6 minutes after **C**. Peak assignments: 1, β-phosphorus of ATP; 2, α-phosphorus of ATP; 3, γ-phosphorus of ATP; 4, phosphocreatine; 5, Pi. Phosphocreatine is used as a major source of energy during exercise. It is hydrolyzed to creatine and Pi. Note that the level of ATP remains relatively constant during exercise because it is produced and used at about the same rate. *After G. Radda, Science* **233**, *641 (1986). Reprinted with permission from the American Association for the Advancement of Science.*

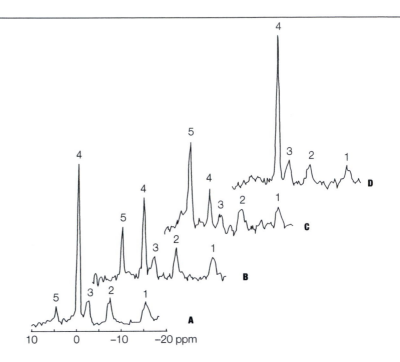

An advantage of NMR is that proteins can be studied in solution in an environment that mimics that in the living cell. One-dimensional NMR, as described to this point, can be applied to structural analysis of smaller molecules. But proteins and other complex biopolymers with large numbers of protons will yield very crowded spectra with many overlapping lines. In multidimensional NMR (2-D, 3-D, 4-D), peaks are spread out through two or more axes to improve resolution. Protein structure determination depends on several homonuclear 2-D ^{1}H NMR experiments: **correlation spectroscopy (COSY), nuclear Overhauser effect spectroscopy (NOESY), 2-D total correlation spectroscopy (TOCSY), foldover-corrected spectroscopy (FOCSY), spin-echo correlation spectroscopy (SECSY),** and others. These computational techniques are based on the observation that nonequivalent protons interact with each other. By using multiple-pulse techniques, it is possible to perturb one nucleus and observe the effects on the spin states of other nuclei. The availability of powerful computers, strong magnets, and Fourier transform (FT) calculations makes it possible to elucidate structures of proteins up to one million daltons, and there is future promise for studies on larger proteins. In addition to structural studies, NMR will also be applied to studies of conformational changes and interactions between ligands and biopolymers.

Many of the instrumental techniques and computational methods for NMR protein structure determination have been developed by Kurt

Wüthrich of the Swiss Federal Institute of Technology (ETH) in Zurich. Beginning in the early 1980's, Wüthrich and his colleagues used NMR to determine the structure of several smaller proteins (below 10,000 daltons). By 1990, NMR structure determination could be applied to proteins of molecular masses up to about 20,000 daltons. One of the largest proteins studied so far by NMR is the GroEL-GroES complex which is about 900,000 daltons (2002). Some of the interesting proteins studied by Wüthrich include tendamistat (an α-amylase inhibitor), rabbit metallothionein, human cyclophilin A-cyclosporin A complex, and the murine prion protein which may be involved in diseases related to bovine spongiform encephalopathy (BSE). Wüthrich was awarded the Nobel Prize in Chemistry in 2002 for this work, which occurred over a period of about 25 years. It is interesting to note that independent studies comparing protein three-dimensional structures derived from X-ray diffraction (on crystals) and NMR (in solution) have yielded identical results. The NMR methods have even been used to correct structural mistakes made in the X-ray diffraction method. Protein NMR is now a very active research area and by the year 2005, over 2700 NMR-derived protein structures had been submitted to the Protein Data Bank (www.rcsb.org/pdb).

The complex principles and techniques behind Wüthrich's studies are best described in his own words (Wüthrich, 2001):

> *"Four principal elements are combined in the NMR method for protein structure determination: (i) the nuclear Overhauser effect (NOE) as an experimentally accessible NMR parameter in proteins that can yield the information needed for* de novo *global fold determination of a polymer chain; (ii) sequence-specific assignment of the many hundred to several thousand NMR peaks from a protein; (iii) computational tools for the structural interpretation of the NMR data and the evaluation of the resulting molecular structures; and, (iv) multidimensional NMR techniques for efficient data collection."*

The preparation of a protein sample for NMR studies is relatively straightforward; however, a rather large amount of protein is required compared to mass spectrometry and X-ray crystallography. For NMR analysis, the protein is usually dissolved in water at concentrations of 3–6 mM. The protein solution may be adjusted to the ionic strength, pH, and temperature found under physiological conditions. Samples as small as 0.5 mL are sufficient for analysis.

D.   MASS SPECTROMETRY

Mass spectrometry (MS) is similar to NMR in that it has historically been of great value in the structure elucidation of relatively small (MW limit around 1000) organic and biomolecules. Only in the last 15 years has the tool of MS been applied to the analysis of biological macromolecules. The MS analysis of proteins and other biopolymers was initially hindered because analytes are usually measured in the gas phase and it was very difficult to vaporize these large molecules. The development of new methods for sample preparation (ionization), multiquadrupole analysis, tandem MS instruments, and powerful computers now makes it possible to study large molecules, especially proteins.

Ionization and Analysis of Proteins

The mass spectrometer generally consists of three components that are coupled together: an **ionization device,** a **mass analyzer,** and an **ion detector** (Figure 7.17). Neutral molecules are ionized and their positively charged ion products are directed through an electric and/or magnetic field where they are separated (analyzed) on the basis of their mass-to-charge ratio (m/z). A detector then records the ions after separation. The "spectrum" generated by MS displays ion intensity as a function of m/z.

Ionization of small organic and biological molecules prior to MS analysis is done by electron impact; however, ionization of nonvolatile, biological macromolecules like peptides, proteins, and nucleic acids requires special

Figure 7.17

The components of a mass spectrometer. The final data output consists of a graph of ion intensity versus m/z. Mass spectrometry, unlike the other techniques described in this chapter, does not involve interaction of molecules with electromagnetic radiation; however, it has traditionally been combined with the spectrophotometric techniques.

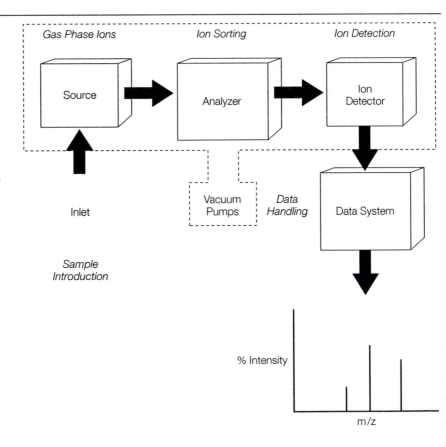

treatment. In the late 1980s, John Fenn, now at Virginia Commonwealth University, and Koichi Tanaka of the Shimadzu instrument company in Kyoto, Japan, introduced methods for the "soft ionization" of macromolecules. These methods are labeled "soft" as they cause minimal degradation to the sample during ionization. Fenn introduced the **electrospray ionization (ESI)** method, in which a solution of the sample is sprayed from a metal needle or capillary tip held at a potential of about +5000 volts. This results in a spray of tiny droplets containing positively charged ions. The solvent evaporates and the ions are directed into the analyzer for separation. Tanaka demonstrated that pulses from a nitrogen laser could be used to ionize proteins from a surface **(soft laser desorption, SLD).** Because ESI and SLD are liquid-phase processes, the samples may be those purified by HPLC or CE. Fenn and Tanaka were awarded the Chemistry Nobel Prize (shared with K. Wüthrich) for this work in 2002.

Laser desorption has now been modified by Franz Hillenkamp of the University of Münster, Germany, to a solid phase method, **matrix-assisted laser desorption ionization (MALDI),** which is the most widely used technique for protein analysis. In MALDI, the sample is placed in a matrix of small organic molecules and irradiated by a laser pulse. The matrices absorb the laser energy causing ionization of the macromolecular sample. A common practice is the use of protein samples obtained from 1-D and 2-D PAGE gels.

Several types of mass analyzers are used for ion separation including time-of-flight (TOF), quadrupoles, and ion traps. In TOF-MS, the most widely used technique, each ion produced has the same initial kinetic energy, but the speed varies with mass. Mass is determined by measurement of an ion's time-of-flight to the detector.

MS Applications in Biochemistry

MS is becoming a standard tool in the analysis of biological molecules and biological processes. Many of the applications of MS described below involve coupling the technique with sample preparation and purification by HPLC, CE, and PAGE (especially 2-DE). (See section on proteomics, Chapter 6, Section B.)

1. **Identification of Peptides and Proteins**   Protein identification is perhaps the most widely used MS application at the present time. The most important ions resulting from the removal of an electron by ionization procedures are the positively charged **molecular ions (M^+)** or **protonated molecular ions $(M + nH)^{n+}$.** Measurement of the molecular mass of these species provides the molecular weight of the original molecule. The accuracy of this method for molecular weight determination is about 0.01%. (Molecular mass measurements using gel filtration are typically no better than 5–10%.) Some molecular ions are unstable and disintegrate to produce fragment ions. These fragmentation processes are useful in structural elucidation of smaller molecules. It should be expected that proteomic studies will be enhanced by coupling SDS-PAGE and 2-DE (for separation and purification) with further size analysis of the proteins by MALDI-TOF.

One MS method for peptide identification is called **peptide mass fingerprinting.** The unknown sample is digested with the proteolytic enzyme trypsin to produce fragments that are analyzed by MALDI-MS. The resulting spectrum displays the masses of the peptide fragments. This may be used as a "fingerprint" of the sample and compared with known fragmented patterns in a sequence database.

2. **Characterization of Post-Translational Modification Processes**
   After translation, proteins in the cell are often modified by covalent attachment of specific functional groups. Such processes might include:

   (a) **phosphorylation** of hydroxyl groups in serine and tyrosine residues, or

   (b) addition of carbohydrate residues **(glycosylation)** to the hydroxyl groups of serine or threonine or to the amide nitrogen of asparagine residues.

   These chemical modifications can easily be detected by MS as the addition of a phosphoryl group ($-PO_3^{2-}$) adds about 80 daltons to the molecular mass of a protein, whereas addition of a glucose residue adds about 180 daltons.

3. **Peptide Sequencing**   Small peptides and proteins may be sequenced by MS procedures. The peptide sample is ionized and fragmented at the peptide bonds by collision-induced dissociation. Size analysis of the series of fragments produced leads to the sequence of amino acids.

4. **Protein–protein Interactions**   One of the most important goals of proteomics is to study how proteins interact with each other in order to initiate biological processes. ESI-MS and MALDI-TOF are currently being used to investigate how proteins associate to form biologically active multisubunit complexes.

E.   X-RAY CRYSTALLOGRAPHY

Historically, **X-ray crystallography** has been the most important tool for studying the three-dimensional structures of biomolecules. In fact, this technique has been the only method available for determining macromolecular structures until the application of NMR in the 1980s. Protein crystals were first studied by X-ray crystallography in 1934 by J. D. Bernal and Dorothy Crowfoot Hodgkin. However, the diffraction patterns were extremely complex and computers were not available for complete analysis of the proteins. John Kendrew, a biochemist working at Cambridge University, announced the first three-dimensional structure of a protein, sperm whale myoglobin, in 1958. Kendrew had determined the complex structure of the protein (153 amino acid residues) after several years of analysis by X-ray crystallography. Other important biomolecules that were studied by X-ray crystallography included DNA (Rosalind Franklin and Maurice Wilkins) and vitamin B_{12} (Dorothy Crowfoot Hodgkin). Analysis of protein structures is now routine. At the beginning of 2005, the Protein Data Bank (www.rcsb.org/pdb) had registered approximately 20,000 protein structures—over 17,000 obtained by X-ray crystallography and about 2700 by NMR. Although the number of structures derived by NMR is growing faster than X-ray, crystallography is still an important source of structural data.

Methodology of X-ray Crystallography

An X-ray crystallography analysis requires three components (see Figure 7.18): (1) a protein crystal, (2) an X-ray source, and (3) a detector (i.e., radiation detector or photographic film).

Protein Crystals

For many proteins, this can be the most difficult step in three-dimensional analysis. The factors important in protein crystallization are not yet fully understood, so the methodology for growing crystals is still an art, and involves much trial and error. Crystallization usually occurs best in a saturated or supersaturated solution of the protein. A common method often used involves changing solution conditions by addition of a precipitant (like ammonium sulfate), changing the pH, or changing the ionic strength. The crystallization process may be hastened by dialysis of the protein against a solution containing the precipitating factor.

Data Collection and Analysis

The theory supporting X-ray crystallography is quite complex and beyond the scope of this text; however, some practical aspects can be described. The crystal is mounted in the diffractometer and bombarded by a collimated beam of X-rays. The crystal is rotated so that it is struck from many different directions. Many of the X-rays passing into the sample are diffracted (scattered) when they encounter electrons associated with atoms. The diffracted beams impinge upon the detector and are recorded. The data at this point are in the form of a regular array of spots called reflections. (Kendrew's pattern of myoglobin had nearly 25,000 reflections, which were analyzed by computer

Figure 7.18

The components required for analysis of protein structure by X-ray crystallography.

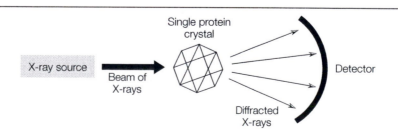

in order to construct the 3-D image of the protein.) The extent of scatter of the reflections depends on the size and position of each atom in the crystal. By extensive computer and mathematical analysis (by Fourier transform) of the angle of scatter and of the pattern collected by the detector, it is possible to construct an **electron density map** of the protein molecule showing the arrangement of the atoms. Using modern equipment and computer software, an analyst can collect data and determine a structure in a day or two.

Although X-ray analysis of proteins is slowly being replaced by NMR, there will continue to be interest in and a need for X-ray analysis of very large and unsymmetrical proteins and other biomolecules. NMR does have some advantages in structure determination—sample preparation is rapid and straightforward, and the sample may be studied under physiological-like conditions. Numerous comparative studies have confirmed that the analysis of an identical protein by X-ray and NMR leads to the same three-dimensional result, even though one is done in the solid state and one in the liquid state.

Study Problems

▶ *1.* For each pair of wavelengths listed below, specify which one is higher in energy.
(a) 1 nm (X-ray) or 10,000 nm (IR)
(b) 280 nm (UV) or 360 nm (VIS)
(c) 200,000 nm (microwave) or 800 nm (VIS)

▶ *2.* In a laboratory experiment, you are asked to determine the molar concentration of a solution of an unknown compound, X. The solution diluted in half by water (1 mL of X and 1 mL of H_2O) has an absorbance at 425 nm of 0.8 and a molar extinction coefficient of $1.5 \times 10^3 \, M^{-1} \, cm^{-1}$. What is the molar concentration of the original solution of X?

▶ *3.* Match the spectral region listed below with the appropriate molecular transition that occurs. The first problem is worked as an example.

Spectral region	Transition
_a_ 1. X-rays	a. inner-shell electrons
___ 2. UV	b. molecular rotations
___ 3. VIS	c. valence electrons
___ 4. IR	d. molecular vibrations
___ 5. microwave	

▶ *4.* Several spectroscopic techniques were studied in this chapter. Which experimental techniques involve an actual measurement of radiation absorbed?
(a) UV-VIS spectroscopy
(b) NMR spectroscopy
(c) MS
(d) Fluorescence spectroscopy

5. Explain the differences between a spectrophotometer that uses a photo-tube for a detector and one that uses a photodiode array detector.

▶ 6. Why can you not use a glass cuvette for absorbance measurements in the UV spectral range?

▶ 7. What is the single structural characteristic that all of the fluorescent molecules in Figure 7.14 possess?

▶ 8. Why must a cuvette with four translucent sides be used for fluorescence measurements?

9. Study the ^{1}H spectrum of valine in Figure 7.15 and match each peak to the corresponding proton in the chemical structure. Explain any spin-spin coupling.

10. Study the ^{13}C spectrum of valine in Figure 7.15. Identify the carbon atom that produces each peak in the spectrum.

Further Reading

J. Berg, J. Tymoczko, and L. Stryer, *Biochemistry,* 5th ed. (2002), Freeman and Company (New York), pp. 89, 107–112, 595–596. Application of NMR, X-ray, and MS to biochemistry.

S. Borman, *Chem. Eng. News,* February 15, 65–67 (1999). "Scaling up Protein NMR."

A. Burlingame, S. Carr, and M. Baldwin, *Mass Spectrometry in Biology and Medicine* (2000), Humana Press (Totowa, NJ).

E. De Hoffmann, and V. Stroobant, *Mass Spectrometry: Principles and Applications,* 2nd ed. (2001), John Wiley & Sons (New York).

E. Eberhardt et al., *Biochem. Mole. Biol. Educ.* **31,** 402–409 (2003). "Preparing Undergraduates to Participate in the Post-Genome Era: A Capstone Laboratory Experience in Proteomics."

F. Finehout and K. Lee, *Biochem. Mole. Biol. Educ.* **32,** 93–100 (2004). "An Introduction to Mass Spectrometry Applications in Biological Research."

R. Garrett and C. Grisham, *Biochemistry,* 3rd ed. (2005), Brooks/Cole (Belmont, CA), pp. 94–95, 125–128, 311–312, 551–552. Introduction to UV-VIS, NMR, and MS.

B. Giles et al., *J. Chem. Educ.* **76,** 1564–1566 (1999). "An In Vivo ^{13}C NMR Analysis of the Anaerobic Yeast Metabolism of 1-^{13}C-Glucose."

R. Hester and R. Girling, Editors, *Spectroscopy of Biological Molecules* (1991), Royal Society of Chemistry Special Publication (Cambridge). Spectroscopy focused on biomolecules.

M. Kinter and N. Sherman, *Protein Sequencing and Identification using Tandem Mass Spectrometry* (2000), Wiley-Interscience (New York).

N. Krishna and L. Berliner, *Protein NMR for the Millenium* (2003), Plenum (New York). Advances in protein NMR.

M. Mann and A. Pandley, *Trends Biochem. Sci.* **26,** 54–61 (2001). "Use of Mass Spectrometry-Derived Data to Annotate Nucleotide and Protein Sequence Databases."

M. Mann et al., *Ann. Rev. Biochem.* **70**, 437–473 (2001). "Analysis of Proteins and Proteomes by Mass Spectrometry."

D. Nelson, and M. Cox, *Lehninger Principles of Biochemistry,* 4th ed. (2005), Freeman (New York), pp. 79–82, 102–104, 136–139, 291–292. Introduction to spectroscopy.

A. Pavia, G. Lampman, and G. Kriz, *Introduction to Spectroscopy,* 3rd ed. (2000), Brooks Cole (Pacific Grove, CA). An introductory textbook.

K. Peterman, K. Lentz, and J. Duncan, *J. Chem. Educ.* **75**, 1283–1284 (1998). "A ^{19}F NMR Study of Enzyme Activity."

N. Price, R. Dwek, M. Wormald, and R. Ratcliffe, *Principles and Problems in Physical Chemistry for Biochemists,* 3rd ed. (2002), Oxford University Press (Oxford). An introduction to spectroscopy with theory and applications.

G. Radda, *Science,* **233**, 640–645 (1986). "The Use of NMR Spectroscopy for the Understanding of Disease."

K. Stewart and R. Ebel, *Chemical Measurements in Biological Systems,* (2000), John Wiley & Sons (New York). Chapter 3 covers the theory of UV-VIS spectroscopy.

I. Tinoco, K. Sauer, J. Wang, and J. Puglisi, *Physical Chemistry: Principles and Applications in Biological Sciences,* 4th ed. (2002), Prentice-Hall (Upper Saddle River, NJ). Chapter 10 covers all areas of spectroscopy.

D. Voet and J. Voet, *Biochemistry,* 3rd ed. (2004), John Wiley & Sons (New York), pp. 172–175, 240–244. NMR, X-ray, and MS in biochemistry.

D. Voet, J. Voet, and C. Pratt, *Fundamentals of Biochemistry,* 2nd ed. (2006), John Wiley & Sons (Hoboken, NJ), pp. 116–118, 144–149, 420–421. NMR, MS, and X-ray in biochemistry.

K. Wilson and J. Walker, Editors, *Principles and Techniques of Practical Biochemistry,* 5th ed. (2000), Cambridge University Press (Cambridge, UK). Chapter 6 covers X-ray diffraction. Chapter 9 covers atomic and molecular electronic spectroscopy. Chapters 10 and 11 cover NMR and MS.

K. Wüthrich, *Nat. Struct. Biol* **8**, 923–925 (2001). "The Way to NMR Structures of Proteins."

K. Wüthrich, *J. Biol. Chem.* **265**, 22059–22062 (1990). "Protein Structure Determination in Solution by NMR Spectroscopy."

http://www.scienceofspectroscopy.info/index.htm
    The Science of Spectroscopy, funded by NASA.

http://www.cis.rit.edu/htbooks/nmr/
    The basics of NMR.

http://www.biophysics.org/btol/index.html
    Biophysics Textbook On Line: Chapters on NMR, sequence analysis, and fluorescence spectroscopy.

http://www.asms.org/
    Web site for the American Society of Mass Spectrometry.

http://www.rcsb.org/pdb

This Web site, the Protein Data Bank, is the single repository for the three-dimensional structures of biomolecules.

http://www.nobel.se

Under the heading "Educational", use "The Virtual Biochemistry Lab" to perform experiments on protein NMR and X-ray diffraction.

BIOMOLECULAR INTERACTIONS: LIGAND BINDING AND ENZYME REACTIONS

Many of the dynamic processes occurring in biological cells and organisms are the result of interactions between molecules. Most often, these interactions involve the binding of a molecule or ion to a specific site (or sites) on a macromolecule, usually a protein or nucleic acid. The binding forces that hold the molecules together are noncovalent and they include hydrogen bonding, van der Waals forces, electrostatic bonds, and hydrophobic interactions. Usually the combination of the two molecules (formation of a complex) will lead to a specific biological action or response. For example, a hormonal response is the result of the hormone molecule interacting with its receptor protein; a chemical transformation in metabolism is the consequence of the initial binding of a substrate to an enzyme, forming an ES complex.

A.   LIGAND-MACROMOLECULE INTERACTIONS (MOLECULAR RECOGNITION)

Throughout our study of biochemistry we will encounter many examples where noncovalent interactions bring together, in specific ways, two different molecules that form a complex. Such molecules display a phenomenon called **molecular recognition.** The importance of these interactions in biology is that the combination of two molecules will lead to some biological action. The action of hormones provides an excellent example. A hormone response is the consequence of a weak, but specific, interaction between the molecule and a receptor protein in the membrane of the target cell. The biochemical response may often be an enhancement in production of an enzyme needed for metabolism **(signal transduction)**. Another example of molecular recognition is the interaction that brings together a substrate molecule with an

---

Table 8.1

Examples of LM Interactions and Their Biological Responses

Type of Interaction (Ligand:Macromolecule)	Biological Significance or Response
Substrate:enzyme	Metabolic reactions
Inhibitor:enzyme	Metabolic regulation
Allosteric effector:enzyme	Metabolic regulation
Coenzyme:enzyme	Metabolic reactions
Hormone:receptor protein	Metabolic regulation; signal transduction
Antigen:antibody	Immune response
Ligand:carrier or transport protein	Storage or transport
Drug:protein or nucleic acid	Disease treatment
Regulatory protein:DNA	Transcription regulation

---

enzyme. Before a metabolic reaction can occur, a substrate molecule must physically interact in a certain well-defined manner with the macromolecular catalyst, an enzyme. The biochemical action of drugs also depends on molecular recognition. A drug is first distributed throughout the body via the bloodstream. Drug molecules in the bloodstream are often bound to plasma proteins (e.g., serum albumin), which act as carriers. When the drug molecules are transported to their site of action, a second molecular interaction is likely to occur. Many drugs elicit their effects by interfering with biochemical processes. This may take the form of enzyme inhibition, where the drug molecule binds to a specific enzyme and prohibits catalytic action. Table 8.1 lists several other molecular recognition events that lead to some dynamic biochemical action. Details of these processes may be found in your biochemistry textbook.

Properties of Noncovalent Binding Interactions

The process of molecular recognition may be defined by a simple, reversible reaction that brings together a **ligand (L)** and a **macromolecule (M)** (R, for receptor, in some textbooks):

$$ L + M \rightleftharpoons LM \longrightarrow \text{Biological response} $$

Equation 8.1

LM represents a complex, held together by noncovalent interactions, that has a specialized biological function. The ligand is usually a relatively small molecule (hormone, substrate, inhibitor, drug, coenzyme, metal ion, etc.), but not always. Examples of large ligands include protein substrates that are cleaved by a protease, a protein molecule that binds to DNA as a transcription regulator, or large antigens that interact with antibodies.

All molecular interactions that are the basis of molecular recognition have at least three common characteristics:

1. **The binding forces that hold the complex together are noncovalent and relatively weak** Four types of noncovalent bonds are important:

   (a) **hydrogen bonds**
   (b) **van der Waals forces**
   (c) **hydrophobic interactions**
   (d) **ionic or electrostatic bond**s

   Properties and examples of these interactions are reviewed in Table 8.2. The strengths of noncovalent binding forces are in the range of 1–30 kJ/mol compared to about 350 kJ/mol for a carbon–carbon single bond, a typical covalent bond. A single, noncovalent bond is usually insufficient to hold two molecules together. Typical biopolymers

Table 8.2

Properties and Examples of Noncovalent Binding Forces

Type	Brief Description and Example	Stabilization Energy (kJ/mol)	Length (nm)
(i) *Hydrogen bonds*	Between a hydrogen atom bonded to an electronegative atom and a second electronegative atom. Between neutral groups	10–30	0.2–0.3
(ii) *van der Waals interactions*	Between molecules with temporary dipoles induced by fluctuating electrons. This may occur between any two atoms in close proximity	1–5	0.1–0.2
(iii) *Hydrophobic interactions*	The presence of water forces nonpolar groups into ordered arrangements to avoid the water	5–30	—
(iv) *Ionic bonds*	Interactions that occur between fully charged atoms or groups. Na^+Cl^- $R-NH_3^+{}^-OOC-R$	20	0.25

that may serve as macromolecules–DNA, RNA, polysaccharides, and proteins–have numerous functional groups that participate in noncovalent interactions. A collection of many of these interactions will lead to highly stabilized complexes.

2. **Reactions that form noncovalent complexes are reversible** Noncovalent interactions are initiated when diffusing (wandering or moving) molecules come into close contact. Diffusion is brought about by thermal motions. An initial close encounter may not always result in the successful formation of a complex. A few weak bonds may form, but be disrupted by thermal motions, causing the molecules to dissociate. Therefore, bonds may constantly form and break until enough bonds have accumulated to result in an intermediate with transient but significant existence. The complex can then initiate a specific biological process. An intermediate rarely lasts longer than a few seconds. Eventually, thermal motions cause the complex to dissociate to the individual molecules. Reversibility is an important characteristic of these interactions so that a static, gridlock situation does not occur. The biological process initiated by the complex LM must have a finite lifetime–a starting and an ending time.

3. **The binding between molecules is specific** Imagine that the interactions to form LM bring together two molecular surfaces. The two surfaces will be held together if noncovalent interactions are established. If on one surface there is a nonpolar molecular group (phenyl ring, hydrophobic alkyl chain), the adjacent region on the other surface must also be hydrophobic and nonpolar. If a positive charge exists on one surface, there may be a neutralizing negative charge on the other surface. A hydrogen-bond donor on one surface can interact favorably with a hydrogen-bond acceptor on the other surface. Simply stated, the two molecules must be compatible or complementary in a chemical sense so the development of stabilizing forces can hold molecules together. In molecular terms, the binding site (or sites) on M displays high specificity; therefore, only certain ligands can bind. L molecules, for a particular M, will be limited to a single, specific molecule or perhaps a group of structurally related molecules.

➤ | **Study Exercise 8.1   Examples of Noncovalent Interactions**

(a) Show how the carboxylate group on the side chain of the amino acid aspartate can interact with a lysyl residue in a protein. Draw structures and define the possible types of noncovalent binding forces.

(b) Show how a peptide carbonyl group can form a hydrogen bond with the free amino acid serine.

Quantitative Characterization of Ligand Binding

A thorough understanding of the biochemical significance of ligand binding to macromolecules comes only from a quantitative analysis of the strength of binding (affinity) between L and M. Binding affinity between two molecules is often expressed as an equilibrium constant, the **formation constant,** K_f, which is derived from the law of mass action. Consider the specific interaction between a small molecule, L (for ligand), and a macromolecule, M (Equation 8.1). These two species combine to form a complex, LM.

K_f, the formation constant for the complex, is defined by Equation 8.2.

>> $$K_f = \frac{[LM]}{[L][M]}$$ **Equation 8.2**

Do not confuse K_f with K_d, the dissociation constant. The relationship between K_f and K_d is defined in Equation 8.3.

>> $$K_d = \frac{[L][M]}{[LM]} = \frac{1}{K_f}$$ **Equation 8.3**

The larger the value of K_f, the greater the strength of binding between L and M. (Large K_f implies a high concentration of LM relative to L and M.) Return to Equation 8.2 and note that in order to determine K_f, a method must be developed to measure equilibrium concentrations of L, M, and the complex LM. In a later section, we will describe experimental techniques that are applied to these measurements of binding constants, but first we must reorganize Equation 8.2 into a form that contains more readily measurable terms. We will begin with the assumption that the macromolecule, M, has several binding sites for L and that these sites do not interact with each other. That is, K_f is identical for all binding sites. The following definitions are necessary for the reorganization of Equation 8.2.

[L] = equilibrium concentration of free or unbound ligand

[M] = equilibrium concentration of macromolecule with no bound L, or the concentration of unoccupied binding sites

[LM] = equilibrium concentration of ligand-macromolecule complex, or the concentration of occupied sites

$[M]_0$ = total or initial concentration of macromolecule, or total concentration of available binding sites

$[L]_0$ = total concentration of bound and unbound ligand; or initial concentration of ligand

v = fraction of available sites on M that are occupied, or the fraction of M that has L in binding site:

>> $$v = \frac{[LM]}{[M]_0}$$ **Equation 8.4**

The term v is particularly significant because it can be considered a ratio of the number of occupied sites to the total number of potential binding sites on M. It can be measured experimentally, but first it must be redefined in the following manner: Since $[M]_0 = [LM] + [M]$, then

$$v = \frac{[LM]}{[LM] + [M]}$$

From Equation 8.2, $[LM] = K_f[L][M]$. Therefore,

$$v = \frac{K_f[L][M]}{K_f[L][M] + [M]}$$

Simplifying,

>> $$v = \frac{K_f[L]}{K_f[L] + 1}$$ *Equation 8.5*

You should recognize the similarity of Equation 8.5 to the Michaelis-Menten equation for enzyme catalysis. A graph of v vs. [L] yields a hyperbolic curve (see Figure 8.1) that approaches a limiting value or saturation level. At this point, all binding sites on M are occupied. Because of the difficulty of measuring the exact point of saturation, this nonlinear curve is seldom used to determine K_f. Linear plots are more desirable, so Equation 8.5 is converted to an equation for a straight line. The equation will first be put into a more general form to account for any number of potential binding sites on M. The

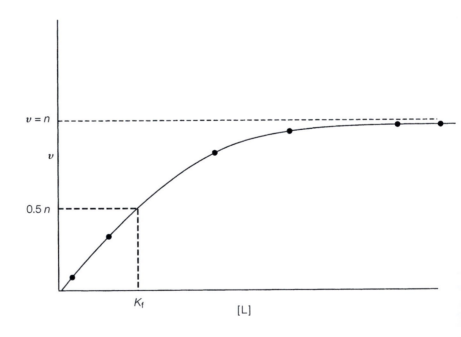

Figure 8.1

A plot illustrating saturation of binding sites with ligand. The n is estimated at the point when all binding sites are occupied by ligand. K_f is represented by the [L] when one-half of the sites are occupied by ligand.

symbol $\bar{v}$ will be used to represent the average number of occupied sites per M, and n will represent the number of potential binding sites per M molecule. Assuming that all the binding sites on M are equivalent, Equation 8.5 becomes

$$\bar{v} = \frac{nK_f[L]}{K_f[L] + 1}$$

Equation 8.6

Scatchard's Equation

Equation 8.6 contains terms such as [L] that are difficult to determine, so it must be converted to a form amenable to experimental measurements.

If $\bar{v}$ is the average number of occupied sites per M molecule, then $n - \bar{v}$ is the average number of unoccupied sites per M molecule.

$$n - \bar{v} = n - \frac{nK_f[L]}{K_f[L] + 1}$$

Simplifying,

$$(n - \bar{v})(K_f[L] + 1) = n(K_f[L] + 1) - n(K_f[L])$$

$$(n - \bar{v}) = \frac{n}{K_f[L] + 1}$$

Equation 8.7

To further simplify Equation 8.7, let the term $\bar{v}/n - \bar{v}$ represent the ratio of occupied sites to nonoccupied sites on M; it can be mathematically represented as

$$\frac{\bar{v}}{n - \bar{v}} = \left(\frac{nK_f[L]}{K_f[L] + 1}\right)\left(\frac{K_f[L] + 1}{n}\right)$$

or

$$\frac{\bar{v}}{n - \bar{v}} = K_f[L]$$

Equation 8.8

A more desirable form for graphical use is known as **Scatchard's equation:**

$$\frac{\bar{v}}{[L]} = K_f(n - \bar{v})$$

Equation 8.9

If a plot of $\bar{v}/[L]$ versus $\bar{v}$ yields a straight line, shown by the solid line in Figure 8.2, then all the binding sites on M are identical and independent, and K_f and n are estimated as shown in the figure.

Cooperative Binding of Ligands

The derivation of Equation 8.9, assumes that K_f is identical for all binding sites; that is, the binding of one molecule of L does not influence the binding of other L molecules to binding sites on M. However, it is common for ligand-macromolecule interactions to display such influences. The binding of one L molecule to M may encourage or inhibit the binding of a second L molecule

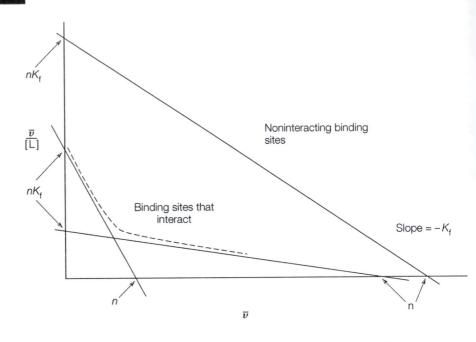

Figure 8.2

Scatchard plot. Two types of Scatchard curves are illustrated. The upper plot (solid line) represents binding to a macromolecule with noninteracting sites. The binding of a ligand molecule at one site is independent of the binding of a second ligand molecule at another site. The plot can be extrapolated to each axis and the n and K_f calculated. The lower, dashed line represents binding of ligand molecules to binding sites that interact. One ligand molecule bound to the macromolecule influences the rate of binding of other ligand molecules. The dashed line is evaluated as shown in the figure. The curved line has two distinct slopes. It can be resolved into two straight lines, each of which may be evaluated for n and K_f. This would indicate that there are two types of binding sites, each with a unique n and K_f.

to M. For example, the binding of oxygen to one of the four subunits of hemoglobin increases the affinity of the other subunits for oxygen. There is said to be **cooperativity** of sequential binding. If the sites do show cooperative binding, the plot is nonlinear, as shown by the dashed line in Figure 8.2. The shape of the nonlinear curve may be used to determine the number of types of binding sites. The dashed line in Figure 8.2 can be resolved into two lines, indicating that two types of binding sites are present on M. The n and K_f for each type of binding site may be estimated by resolving the smooth curve into straight lines as shown in Figure 8.2. K_f and n can be estimated by extrapolating the two straight lines to the axes.

Experimental Measurement of Ligand-Binding Interactions

To analyze ligand-macromolecule interactions quantitatively (to use Equation 8.9), one must be able to distinguish experimentally between bound ligand (LM) and free ligand (L). Many techniques have been developed for measuring the dynamics of LM interactions. Widely used techniques include equilibrium dialysis (Chapter 3, Section D), ultra-filtration (Chapter 3, Section D), and spectroscopic (especially fluorescence) measurements (Chapter 7). Many of these methods require specialized and expensive equipment, sometimes cumbersome procedures, and the use of radiolabeled ligands.

> **Study Exercise 8.2   Using Equilibrium Dialysis**   Ligand binding to a protein can be measured by the procedure of equilibrium dialysis (see Chapter 3, Section D). A solution of the protein (M) is sealed in a dialysis bag which is then placed in a large volume of solution containing the ligand (L). The pore size of the dialysis bag allows the passage of ligand molecules, but not protein molecules. After equilibrium is reached several hours later, the bag is opened and the concentrations of free ligand inside and outside are measured. The difference between these two values is the amount of ligand bound to the protein. The following experimental data are collected:
>
> Concentration of free ligand, $[L] = 1 \times 10^{-5}$ mole/liter
>
> Concentration of bound ligand, $[LM] = 5 \times 10^{-6}$ mole/liter
>
> K_f for the reaction $L + M \rightleftharpoons LM = 1 \times 10^5$ moles/liter
>
> Calculate the concentration of the protein, M, in moles per liter. Assume that M has only one binding site for the ligand.
>
> **Solution:** Use Equation 8.2 to calculate [M]. The concentration of $M = 5 \times 10^{-6}$ moles/liter.

One of the simplest and most convenient methods for monitoring ligand binding is the differential method, which detects and quantifies some measurable change in spectral absorption or fluorescence, in the UV-VIS regions, that accompanies ligand binding (see next section). A ligand may show enhanced fluorescence when bound to a macromolecule or amino acid residues (i.e. Trp) in a protein may show enhanced fluorescence when a ligand molecule is bound.

The Bradford Protein Assay as an Example of Ligand Binding

When Coomassie Blue dye binds to proteins, the dye undergoes a significant spectral change in the visible region (Figure 8.3). The spectrum of free dye displays a minimum in the range 575–625 nm. When dye and a protein (ovalbumin) are mixed, a new absorption band appears in the region of 595 nm. The increase in absorption at 595 nm is directly related to the concentration of protein, which is the basis of the Bradford protein assay (Chapter 3, Section B). The new absorption band is due to the presence of dye-ovalbumin complex.

The spectral changes occurring during the binding of Coomassie Blue dye to ovalbumin. **A** free dye in solution, **B** dye plus 0.5 mg of ovalbumin, **C** dye plus 1.0 mg of ovalbumin. The dye concentration is identical in all three curves.

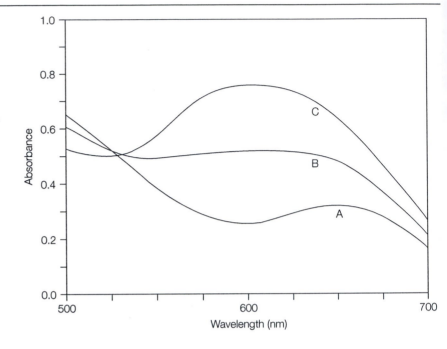

Note that the new band is rather broad and that a range of wavelengths may actually be used, but 595 nm is usually chosen. In this section, we will use the Bradford assay to illustrate an alternate method for using Scatchard's equation.

Equation 8.6 for evaluating protein-ligand interactions is easily adaptable to the spectroscopic differential method, but it must be converted into a linear form. Absorption data can be expressed in terms of [bound]/[free] ligands or [occupied]/[unoccupied] sites on the macromolecule. The data analysis described here assumes that L and LM, but not M, contribute to the measured absorbance. The measured absorbance change is proportional to the ratio of bound L to unbound L.

The absorbance observed from a mixture of L and M, A_{obs}, is

$$A_{obs} = A_L + A_{LM}$$

where

$$A_L = \text{absorbance of free L}$$
$$A_{LM} = \text{absorbance of bound L}$$

Absorbance data may be used directly to calculate the fraction of ligand bound:

>>   Fraction of ligand bound $= f = \dfrac{A_{obs} - A_L}{A_{max} - A_L}$ **Equation 8.10**

where

> A_{max} = total absorbance of LM when all the L molecules in solution are bound to M. This number is experimentally determined in Chapter 8, Section B.

The term $A_{max} - A_L$ is proportional to the total number of binding sites, and $A_{obs} - A_L$ represents the average number of occupied sites on M caused by a certain concentration of L.

For graphical analysis of the experimental data, $\bar{v}$ for Equation 8.6 can be defined in terms that can be measured experimentally:

>> $$\bar{v} = f\frac{[L]_0}{[M]_0}$$ *Equation 8.11*

If Equation 8.11 is combined with Equation 8.6, an equation for a straight line can be derived in terms that are readily measured.

$$f\frac{[L]_0}{[M]_0} = \frac{nK_f[L]}{K_f[L] + 1}$$

Take the reciprocal of each side and solve for $[M]_0/f$:

$$\frac{[M]_0}{f[L]_0} = \frac{K_f[L] + 1}{nK_f[L]}$$

$$\frac{[M]_0}{f} = \frac{[L]_0}{n} + \frac{[L]_0}{nK_f[L]}$$

The term $[L]/[L]_0 = 1 - f$; therefore:

$$\frac{[M]_0}{f} = \frac{[L]_0}{n} + \frac{1}{nK_f(1 - f)}$$

Rearrange to the form for a straight line, $y = mx + b$:

>> $$\frac{[M]_0}{f} = \frac{1}{nK_f(1 - f)} + \frac{[L]_0}{n}$$ *Equation 8.12*

This is an alternate form of the Scatchard equation. The values n and K_f can be calculated from a plot of $1/(1 - f)$ vs. $[M]_0/f$. The slope of the line is represented by $1/nK_f$, whereas the y intercept is $[L]_0/n$. If the data points fall on a straight line as in Figure 8.4, then all the dye binding sites on the protein M are identical and independent.

The derivation of Equation 8.12 assumes that K_f is identical for all binding sites; that is, the binding of one molecule of L does not influence the binding of other L molecules to binding sites on M. However, it is common for ligand-macromolecule interactions to display such influences. The binding of one L molecule to M may encourage or inhibit the binding of a second L molecule to M. For example, the binding of oxygen to one of the

Figure 8.4

Scatchard plot for typical binding data. See text for details.

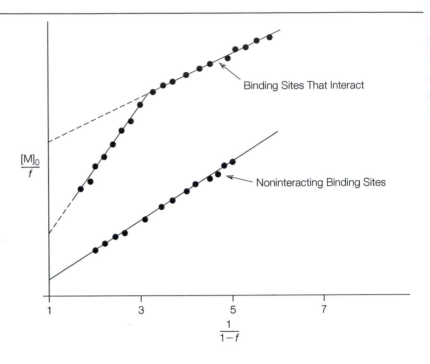

four subunits of hemoglobin increases the affinity of the other subunits for oxygen. There is said to be cooperativity of sequential binding. If the sites do show cooperative binding, the plot becomes nonlinear as shown by the curved line in Figure 8.4. This line is resolved into two lines, indicating that two types of binding sites are present on M. The n and K_f for each type of binding may be estimated by resolving the smooth curve into two straight lines. K_f and n may be calculated from the intercepts and slopes.

Computer Software for Analysis of LM Binding

Numerous computer programs exist that may be used for graphical analysis of receptor binding data. Most programs perform a linear or nonlinear least-squares regression analysis of experimental data. Many are available as free-ware on the Internet. Some popular software programs include SYSTAT (www.systat.com), Quattro.Pro (Borland International), CurveExpert (www.curveexpert.webhop.biz/), and DNRPEASY (www.biosci.uq.edu.au/~duggleby/rgd3a.htm). Additional Web sites or references are available at the end of this chapter and in Appendix I. One very useful program is DynaFit (www.biokin.com), which may be applied to the analysis of ligand binding data or enzyme kinetic data. The program has been designed by Dr. Petr Kuzmic, President and CEO of Biokin Ltd. DynaFit is available free for academic users (students, professors, university researchers), but one must register for an academic license at the Web site.

B.   BIOLOGICAL CATALYSIS (ENZYMES)

Enzyme-catalyzed reactions may also be described in terms of ligand-macromolecule interactions. Enzyme reactions are initiated by the interaction of a substrate molecule with an enzyme molecule:

$$E + S \rightleftharpoons ES$$

Equation 8.13

Enzymes are biological catalysts. Without their presence in a cell, most biochemical reactions would not proceed at the required rate. The chemical and biological properties of enzymes have been investigated since the early 1800s. The unrelenting interest in enzymes is due to several factors–their dynamic and essential roles in the cell, their extraordinary catalytic power, and their selectivity.

Enzyme-catalyzed reactions proceed through an ES complex, as shown in Equation 8.14. Individual rate constants k are assigned to each reaction step.

$$E + S \underset{k_2}{\overset{k_1}{\rightleftharpoons}} ES \underset{k_4}{\overset{k_3}{\rightleftharpoons}} E + P$$

Equation 8.14

E represents the enzyme, S the substrate or reactant, and P the product. For a specific enzyme, only one or a few different substrate molecules can bind in the proper manner and produce a functional ES complex. The substrate must have a size, shape, and polarity compatible with the active site of the enzyme. Some enzymes catalyze the transformation of many different molecules as long as there is a common type of chemical linkage in the substrate. Others have absolute specificity and can form reactive ES complexes with only one molecular structure. In fact, some enzymes are able to differentiate between D and L isomers of substrates.

Classes of Enzymes

Thousands of enzymes have now been isolated and studied; confusion would reign without some system for nomenclature and classification. Common names for enzymes are usually formed by adding the suffix **–ase** to the name of the substrate. The enzyme tyrosinase catalyzes the oxidation of tyrosine; cellulase catalyzes the hydrolysis of cellulose to produce glucose. Common names of this type define the substrate, but do not describe the chemistry of the reaction. Some very early names, such as catalase, trypsin, and pepsin, are even less descriptive and give no clue to their function or substrates. To avoid such confusion, enzymes now have official names that reflect the reactions they catalyze. All known enzymes can be classified into one of six categories (Table 8.3). Each enzyme has an official international name ending in *–ase* and a classification number. The number consists of four digits, each referring to a class and subclass of reaction. Table 8.3 shows an example from each class of enzyme.

Table 8.3

An Example of Each Class of Enzyme

1. Oxidoreductase

$$CH_3 - \underset{\underset{OH}{|}}{CH}COO^- \quad \underset{NAD^+ \quad NADH + H^+}{\overset{NAD^+ \quad NADH + H^+}{\rightleftharpoons}} \quad CH_3\underset{\underset{O}{\|}}{C}COO^-$$

Lactate   Pyruvate

Common name: Lactate dehydrogenase
Official name: L-Lactate: NAD^+ oxidoreductase
Official number: 1.1.2.3

2. Transferase

$$(d\text{-}NMP)_n + d\text{-}NTP \rightleftharpoons (d\text{-}NMP)_{n+1} + PP_i$$

$(d\text{-}NMP)_n$ = DNA with n nucleotides
$(d\text{-}NMP)_{n+1}$ = DNA with $n + 1$ nucleotides
PP_i = Pyrophosphate
Common name: DNA polymerase
Official name: Deoxynucleoside triphosphate: DNA deoxynucleotidyltransferase (DNA directed)
Official Number: 2.7.7.7

3. Hydrolase

$$H_3C - \underset{\underset{O}{\|}}{C} - O - CH_2 - CH_2 - \overset{+}{N}(CH_3)_3 + H_2O \rightleftharpoons CH_3\underset{\underset{O}{\|}}{C} - O^- + CH_2 - CH_2 - \overset{+}{N}(CH_3)_3$$
$$\underset{OH}{|}$$

Acetylcholine   Acetate   Choline

Common name: Acetylcholinesterase
Official name: Acetylcholine acetylhydrolase
Official number: 3.1.1.7

4. Lyase

$$CO_2 + H_2O \rightleftharpoons H_2CO_3$$
Carbonic acid

Common name: Carbonic anhydrase
Official name: Carbonate hydrolyase
Official number: 4.2.1.1

5. Isomerase

$$\begin{array}{ccc} CH_2OPO_3^{2-} & & CH_2OPO_3^{2-} \\ | & & | \\ C=O & \rightleftharpoons & CHOH \\ | & & | \\ CH_2OH & & C=O \\ & & \underset{H}{} \end{array}$$

Dihydroxyacetone   Glyceraldehyde
phosphate   3-phosphate

Common name: Triose phosphate isomerase
Official name: D-Glyceraldehyde-3-phosphate ketoisomerase
Official number: 5.3.1.1

Table 8.3

continued

6. **Ligase**

$$CH_3C - COO^- + CO_2 \underset{}{\overset{ATP}{\rightleftharpoons}} {}^-OOC - CH_2CCOO^-$$

Pyruvate   Oxaloacetate

Common name: Pyruvate carboxylase
Official name: Pyruvate: CO_2 ligase (ADP-forming)
Official number: 6.4.1.1

Kinetic Properties of Enzymes

The **initial reaction velocity, v_0,** of an enzyme-catalyzed reaction varies with the substrate concentration, [S], as shown in Figure 8.5. The Michaelis-Menten equation has been derived to account for the kinetic properties of enzymes. (Consult a biochemistry textbook for a derivation of this equation and for a discussion of the conditions under which the equation is valid.) The common form of the equation is

>> $$v_0 = \frac{V_{max}[S]}{K_M + [S]}$$ **Equation 8.15**

where

v_0 = initial reaction velocity

V_{max} = maximal reaction velocity; attained when all enzyme active sites are filled with substrate molecules

[S] = substrate concentration

K_M = Michaelis constant = $\dfrac{k_2 + k_3}{k_1}$

Figure 8.5

Michaelis-Menten plot for an enzyme-catalyzed reaction.

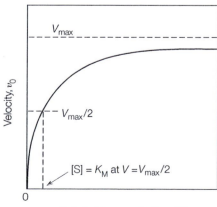

The important kinetic constants, V_{max} and K_M, can be graphically determined as shown in Figure 8.5. Equation 8.15 and Figure 8.5 have all of the disadvantages of nonlinear kinetic analysis. V_{max} can be estimated only because of the asymptotic nature of the line. The value of K_M, the substrate concentration that results in a reaction velocity of $V_{max}/2$, depends on V_{max}, so both are in error. By taking the reciprocal of both sides of the Michaelis-Menten equation, however, it is converted into the Lineweaver-Burk relationship (Equation 8.16).

>>
$$\frac{1}{v_0} = \frac{K_M}{V_{max}} \cdot \frac{1}{[S]} + \frac{1}{V_{max}}$$
Equation 8.16

This equation, which is in the form $y = mx + b$, gives a straight line when $1/v_0$ is plotted against $1/[S]$ (Figure 8.6). The intercept on the $1/v_0$ axis is $1/V_{max}$ and the intercept on the $1/[S]$ axis is $-1/K_M$. A disadvantage of the Lineweaver-Burk plot is that the data points are compressed in the high substrate concentration region.

Significance of Kinetic Constants

The **Michaelis constant,** K_M, for an enzyme-substrate interaction has two meanings: (1) K_M is the substrate concentration that leads to an initial reac-

Figure 8.6

Lineweaver-Burk plot for an enzyme-catalyzed reaction.

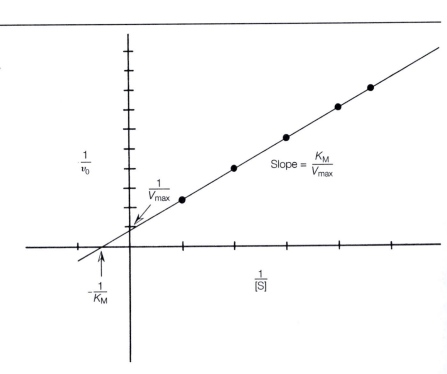

tion velocity of $V_{max}/2$ or, in other words, the substrate concentration that results in the filling of one-half of the enzyme active sites, and (2) $K_M = (k_2 + k_3)/k_1$. The second definition of K_M has special significance in certain cases. When $k_2 \gg k_3$, then $K_M = k_2/k_1$ so K_M is equivalent to the dissociation constant of the ES complex. When $k_2 \gg k_3$, a large K_M implies weak interaction between E and S, whereas a small K_M indicates strong binding between E and S.

V_{max} is important because it leads to the analysis of another kinetic constant, k_3, **turnover number.** The analysis of k_3 begins with the basic rate law for an enzyme-catalyzed process (Equation 8.17), which is derived from Equation 8.14.

▶▶ $$v_0 = k_3[ES]$$ *Equation 8.17*

If all of the enzyme active sites are saturated with substrate, then [ES] in Equation 8.17 is equivalent to $[E_T]$, the total concentration of enzyme, and v_0 becomes V_{max}; hence

▶▶ $$V_{max} = k_3[E_T]$$ *Equation 8.18*

or

$$k_3 = \frac{V_{max}}{[E_T]}$$

For an enzyme with one active site per molecule, the turnover number, k_3, is the number of substrate molecules transformed to product by one enzyme molecule per unit time, usually in minutes or seconds. The turnover number is a measure of the efficiency of an enzyme. K_M and k_3 values for some enzymes are listed in Table 8.4.

Table 8.4

K_M and k_3 Values for Some Enzyme:Substrate Systems

Enzyme	Substrate	K_M (mM)	k_3 (sec^{-1})
Catalase	H_2O_2	0.001	40,000,000
Carbonic anhydrase	HCO_3^-	9	400,000
Acetylcholinesterase	Acetylcholine	0.09	25,000
Penicillinase	Benzylpenicillin	0.050	2,000
Chymotrypsin	Glycyltyrosinylglycine	108	100
DNA polymerase	DNA	–	15
Ribulose-1,5-bis-phosphate carboxylase	Ribulose-1,5-bisphosphate CO_2	0.028 0.009	3.3

➤ | **Study Exercise 8.3   Enzyme Kinetics**   An enzyme displays the following reaction kinetics:

$$V_{max} = 200 \ \mu mol/min$$
Initial rate $(v_0) = 75 \ \mu mol/min$
$$[S] = 10 \ \mu M$$

Calculate K_M.

Inhibition of Enzyme Activity

Nonsubstrate molecules may interact with enzymes, leading to a decrease in enzymatic activity. Enzyme inhibition, the study of molecules that interfere with enzymes, is of interest because it often provides clues about the mechanism of enzyme action and also reveals information about metabolic control and regulation. In addition, many toxic substances, including drugs, express their action by enzyme inhibition. Inhibition of enzymes by reversible pathways will be reviewed here. For more details on enzyme inhibition, including irreversible processes, see your biochemistry course textbook.

The process of reversible inhibition is described by an equilibrium interaction between enzyme (E) and inhibitor molecule (I):

>> $E + I \rightleftharpoons EI$ *Equation 8.19*

Most inhibition processes can be classified as **competitive, mixed,** or **uncompetitive,** depending on how the inhibitor interferes with enzyme action.

1.   **Competitive inhibition**   A competitive inhibitor is usually similar in structure to the substrate and is capable of reversible binding to the enzyme active site. In contrast to the substrate molecule, the inhibitor molecule cannot undergo chemical transformation to a product; however, it does interfere with substrate binding. The binding of substrate and competitive inhibitor is a mutually exclusive process: When inhibitor is bound to enzyme, substrate is unable to bind and vice versa. The kinetic scheme for competitive inhibition is as follows:

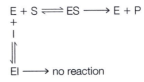

2.   **Mixed inhibition**   A mixed inhibitor does not bind in the active site of the enzyme but binds to some other region of the enzyme molecule. The presence of inhibitor may or may not affect substrate binding, but

it does interfere with the catalytic functioning of the enzyme. [A non-competitive inhibitor, a special type of mixed inhibitor, does not affect substrate binding (no change in K_M), but does affect the catalytic efficiency of the enzyme (lowers V_{max})]. The enzyme with mixed inhibitor bound may be converted to a nonfunctional conformational state. The kinetic scheme for noncompetitive inhibition is as follows (note that the inhibitor may bind to the free enzyme, E, to EI, or to ES):

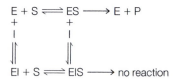

3. **Uncompetitive inhibitor**   An uncompetitive inhibitor is similar to a noncompetitive inhibitor—it allows substrate to bind to the active site. It differs, however, because an uncompetitive inhibitor binds only to the ES complex:

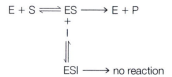

Because the inhibitor combines only with ES and not free enzyme, it will influence the activity of the enzyme only when substrate concentrations and, in turn, ES concentrations are high.

Any complete inhibition study requires the experimental differentiation among the three types of inhibition. The three inhibitory processes are kinetically distinguishable by application of the Lineweaver-Burk equation. For each inhibitor studied, at least two sets of rate experiments are completed. In all sets, the enzyme concentration is identical. In set 1, the substrate concentration is varied and no inhibitor is added. In set 2, the same variable substrate concentrations are used as in set 1, and a constant amount of inhibitor is added to each assay. If additional data are desired, more sets are prepared with variable substrate concentrations as in set 2; but a constant, and different, concentration of inhibitor is present. These data, when plotted on a Lineweaver-Burk graph, lead to three lines as shown in Figure 8.7. In competitive inhibition, all three lines intersect at the same point on the $1/v_0$ axis; hence, V_{max} is not altered by the presence of the competitive inhibitor. If enough substrate is added, the competitive inhibition may be overcome. The apparent K_M value (measured on the $1/[S]$ axis) changes with each increase in inhibitor concentration. In mixed (noncompetitive) inhibition, the family of lines has a common intercept on the $1/[S]$ axis (unchanged K_M for the lines). For uncompetitive inhibition, parallel lines are obtained indicating that both V_{max} and K_M are changed. The kinetic characteristics of reversible inhibition are compared in Table 8.5.

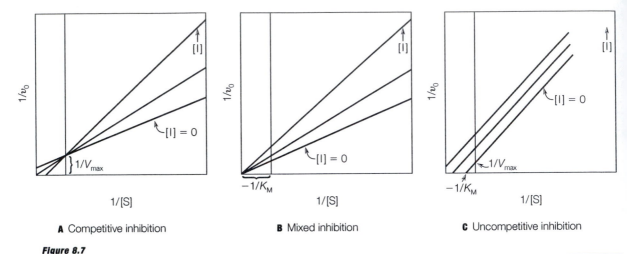

A Competitive inhibition   **B** Mixed inhibition   **C** Uncompetitive inhibition

Figure 8.7

Lineweaver-Burk plots to determine type of inhibition. Each set of three lines includes a line representing no inhibitor present, and two lines showing two different concentrations of inhibitor. **A** competitive inhibition: The lines intersect on the $1/v_0$ axis, where V_{max} can be calculated. **B** Mixed (noncompetitive) inhibition: The lines intersect on the $1/[S]$ axis where K_M can be calculated. **C** Uncompetitive inhibition: The lines are parallel. The V_{max} and K_M can be calculated for each line as shown.

Units of Enzyme Activity

The actual molar concentration of an enzyme in a cell-free extract or purified preparation is seldom known. Only if the enzyme is available in a pure crystalline form, carefully weighed, and dissolved in a solvent can the actual molar concentration be accurately known. It is, however, possible to develop a precise and accurate assay for enzyme activity. Consequently, the amount of a specific enzyme present in solution is most often expressed in units of activity. Three units are in common use, the *international unit* (IU), the *katal*, and *specific activity*. The International Union of Biochemistry Commission on Enzymes has recommended the use of a standard unit, the

Table 8.5

Kinetic Characteristics of Reversible Inhibition

Type of inhibition	Effect on Inhibited Reaction[1]		
	K_m	V_{max}	K_m/V_{max} (Slope)
Competitive	Higher	Same	Increase
Mixed	Higher	Lower	Increase
Noncompetitive	Same	Lower	Increase
Uncompetitive	Lower	Lower	Same

[1]*Relative to uninhibited reaction.*

international unit, or just **unit,** of enzyme activity. One IU of enzyme corresponds to the amount that catalyzes the transformation of 1 μmole of substrate to product per minute under specified conditions of pH, temperature, ionic strength, and substrate concentration. If a solution containing enzyme converts 10 μmoles of substrate to product in 5 minutes, the solution contains 2 enzyme units. A new unit of activity, the katal, has been recommended. One **katal** of enzyme activity represents the conversion of 1 mole of substrate to product in 1 second. One international unit is equivalent to 1/60 μkatal or 0.0167 μkatal. One katal, therefore, is equivalent to 6×10^7 international units. It is convenient to represent small amounts of enzyme in millikatals (mkatals), microkatals (μkatals), or nanokatals (nkatals). The enzyme activity in the above example was 2 units, which can be converted to katals as follows: Since one katal is 6×10^7 units, 2 units are equivalent to $2/6 \times 10^7$ or 33 nkatals (0.033 μkatals). If the enzyme is an impure preparation in solution, the activity is most often expressed as units/mL or katals/mL.

Specific Activity

Another useful quantitative definition of enzyme efficiency is **specific activity.** The specific activity of an enzyme is the number of enzyme units or katals per milligram of protein. This is a measure of the purity of an enzyme. If a solution contains 20 mg of protein that express 2 units of activity (33 nkatals), the specific activity of the enzyme is 2 units/20 mg = 0.1 units/mg or 33 nkatals/20 mg = 1.65 nkatals/20 mg. As an enzyme is purified, its specific activity increases. That is, during purification, the enzyme concentration increases relative to the total protein concentration until a limit is reached. The maximum specific activity is attained when the enzyme is homogeneous or in a pure form.

The activity of an enzyme depends on several factors, including substrate concentration, cofactor concentration, pH, temperature, and ionic strength. The conditions under which enzyme activity is measured are critical and must be specified when activities are reported.

Design of an Enzyme Assay

Whether an enzyme is obtained commercially or prepared in a multistep procedure, an experimental method must be developed to detect and quantify the specific enzyme activity. During isolation and purification of an enzyme, the assay is necessary to determine the amount and purity of the enzyme and to evaluate its kinetic properties. An assay is also essential for a further study of the mechanism of the catalyzed reaction.

The design of an assay requires certain knowledge of the reaction:

1. The complete stoichiometry.
2. What substances are required (substrate, metal ions, cofactors, etc.) and their kinetic dependence.
3. Effect of pH, temperature, and ionic strength.

The most straightforward approach to the design of an enzyme assay is to measure the change in substrate or product concentration during the reaction.

If an enzyme assay involves continuous monitoring of substrate or product concentration, the assay is said to be **kinetic.** If a single measurement of substrate or product concentration is made after a specified reaction time, a **fixed-time assay** results. The kinetic assay is more desirable because the time course of the reaction is directly observed and any discrepancy from linearity can be immediately detected.

Kinetic versus Fixed-time Assay

Figure 8.8 displays the kinetic progress curve of a typical enzyme-catalyzed reaction and illustrates the advantage of a kinetic assay. The rate of product formation decreases with time. This may be due to any combination of factors such as decrease in substrate concentration, denaturation of the enzyme, and product inhibition of the reaction. The solid line in Figure 8.8 represents the continuously measured time course of a reaction (kinetic assay). The true rate of the reaction is determined from the slope of the dashed line drawn tangent to the experimental result. From the data given,

Figure 8.8

Kinetic progress of an enzyme-catalyzed reaction. See text for details.

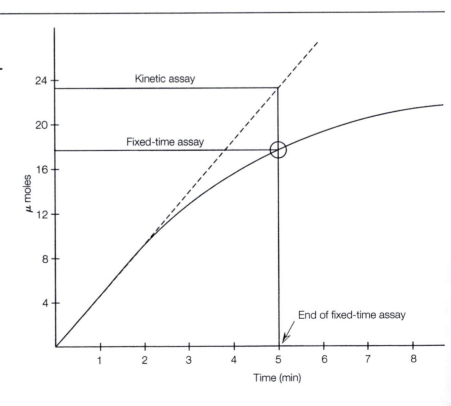

the rate is 5 μmoles of product formed per minute. Data from a fixed-time assay are also shown on Figure 8.8. If it is assumed that no product is present at the start of the reaction, then only a single measurement after a fixed period is necessary. This is shown by a circle on the experimental rate curve. The measured rate is now 16 μmoles of product formed every 5 minutes or about 3 μmoles/minute, considerably lower than the rate derived from the continuous, kinetic assay. Which rate measurement is correct? Obviously, the kinetic assay gives the true rate because it corrects for the decline in rate with time. The fixed-time assay can be improved by changing the time of the measurement, in this example, to 2 minutes of reaction time, when the experimental rate is still linear. It is possible to obtain true rates with the fixed-time assay, but one must choose the time period very carefully. In the laboratory, this is done by removing aliquots of a reaction mixture at various times and measuring the concentration of product formed in each aliquot. Figure 8.8 reinforces an assumption used in the derivation of the Michaelis-Menten equation. Only measurements of initial velocities lead to true reaction rates. This avoids the complications of enzyme denaturation, decrease of [S], product inhibition, and reversion of the product to substrate.

Several factors must be considered when the experimental assay conditions are developed. The reaction rate depends on the concentrations of substrate, enzyme, and necessary cofactors. In addition, the reaction rate is under the influence of environmental factors such as pH, temperature, and ionic strength. Enzyme activity increases with increasing temperature until the enzyme becomes denatured. The enzyme activity then decreases until all enzyme molecules are inactivated by denaturation. During kinetic measurement, it is essential that the temperature of all reaction mixtures be maintained constant.

Ionic strength and pH should also be monitored carefully. Although some enzymes show little or no change in activity over a broad pH range, most enzymes display maximum activity in a narrow pH range. Any assay developed to evaluate the kinetic characteristics of an enzyme must be performed in a buffered solution, preferably at the optimal pH.

Applications of an Enzyme Assay

The conditions used in an enzyme assay depend on what is to be accomplished by the assay. There are two primary applications of an enzyme assay procedure. First, it may be used to measure the concentration of active enzyme in a preparation. In this circumstance, the measured rate of the enzyme-catalyzed reaction must be proportional to the concentration of enzyme; stated in more kinetic terms, there must be a linear relationship between initial rate and enzyme concentration (the reaction is first order in enzyme concentration). To achieve this, certain conditions must be met: (1) the concentrations of substrates(s), cofactors, and other requirements must be in excess; (2) the reaction mixture must not contain inhibitors of the enzyme; and (3) all environmental factors such as pH, temperature, and ionic strength should be controlled. Under these conditions, a plot of enzyme activity (μmole product formed/minute) vs. enzyme concentration is a straight line and can be used to estimate the concentration of active enzyme in solution.

Second, an enzyme assay may be used to measure the kinetic properties of an enzyme such as K_M, V_{max}, and inhibition characteristics. In this situation, different experimental conditions must be used. If K_M for a substrate is desired, the assay conditions must be such that the measured initial rate is first order in substrate. To determine K_M of a substrate, constant amounts of enzyme are incubated with varying amounts of substrate. A Lineweaver-Burk plot ($1/v_0$ vs. $1/[S]$) may be used to determine K_M and V_{max}. If a reaction involves two or more substrates, each must be evaluated separately. The concentration of only one substrate is varied, while the other is held constant at a saturating level. The same procedure holds for determining the kinetic dependence on cofactors. Substrate(s) and enzyme are held constant and the concentration of the cofactor is varied.

Inhibition kinetics are included in the second category of assay applications. An earlier discussion outlined the kinetic differentiation among competitive, uncompetitive, and mixed inhibition. In summary, a study of enzyme kinetics is approached by measuring initial reaction velocities under conditions where only one factor (substrate, enzyme, cofactor) is varied and all others are held constant.

➤ | **Study Exercise 8.4   Converting Rate Data from ΔA/min Units to Concentration Units**   Because most enzyme assays are completed with the use of spectroscopic methods, initial rate data are in the form of ΔA/min. The change of color intensity per time interval is not a useful unit for kinetics. A more valuable and widely used unit that actually defines the **amount of product formed per time interval** is **mmol/min** or **μmol/min.** The conversion of ΔA units to concentration units may be done using Beer's Law:

$$c = A/El$$

where

    A = the absorbance

    E = the molar absorption coefficient of the substrate or product

    l = path length of the cuvette, usually 1 cm

    c = concentration of substrate or product

For our example, let us assume that we are measuring the appearance of product that has an E value of 3600 M^{-1}cm^{-1}. According to Beer's Law, the absorbance change, in a 3 mL cuvette, during 1 minute for production of 1 molar product is 3600:

$$1 \text{ molar} = \frac{A}{3600 \ M^{-1} \text{cm}^{-1} \times 1 \text{ cm}}$$

$$A = 3600$$

To convert the raw data to moles of product formed per minute per liter, each $\Delta A/\text{min}$ is divided by 3600. For example, a $\Delta A/\text{min}$ of 0.1 is converted to μmoles of product formed (Δc) in the following way:

$$\Delta c = \frac{\Delta A/\text{min}}{El} = \frac{0.1/\text{min}}{3600 \ M^{-1} \text{cm}^{-1} \text{cm}} = 2.78 \times 10^{-5} \frac{M}{\text{min}}$$

Now convert to moles/min:

$$2.78 \times 10^{-5} \frac{\text{moles}}{\text{liter} \times \text{min}} \times 0.003 \text{ liters} = 8.33 \times 10^{-8} \frac{\text{moles}}{\text{min}}$$

Now convert to μmoles/min:

$$\Delta c = 8.33 \times 10^{-2} \frac{\mu\text{moles}}{\text{min}}$$

Computer Software for Analysis of Enzyme Kinetic Data

Many of the computer programs listed for use in ligand binding analysis (Chapter 8, Section A) may also be used for enzyme kinetic data. A widely used program is DynaFit (www.biokin.com), which is available free to academic users. The purpose of DynaFit is to aid the graphical analysis of enzyme kinetic data. The experimental data may be in the form of initial reaction rates with dependence on the concentration of substrate or inhibitor. Reaction progress curves (time vs. absorbance) may also be analyzed.

Another useful program for enzyme kinetic analysis is Leonora. This program is associated with the book *Analysis of Enzyme Kinetic Data* by Cornish-Bowden (see Table 8.6).

Many Internet databases, in addition to kinetic analysis, also contain general information on enzyme names, EC numbers, catalytic mechanisms, cofactors, structures, reactions, kinetics, associated diseases, and other facts. The most popular, current ones are the ENZYME databases, part of the Expert Protein Analysis System (ExPASy), BRENDA, LIGAND, EMP, and the Enzyme Structure Database (Table 8.6 and Appendix I).

Performing enzyme kinetic/inhibition experiments in the laboratory and analyzing data may be too complex and not appropriate for some levels of students. In addition, equipment and reagents can be quite expensive and difficult to set up. Computer simulation of enzyme kinetic experiments may be an appropriate alternative (Gonzalez-Cruz et al., 2003).

Table 8.6

Databases for Searches of General Enzyme Information and Analysis of Kinetic Data

Name	Information	Web site
DynaFit	Graphical analysis of enzyme kinetic and ligand binding data.	www.biokin.com
Leonora	Analysis of enzyme kinetic data.	http://ir2lcb.cnrs-mrs.fr/athel/leonora0.htm
ENZYME	General enzyme information from ExPASy. Searched by EC number or enzyme name.	http://www.expasy.ch/enzyme
BRENDA	The comprehensive enzyme information system. Searched by EC number and enzyme name.	http://www.brenda.uni-koeln.de/
EMP	The Enzymology Database. Data that have been published in the literature.	http://wit.mcs.anl.gov/
Enzyme structure database	Contains the known enzyme structures in Protein Data Bank.	http://www.biochem.ucl.ac.uk/bsm/enzymes/index.html

Study Problems

> 1. A drug, X, was studied for its affinity for serum albumin. When X was bound to albumin, an increase in absorbance was noted. The $\bar{v}$ values were determined from these absorbance measurements. Use the data below to determine K_f and n for the interaction between X and albumin. Prepare two types of graphs and compare the results. In one graph plot $\bar{v}$ vs. [X], and in the second plot $\bar{v}/[X]$ vs. $\bar{v}$. Which is the better method? Why? You may also want to try Dynafit.

[X]	$\bar{v}$
0.36	0.43
0.60	0.68
1.2	1.08
2.4	1.63
3.6	1.83
4.8	1.95
6.0	1.98

> 2. You are attempting to develop a colorimetric probe for use in binding studies. List several requirements that the probe must meet in order to be effective. For example, it must bind at specific locations of the macromolecule. Can you think of other requirements?

▶ 3. A research project you are working on involves the study of sugar binding to human albumin. The sugars to be tested are not fluorescent, and you do not wish to use a secondary probe such as a dye. Human albumin has only one tryptophan residue, and you know that this amino acid is fluorescent. You find that the tryptophan fluorescence spectrum of human albumin undergoes changes when various sugars are added. Can you explain the results of this experiment and discuss the significance of the finding?

4. Equation 8.6 in this chapter can be used to determine K_f values, but hyperbolic plots are obtained. Convert Equation 8.6 into an equation that will yield a linear plot without going through all the changes necessary for the Scatchard equation. Hint: See the Michaelis-Menten equation and the Lineweaver-Burk equation.

5. One of the many straight-line modifications of the Michaelis-Menten equation is the Eadie-Hofstee equation:

$$v_o = -K_M \frac{v_0}{[S]} + V_{max}$$

Beginning with the Lineweaver-Burk equation, derive the Eadie-Hofstee equation. Explain how it may be used to plot a straight line from experimental rate data. How are V_{max} and K_M calculated?

▶ 6. The table below gives initial rates of an enzyme-catalyzed reaction along with the corresponding substrate concentration. Use any graphical method or computer program to determine K_M and V_{max}.

v_0 (μM/min)	[S] (mole/liter)
130	65×10^{-5}
116	23×10^{-5}
87	7.9×10^{-5}
63	3.9×10^{-5}
30	1.3×10^{-5}
10	0.37×10^{-5}

▶ 7. The enzyme concentration used in each assay in Problem 6 is 5×10^{-7} mole/liter. Calculate K_3, the turnover number, in units of sec^{-1}.

8. Using the data in Problems 6 and 7, calculate the specific activity of the enzyme in units/mg and katal/mg. Assume the enzyme has a molecular weight of 55,000 and the reaction mixture for each assay is contained in a total volume of 1.00 mL.

▶ 9. Compound X was tested as an inhibitor of the enzyme in Problem 6. Use the rate data in Problem 6 and the following inhibition data to evaluate compound X. Is it a competitive or noncompetitive inhibitor?

$[X] = 3.7 \times 10^{-4} M$ v_0 (μM/min)	[S] (mole/liter)	$[X] = 1.58 \times 10^{-3} M$ v_0 (μM/min)
125	6.5×10^{-4}	110
102	2.3×10^{-4}	80
70	7.9×10^{-5}	40
45	3.9×10^{-5}	20
20	1.3×10^{-5}	—
5	0.37×10^{-5}	—

10. Show the mathematical steps required to derive the Lineweaver-Burk equation beginning with the Michaelis-Menten equation.

▶ 11. Figure 8.8 displays the kinetic progress of an enzyme-catalyzed reaction. What time limit must be imposed on rate measurements taken with the use of the fixed-time assay? Why?

12. In your biochemistry research project, you find that the binding of a ligand to a protein decreases if the ionic strength of the buffer solvent is increased. What type of noncovalent bonding might be involved in the ligand-protein complex?

13. The following data were collected in a ligand binding experiment. The protein concentration was 7.5×10^{-6}. Use a graphical analysis program to determine n and K_f.

Reaction No.	L added (μM)	L bound (μM)
1	20	11
2	50	26
3	100	44
4	150	55
5	200	60
6	400	70

Further Reading

A. Attie and R. Raines, *J. Chem. Educ.* **72,** 119–124 (1995). "Analysis of Receptor-Ligand Interactions."

M. Benore-Parsons and K. Sufka, *Biochem. Mol. Biol. Educ.* **31,** 85–92 (2003). "Teaching Receptor Theory to Biochemistry Undergraduates."

J. Berg, J. Tymoczko, and L. Stryer, *Biochemistry,* 5th ed. (2002), Freeman (New York), pp. 8–11. Noncovalent bonding.

A. Bordbar, A. Saboury, and A. Moosavi-Movahedi, *Biochem. Educ.* **24,** 172–175 (1996). "The Shapes of Scatchard Plots for Systems with Two Sets of Binding Sites."

R. Boyer, *Concepts in Biochemistry,* 2nd ed. (2002), John Wiley & Sons (New York), pp. 49–51, 126–154. Introduction to ligand binding and enzyme kinetics.

M. Bradford, *Anal. Biochem.* **72**, 248–254 (1976). "A Rapid and Sensitive Method for the Quantitation of Microgram Quantities of Protein Utilizing the Principle of Protein-Dye Binding."

B. Cantor and P. Schimmel, *Biophysical Chemistry,* Part III (1980), Freeman (San Francisco), pp. 849–886. "Ligand Interactions at Equilibrium."

A. Cornish-Bowden, *Analysis of Enzyme Kinetic Data* (1995), Oxford University Press (Oxford).

R. Eisenthal and M. Danson, *Enzyme Assays: A Practical Approach* (1992), IRL Press (Oxford).

D. Fell, *Understanding the Control of Metabolism* (1997), Portland Press (London).

R. Garrett and C. Grisham, *Biochemistry,* 3rd ed. (2005), Brooks Cole (Belmont, CA), pp. 404–441. Enzyme kinetics.

J. Gonzalez-Cruz, R. Rodriquez-Sotres, and M. Rodriquez-Penagos, *Biochem. Mol. Biol. Educ.* **31,** 93–101 (2003). "On the Convenience of Using Computer Simulation to Teach Enzyme Kinetics to Undergraduate Students with Biological Chemistry-related Curricula."

P. Kuzmic, *Anal. Biochem.* **237,** 260–273 (1996). "Program DYNAFIT for the Analysis of Enzyme Kinetic Data: Application to HIV Protease."

V. Leskovac, *Comprehensive Enzyme Kinetics* (2003), Kluwer Academic/Plenum Publishers (New York).

A. Marangoni, *Enzyme Kinetics: A Modern Approach* (2002), John Wiley & Sons (New York).

M. Moller and A. Denicola, *Biochem. Mol. Biol. Educ.* **30,** 309–312 (2002). "Study of Protein-Ligand Binding by Fluorescence."

D. Nelson and M. Cox, *Lehninger Principles of Biochemistry,* 4th ed. (2005), Freeman (New York), p. 439. Scatchard analysis and noncovalent bonding; pp. 190–237; enzyme kinetics.

N. Price, R. Dwek, M. Wormald, and R. Ratcliffe, *Principles and Problems in Physical Chemistry for Biochemists,* 3rd ed. (2002), Oxford University Press (Oxford). Chapter 4 covers ligand binding and Chapters 9–13 cover enzyme reactions.

F. Ranaldi, P. Vanni, and E. Gischetti, *Biochem. Educ.* **27,** 87–91 (1999). "What Students Must Know about the Determination of Enzyme Kinetic Parameters."

I. Schomburg et al., *Trends Biochem. Sci.* **27,** 54–56 (2002). "BRENDA: A Resource for Enzyme Data and Metabolic Information."

D. Sheehan, *Physical Biochemistry: Principles and Applications* (2001), John Wiley & Sons (New York).

J. Sohl and A. Splittgerber, *J. Chem. Educ.* **68,** 262–264 (1991). "The Binding of Coomassie Brilliant Blue to Bovine Serum Albumin—A Physical Biochemistry Experiment."

I. Tinoco, K. Sauer, J. Wang, and J. Puglisi, *Physical Chemistry: Principles and Applications in Biological Sciences,* 4th ed. (2002). Chapter 7 covers enzyme reactions and Chapter 9 covers ligand binding.

B. Tsai, *Computational Biochemistry* (2002), John Wiley & Sons (New York), pp. 107–146. Ligand interactions and enzyme kinetics.

D. Voet and J. Voet, *Biochemistry,* 3rd ed. (2004), John Wiley & Sons, (Hoboken, NJ), pp. 258–265; 660. Hormone binding and noncovalent interactions, pp. 459–495; enzyme kinetics.

D. Voet, J. Voet, and C. Pratt, *Fundamentals of Biochemistry,* 2nd ed. (2006), John Wiley & Sons (Hoboken, NJ), pp. 25–26; 358–394. Noncovalent interactions and enzyme kinetics.

J. Watson et al., *Molecular Biology of the Gene,* 5th ed. (2004), Benjamin/Cummings (San Francisco). Chapters 3 and 4.

K. Wilson and J. Walker, *Principles and Techniques of Practical Biochemistry*, 5th ed. (2000), Cambridge University Press (Cambridge), pp. 357–452. Enzyme kinetics and ligand interactions.

C. Whiteley, *Biochem. Educ.* **27,** 15–18 (1999). "Enzyme Kinetics: Partial and Complete Non-Competitive Inhibition."

C. Whiteley, *Biochem. Educ.* **28,** 144–147 (2000). "Enzyme Kinetics: Partial and Complete Uncompetitive Inhibition."

http://www.biophysics.org/btol/index.html
Biophysics textbook on line. See chapter on Intermolecular Forces.

http://www.scripps.edu/pub/olson-web/people/gmm/
Study Ligand-Protein Docking for information about drug design. Review movies showing interactions with a variety of ligand-protein systems.

http://www.science.smith.edu/Biochem/Chm_357/design.htm
Introduction to Computational Drug Design.

http://www.graphpad.com/www/radiolig/radiolig1.htm
GraphPad Software for analyzing binding data. Provides an introduction to analysis of ligand-protein interactions.

http://www.curvefit.com/index.htm
The CurveFit Web site by GraphPad.

http://www.chem.qmw.ac.uk/iubmb/enzyme/
The Enzyme List maintained by the International Union of Biochemistry and Molecular Biology.

http://www.biokin.com
Contains the computer program DynaFit, which may be used for the graphical analysis of ligand-receptor binding and enzyme kinetic data.

http://www.curveexpert.webhop.biz
The nonlinear regression program, CurveExpert.

http://www.systat.com
   Computer programs for curve fitting.

http://lab3d.chem.virginia.edu
   Virtual biochemistry lab experiment on enzyme kinetics.

MOLECULAR BIOLOGY I: STRUCTURES AND ANALYSIS OF NUCLEIC ACIDS

In this chapter we shall focus on the **nucleic acids,** biomolecules important for their cellular roles in the storage, transfer, and expression of genetic information. Two fundamental types of nucleic acids participate as genetic molecules: **(1) deoxyribonucleic acid (DNA) and (2) ribonucleic acid (RNA).** We begin this chapter with a review of the chemical and biological properties of the nucleic acids and then turn to a discussion of laboratory methods for their isolation and characterization. Several procedures are described here that provide information about the structure and function of DNA and RNA. These include the isolation of chromosomal and plasmid DNA, isolation of RNA, ultraviolet absorption of the nucleic acids, thermal denaturation curves, ethidium bromide binding and fluorescence, agarose gel electrophoresis, and sequencing DNA molecules. Early studies on DNA and RNA delved into understanding the molecular details of DNA replication, RNA transcription, and translation to produce proteins. More recently, research on the nucleic acids has concentrated on the biotechnological development of **recombinant DNA** for molecular cloning procedures, which will be discussed in Chapter 10.

A.   INTRODUCTION TO THE NUCLEIC ACIDS

Chemical Components of DNA and RNA

In chemical terms, the nucleic acids are linear polymers made up of monomeric units called **nucleotides.** A nucleotide is comprised of three chemical entities (see Figure 9.1):

1. A purine or pyrimidine nitrogen-containing, heterocyclic base (Figure 9.2).

Figure 9.1

General structure of a nucleotide showing the three fundamental components: a purine or pyrimidine base, a ribose (or deoxyribose), and phosphate.

2. A five-carbon carbohydrate (aldopentose), **β-D-ribose** or **β-D-2-deoxyribose** (Figure 9.3).

3. One, two, or three phosphate groups.

The predominant nitrogen bases in DNA include the purines, **adenine (A)** and **guanine (G)** and the pyrimidines, **thymine (T)** and **cytosine (C)**. In RNA, the purines include **A** and **G** and the pyrimidines, **C** and **uracil (U)**. The nucleic acids, especially RNA, contain small quantities of methylated nitrogen bases (Figure 9.2).

Major Bases

Minor Bases

Adenine   Guanine   2-Methyladenine   1-Methylguanine

Purines

Thymine (DNA)   Cytosine   Uracil (RNA)   5-Methylcytosine   5-Hydroxymethylcytosine

Pyrimidines

Figure 9.2

The major and some minor heterocyclic bases in DNA and RNA. All are derived from purine or pyrimidine.

Figure 9.3

The aldopentoses in
RNA and DNA: **A** β-D-ribose,
B β-D-2-deoxyribose.

A β-D-Ribose   **B** β-D-2-Deoxyribose

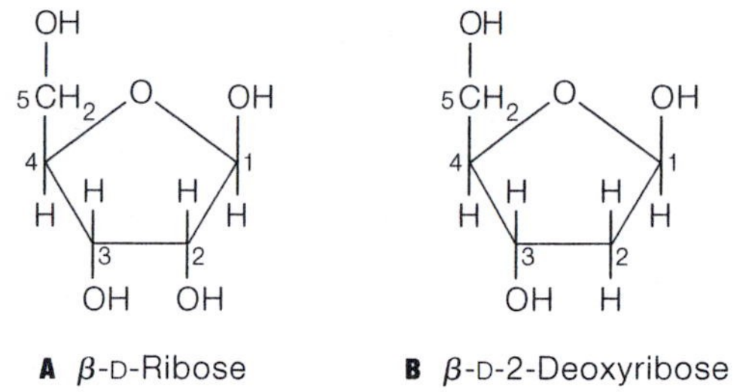

Figure 9.4

A nucleoside consists of a
purine or pyrimidine base linked
to a ribose or deoxyribose by an
N-glycosidic bond. Two
numbering systems (primed and
unprimed) are necessary to
distinguish the two rings.

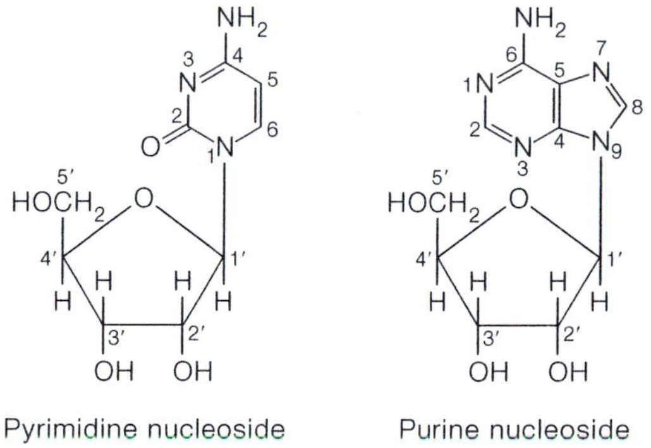

Pyrimidine nucleoside   Purine nucleoside

When a nitrogen base and aldopentose are combined via an N-glycosidic bond, the product is a **nucleoside** (Figure 9.4). Addition of a phosphoryl group ($-PO_3^{2-}$) to a hydroxyl group on the carbohydrate leads to a **nucleotide.** The most common site for phosphorylation is the 5′-hydroxyl group. The names of 5′ nucleotides derived from the common nitrogen bases are listed in Table 9.1.

Table 9.1

Nomenclature for Nucleosides and Nucleotides in DNA and RNA

Base	Nucleoside	Nucleotide (Abbreviation)	Nucleic Acid
Purine			
Adenine	Adenosine, deoxyadenosine	Adenosine 5′-monophosphate (5′-AMP)	RNA
		Deoxyadenosine 5′-monophosphate (5′-dAMP)	DNA
Guanine	Guanosine, deoxyguanosine	Guanosine 5′-monophosphate (5′-GMP)	RNA
		Deoxyguanosine 5′-monophosphate (5′-dGMP)	DNA
Pyrimidine			
Cytosine	Cytidine, deoxycytidine	Cytidine 5′-monophosphate (5′-CMP)	RNA
		Deoxycytidine 5′-monophosphate (5′-dCMP)	DNA
Thymine	Deoxythymidine	Deoxythymidine 5′-monophosphate (5′-dTMP)	DNA
Uracil	Uridine	Uridine 5′-monophosphate (5′-UMP)	RNA

DNA Structure and Function

DNA in all forms of life is a polymer made up of nucleotides containing four major types of heterocyclic nitrogen bases. The nucleotides are held together by **3′,5′-phosphodiester bonds** (Figure 9.5). The quantitative ratio and sequence of bases vary with the source of the DNA. The covalent backbone of DNA (and RNA) consists of alternating deoxyriboses (or riboses) and phosphate groups. This feature, which is defined as a **common, invariant region,** is found in all nucleic acids. The **variable region** of DNA and RNA displays the sequence of the four kinds of bases in the nucleic acid. The purine/pyrimidine nitrogen bases protrude from the backbone-like side chains. The sequence of bases carries the specific genetic message. The DNA or RNA chain also displays **directionality**—one end of the chain has a 3′-hydroxyl (or phosphate) group and the other end has a 5′-hydroxyl (or phosphate) group.

Native DNA exists as two complementary, antiparallel strands arranged in a **double helix** held together by noncovalent bonding. DNA in most prokaryotic cells (simple cells with no major organelles and a single chromosome) exists as a single molecule in a circular, double-stranded form with a molecular weight of at least 2×10^9 amu. Eukaryotic cells (cells with major organelles) contain several chromosomes and, thus, several very large DNA molecules (Table 9.2).

DNA was first isolated from biological material in 1869, but its participation in the transfer of genetic information was not recognized until the mid-1940's. Since that time, DNA has been the subject of thousands of physical, chemical, and biological investigations. A landmark discovery was the elucidation of the three-dimensional structure of DNA by X-ray diffraction analysis by Watson and Crick in 1953. The double helix as envisioned by Watson and Crick is now recognized as a significant form of native DNA (Figure 9.6). The most important structural and functional feature of the double helix is the **complementary base pairing.** This not only holds the two strands of the double helix together, but it also allows the DNA to function in the storage and transfer of genetic information.

The double helix is stabilized by two types of noncovalent forces (review Table 8.2):

1. **Hydrogen bonding between pairs of complementary bases: A:T and G:C** This combination leads to the maximum number of hydrogen bonds for stability. Each base pair consists of a pyrimidine and a purine base. The distance spanned across the double helix is just right for a purine:pyrimidine pair. Two pyrimidines are too small to form strong hydrogen bonds, and two purines are too sterically crowded.

2. **Hydrophobic interactions and van der Waals forces between "stacked bases" (see Chapter 8, Section A)** The planes of the nitrogen bases are nearly perpendicular to the common axis of the helix and take on a stacking arrangement, bringing the purine and pyrimidine rings close together. This allows for favorable hydrophobic interactions between nonpolar regions of the bases and interactions between π electrons in the sp^2 hybrid orbitals of the aromatic rings.

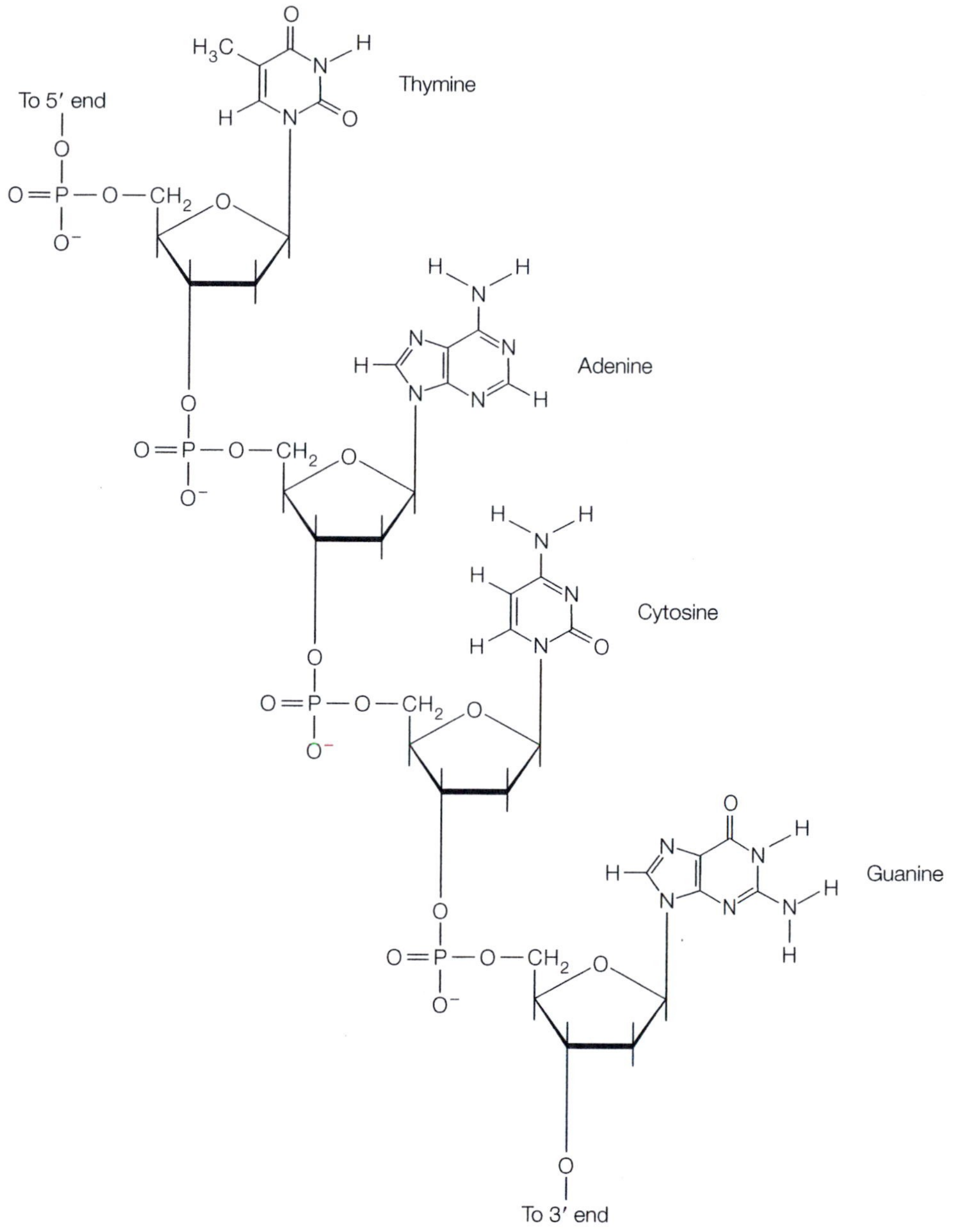

Figure 9.5

The covalent structure of DNA showing the phosphodiester backbone linking deoxyribose through the 3' and 5' hydroxyl groups.

Table 9.2

Comparison of DNA from Different Species

Organism	Number of Base Pairs	Length (μm)	Conformation
Viruses			
SV40	5100	1.7	Circular
Adenovirus	36,000	12	Linear
λ phage	48,600	17	Circular
Bacteria			
E. coli	4,700,000	1400	Circular
Eukaryotes			
Yeast	13,500,000	4600	Linear
Fruit fly	165,000,000	56,000	Linear
Human	3,200,000,000	$1-2 \times 10^6$	Linear

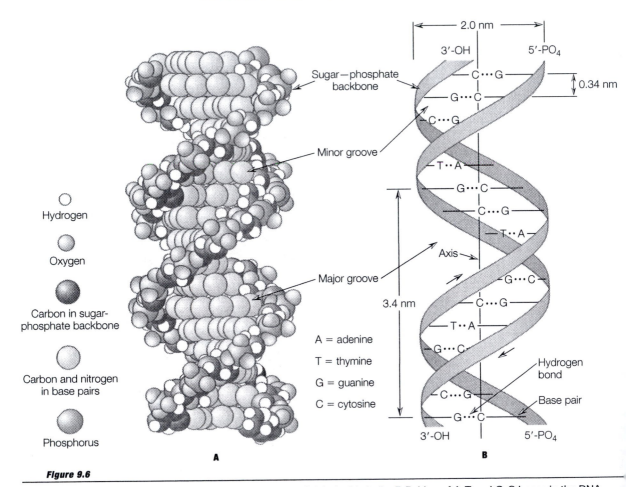

Hydrogen

Oxygen

Carbon in sugar-
phosphate backbone

Carbon and nitrogen
in base pairs

Phosphorus

A = adenine
T = thymine
G = guanine
C = cytosine

Figure 9.6

A The Watson-Crick double helix. **B** Pairing of A-T and G-C bases in the DNA double helix. Note that the two strands are antiparallel; their 3′ and 5′ phosphodiester bonds run in opposite directions.

RNA Structure and Function

Unlike DNA which is primarily a homogeneous molecule in the cell, RNA exists in three major forms (Table 9.3). **Ribosomal RNA (rRNA),** the most abundant form, is found associated with the ribosomes, the protein-synthesizing organelles. The three sizes of prokaryotic rRNA, 5S, 16S, and 23S, can be separated by centrifugation (Chapter 4). **Messenger RNA (mRNA)** carries the transient message for protein synthesis from nuclear DNA to the ribosomes. The smallest RNA molecules, **transfer RNA (tRNA),** select, bind, and activate amino acids for use in protein synthesis. Because RNA molecules are heterogeneous and short lived in the cell, it has been a challenge to isolate them in an intact form in order to study their structures and properties. RNA is usually single stranded, but it does have important structural features. tRNA molecules have been purified and crystallized for X-ray crystallography. The basic two-dimensional structure of all tRNAs is often shown in a cloverleaf pattern in order to display its single-stranded and double-stranded regions. mRNA and rRNA molecules recently isolated and crystallized display the structural elements of hairpin turns, right-handed double helixes, and internal loops. Complementary bases in RNA are A:U and G:C.

Now that some of the major players in the flow of biological information have been identified, it is possible to provide a schematic outline for the process of protein synthesis (see Figure 9.7). The general direction of information flow is:

$$DNA \longrightarrow RNA \longrightarrow Proteins \longrightarrow Cell\ Structure/Function$$

Note that DNA is the original source of genetic information. DNA is faithfully copied by the process of **replication.** In **transcription,** a smaller region of the DNA (gene) is converted into the language of mRNA. In **translation,** proteins are synthesized from amino acids. After synthesis, many proteins are not yet biologically active and must go through **posttranslational**

Table 9.3				
RNA Molecules in *E. coli*				
Type	Relative Amount (%)	S^a Value	MW	Average Number of Nucleotides
Ribosomal RNA (rRNA)	80	23S	1.2×10^6	3700
		16S	0.55×10^6	1700
		5S	36,000	120
Transfer RNA (tRNA)	15	4S	25,000	74–93
Messenger RNA (mRNA)	5	4S	Heterogeneous mixture	

^{a}*The S value refers to the sedimentation coefficient and is related to the size of the molecule.*

Figure 9.7

The storage and replication of biological information in DNA and its transfer via RNA to synthesize proteins that direct and maintain cellular structure and function.

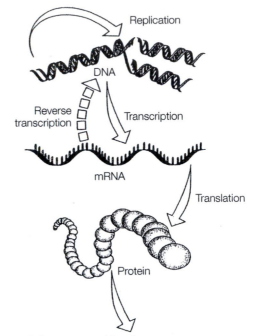

Cell structure and function
- Energy metabolism
- Synthesis and breakdown of biomolecules
- Storage and transport of biomolecules
- Muscle contraction
- Cellular communication (signal transduction)

processing before they are functional. Changes that are made in this process include protein folding into the native conformation (assisted by chaperones), protein shortening by proteases, amino acid residue modification (i.e., phosphorylation of seryl hydroxyl groups), attachment of carbohydrate residues, and addition of prosthetic groups. Finished and biologically active proteins play essential roles in the general maintenance, functioning, and structure of the cell.

The nucleic acids are among the most complex molecules that you will encounter in your biochemical studies. When the dynamic roles that are played by DNA and RNA in the life of the cell are realized, the complexity is understandable. The discovery of the structures and functions of the nucleic acids will always be considered as some of the most significant breakthroughs in our understanding of the chemistry of life.

This section has been a brief review of many important concepts in molecular biology. For more details, refer to one of the standard biochemistry textbooks listed in the Further Reading section at the end of the chapter.

➤ **Study Exercise 9.1   Complementary Bases in DNA**   A short polynucleotide strand of DNA has the following sequence of bases. Write the sequence of the complementary strand.

   5′ AGCTTACGTCC 3′

➤ **Study Exercise 9.2   Complementary Bases in RNA**   A short polynucleotide strand of double-stranded RNA has the following sequence of bases. Write the sequence of the complementary strand of RNA.

   5′ UAGGUACUUCG 3′

B.   LABORATORY METHODS FOR INVESTIGATION OF DNA AND RNA

Isolation of Chromosomal DNA

Because of the large size and the fragile nature of chromosomal DNA, it is difficult to isolate in a completely intact, undamaged form. Several isolation procedures have been developed that provide DNA in a biologically active form, but this does not mean it is completely undamaged. These preparations yield DNA that is stable, of high molecular weight, and relatively free of RNA and proteins. Here a general method will be described for the isolation of chromosomal DNA, in a relatively pure form, from microorganisms.

Designing an isolation procedure for DNA requires extensive knowledge of the chemical stability of DNA as well as its condition in the cellular environment. Figures 9.5 and 9.6 illustrate several chemical bonds in DNA that may be susceptible to cleavage during the extraction process. The experimental factors that must be considered and their effects on various structural aspects of intact DNA are outlined below.

1. **pH**
   (a) Hydrogen bonding between the complementary strands is stable between pH 4 and 10.
   (b) The phosphodiester linkages in the DNA backbone are stable between pH 3 and 12.
   (c) *N*-glycosidic bonds to purine bases (adenine and guanine) are hydrolyzed at pH values of 3 and less.

2. **Temperature**
   (a) There is considerable variation in the temperature stability of the hydrogen bonds in the double helix, but most DNA begins to unwind in the range of 80–90°C.

(b) Phosphodiester linkages and *N*-glycosidic bonds are stable up to 100°C.

3. **Ionic strength**

   (a) DNA is most stable and soluble in salt solutions. Salt concentrations of less than 0.05 *M* weaken the hydrogen bonding between complementary strands.

4. **Cellular conditions**

   (a) Before the DNA can be released, the cell wall of the organism must by lysed. The ease with which the cell wall is disrupted varies from organism to organism. In some cases (yeast), extensive grinding or sonic treatment is required, whereas in others (*Bacillus subtilis*), enzymatic hydrolysis of the cell wall is possible.

   (b) Several enzymes are present in the cell that may act to degrade DNA, but the most serious damage is caused by the deoxyribonucleases. These enzymes catalyze the hydrolysis of phosphodiester linkages.

   (c) Native DNA is present in the cell as DNA-protein complexes. Basic proteins called **histones** must be dissociated from the DNA during the extraction process.

5. **Mechanical stress on the DNA**

   (a) Gentle manipulations may not always be possible during the isolation process. Grinding, shaking, stirring, and other disruptive procedures may cause cleavage (shearing or scission) of the DNA chains. This usually does not cause damage to the secondary structure of the DNA, but it does reduce the length of the molecules.

Now that these factors are understood, a general procedure of DNA extraction from bacteria will be outlined.

Step 1. Disruption of the cell wall and release of the DNA into a medium in which it is soluble and protected from degradation The isolation procedure described here calls for the use of an enzyme, lysozyme, to disrupt the cell wall. Lysozyme catalyzes the hydrolysis of glycosidic bonds in cell wall peptidoglycans, thus causing destruction of the cell wall and allowing the release of DNA and other cellular components. The medium for solution of DNA is a buffered saline solution containing EDTA. DNA, which is ionic, is more soluble and stable in salt solution than in distilled water. The EDTA serves at least two purposes. First, it binds divalent metal ions (Ca^{2+}, Mg^{2+}, Mn^{2+}) that could form salts with the anionic phosphate groups of the DNA. Second, it inhibits deoxyribonucleases that have a requirement for Mg^{2+} or Mn^{2+}. The mildly alkaline medium (pH 8) acts to reduce electrostatic interaction between DNA and the basic histones and the polycationic amines, spermine and spermidine. The relatively high pH also tends to diminish nuclease activity and denature other proteins.

Step 2. Dissociation of the protein-DNA complexes Detergents are used at this stage to solubilize the inner membrane and disrupt the ionic interactions between positively charged histones and the negatively charged backbone of DNA. Sodium dodecyl sulfate (SDS), an anionic detergent, binds to proteins and gives them extensive anionic character. A secondary action of SDS is to act as a denaturant of deoxyribonucleases and other proteins. Also favoring dissociation of protein-DNA complexes is the alkaline pH, which reduces the positive character of the histones. To ensure complete dissociation of the DNA-protein complex and to remove bound cationic amines, a high concentration of a salt (NaCl or sodium perchlorate) is added. The salt acts by diminishing the ionic interactions between DNA and cations.

Step 3. Separation of the DNA from other soluble cellular components Before DNA is precipitated, the solution must be deproteinized. This is brought about by treatment with chloroform–isoamyl alcohol followed by centrifugation. Upon centrifugation, three layers are produced: an upper aqueous phase, a lower organic layer, and a compact band of denatured protein at the interface between the aqueous and organic phases. Chloroform causes surface denaturation of proteins. Isoamyl alcohol reduces foaming and stabilizes the interface between the aqueous phase and the organic phases where the protein collects.

The upper aqueous phase containing nucleic acids is then separated and the DNA precipitated by addition of ethanol. Because of the ionic nature of DNA, it becomes insoluble if the aqueous medium is made less polar by addition of an organic solvent. The DNA forms a threadlike precipitate that can be collected by "spooling" onto a glass rod. The isolated DNA may still be contaminated with protein and RNA. Protein can be removed by dissolving the spooled DNA in saline medium and repeating the chloroform–isoamyl alcohol treatment until no more denatured protein collects at the interface.

RNA does not normally precipitate like DNA, but it could still be a minor contaminant. RNA may be degraded during the procedure by treatment with ribonuclease after the first or second deproteinization step. Removal of RNA sometimes makes it possible to denature more protein using chloroform–isoamyl alcohol. If DNA in a highly purified state is required, several deproteinization and alcohol precipitation steps may be carried out. It is estimated that up to 50% of the cellular DNA is isolated by this procedure. The average yield is 1 to 2 mg per gram of wet packed bacterial cells.

Isolation of Plasmid DNA

Many bacterial cells contain self-replicating, extrachromosomal DNA molecules called **plasmids.** This form of DNA is closed circular, double-stranded, and much smaller than chromosomal DNA; its molecular weight ranges from 2×10^6 to 20×10^6, which corresponds to between 3000 and 30,000 base pairs. Plasmids are widely used as cloning vehicles in the preparation of recombinant DNA (see Chapter 10, Section A). Bacterial plasmids normally contain genetic information for the translation of proteins that confer a specialized and sometimes protective characteristic (phenotype) on the organism. Examples of these characteristics include enzyme systems that degrade antibiotics, and enzymes necessary for the production of antibodies and toxins. Plasmids are replicated in the cell by one of two possible modes. **Stringent replicated plasmids** are present in only a few copies and **relaxed replicated plasmids** are present in many copies, sometimes up to 200. Some relaxed plasmids continue to be produced even after the antibiotic chloramphenicol is used to inhibit chromosomal DNA synthesis in the host cell. Under those conditions, many copies of the plasmid DNA may be produced (up to 2000 or 3000), and may accumulate to 30 to 40% of the total cellular DNA.

The extensive use of plasmid DNA as a cloning vehicle often requires the isolation and characterization of plasmids. Characterization of plasmids might be necessary for any of the following reasons (see Chapter 10):

1. Construction of new recombinant DNA.

2. Analysis of molecular size by agarose gel electrophoresis.

3. Electrophoretic analysis of restriction enzyme digests and construction of a restriction enzyme map.

4. Sequence analysis of nucleotides by the Sanger or Maxam-Gilbert method.

Several methods for isolating plasmid DNA have been developed; some lead to a more highly purified product than others. All isolation methods have the same objective—separation of plasmid DNA from chromosomal DNA. Plasmid DNA has two major structural differences from chromosomal DNA.

1. Plasmid DNA is almost always extracted in a covalently closed circular form, whereas isolated chromosomal DNA usually consists of sheared linear fragments, and,

2. Plasmids are much smaller than chromosomal DNA.

The structural differences cause physicochemical differences that can be exploited to separate the two types of DNA molecules. Methods for isolating plasmid DNA fall into three major categories:

1. Methods that rely on specific interaction between plasmid DNA and a solid support. Examples are adsorption to nitrocellulose microfilters and hydroxyapatite columns.

2. Methods that cause selective precipitation of chromosomal DNA by various agents. These methods exploit the relative resistance of covalently closed circular DNA to extremes of pH, temperature, or other denaturing agents.

3. Methods based on differences in sedimentation behavior between the two types of DNA. This is the approach of choice if highly purified plasmid DNA is required.

Two widely used isolation procedures based on Method 2 are described here. Both procedures yield plasmid DNA that is sufficiently pure for size analysis by agarose gel electrophoresis and for digestion by restriction enzymes.

Method A: Separation of Plasmid DNA by Boiling (Holmes and Quigley)

The total cellular DNA must first be released by lysis of the bacterial cells. This is brought about by incubation with the enzyme lysozyme in the presence of reagents that inhibit nucleases. Chromosomal DNA is then separated from the plasmid DNA by boiling the lysis mixture for a brief period, followed by centrifugation. In contrast to closed circular plasmid DNA, linear chromosomal DNA becomes irreversibly denatured by heating and forms an insoluble gel, which sediments during centrifugation. Even though plasmid DNA may become partially denatured during boiling, the closed circular helix reforms upon cooling. The boiling serves a second purpose, that of denaturing deoxyribonucleases and other proteins.

Method B: Microscale Isolation of Plasmids by Alkaline Lysis (Birnboim)

Often it is necessary to detect and analyze plasmid DNA in a large number of small bacterial samples. One widely used microscale method is the alkaline lysis procedure. For this procedure, host bacterial cells harboring the plasmids are grown in small culture volumes (1–5 mL) or in single colonies on agar plates. The cells are lysed and their contents denatured by alkaline sodium dodecyl sulfate (SDS). Proteins and high-molecular-weight chromosomal DNA denatured under these conditions precipitate as a gel that can be centrifuged from the supernatant, which contains plasmid DNA and bacterial RNA.

Characterization of DNA

Ultraviolet Absorption

A complete understanding of the biochemical functions of DNA requires a clear picture of its structural and physical characteristics. DNA has significant absorption in the UV range because of the presence of the aromatic bases adenine, guanine, cytosine, and thymine. This provides a useful probe into DNA structure because structural changes such as helix unwinding affect the extent of absorption. In addition, absorption measurements are used as an indication of DNA purity. The major absorption band for purified DNA peaks at about 260 nm. Protein material, the primary contaminant in DNA, has a peak absorption at 280 nm. The ratio A_{260}/A_{280} is often used as a relative measure of the nucleic acid/protein content of a DNA sample. The typical A_{260}/A_{280} for isolated DNA is about 1.8. A smaller ratio indicates increased contamination by protein.

Thermal Denaturation

If DNA solutions are treated with denaturing agents (heat, alkali, organic solvents) their ultraviolet-absorbing properties are strikingly increased. Figure 9.8 shows the effect of temperature on the UV absorption of DNA. The curve is obtained by plotting $A_{260(T)}/A_{280(25°)}$ vs. temperature (T). Heating through the temperature range of 25°C to about 80°C results in only minor increases in absorption. However, as the temperature is further increased, there is a sudden increase in UV absorption followed by a constant A_{260}. The total increase in absorption is usually on the order of 40% and occurs over a small temperature range. Figure 9.8 is called a **thermal denaturation curve,** a **temperature profile,** or a **melting curve.** The temperature corresponding to the midpoint of each absorption increase is defined as T_m, the **transition temperature** or **melting temperature.** (This should not be confused with melting point, transformation of a substance from solid to liquid, as was routinely studied in organic laboratory.) Each species of DNA

Figure 9.8

A typical thermal denaturation curve for DNA. T_m is measured as the midpoint of the absorbance increase.

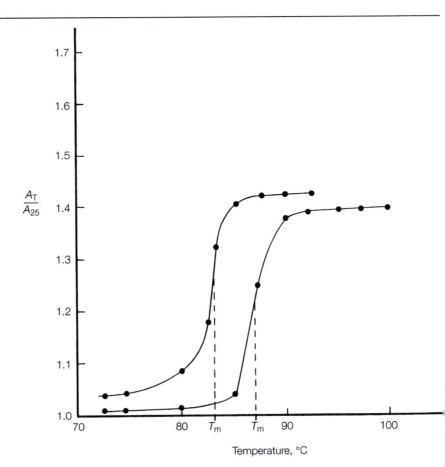

has a characteristic T_m value that can be used for identification and characterization purposes.

The origin of the absorption increase, called a **hyperchromic effect,** is well understood. The absorption changes are those that result from the transition of an ordered double-helix DNA structure to a denatured state or random, unpaired DNA strands. Native DNA in solution exists in the double helix held together primarily by hydrogen bonding between complementary base pairs on each strand (see Figure 9.6). Hydrophobic and π-π interactions between stacked base pairs also strengthen the double helix. Agents that disrupt these forces (hydrogen bonding, hydrophobic and π-π interactions) cause dissociation or unwinding of the double helix. In a random coil arrangement, base-base interactions are at a minimum; this alters the resonance behavior of the aromatic rings, causing an increase in absorption. The process of DNA dissociation can, therefore, be characterized by monitoring the UV-absorbing properties of DNA under various conditions.

Ethidium Bromide Binding and Fluorescence

Fluorescence assays are considered among the most convenient, sensitive, and versatile of all laboratory techniques (Chapter 7, Section B). However, the purine and pyridines of the nucleic acids yield only weak fluorescence spectra. The fluorescence of the dye, ethidium bromide, is enhanced about 25-fold when it interacts with DNA. Ethidium bromide, which is a relatively small planar molecule (Figure 9.9), binds to DNA by insertion between stacked base pairs (intercalation). The process of intercalation is especially significant for aromatic dyes, antibiotics, and other drugs. Some dyes, when intercalated into DNA, show enhanced fluorescence that can be used to detect DNA molecules after gel electrophoresis measurements (Chapter 6, Section C).

The concentration of RNA and DNA solutions may be determined using ethidium bromide binding. The spectrophotometric assay for DNA/RNA (Chapter 3, Section C) measures both double-stranded and single-stranded DNA because it measures the UV absorption by purine and pyrimidine bases. Ethidium bromide interaction with single-stranded DNA does not lead to increased fluorescence, so duplex DNA can be quantified in the presence of dissociated DNA. Solutions of purified DNA are commonly contaminated with RNA. Because single-stranded RNA can form hairpin loops with base pairing and duplex formation (as in tRNA), ethidium bromide also binds with enhanced fluorescence to these duplex regions of RNA. Addition of ribonuclease A results in digestion of RNA and loss of fluorescence due to ethidium binding of RNA. The amount of fluorescence lost is proportional to the concentration of RNA. The ethidium fluorescence remaining after ribonuclease treatment is directly proportional to the concentration of duplex DNA.

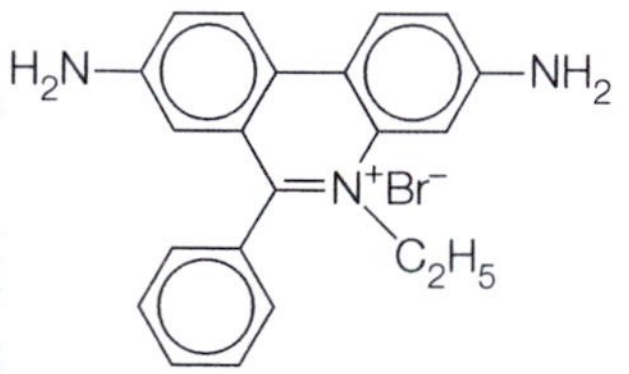

Figure 9.9

Structure of the fluorescent intercalation dye ethidium bromide.

$$F(\text{total}) = F(\text{DNA}) + F(\text{RNA})$$

where

F = fluorescence yield due to each type of nucleic acid

When the F (RNA) term is reduced to zero, the total fluorescence is a direct measurement of concentration of double-stranded DNA. The actual concentration of DNA in solution can be calculated by using a standard solution of DNA or from a standard curve.

➤ | **Study Exercise 9.3 Ethidium Bromide–Fluorescence Assay for DNA**
Solution A of DNA (unknown concentration) is known to contain double-stranded DNA, damaged single-stranded DNA, and RNA. The ethidium bromide fluorescence assay was done on the solution and the following data were collected. The fluorescence intensity values (F) were obtained on an instrument with a standardized scale of 0–100.

F of solution A = 87

F after treatment of A with ribonuclease = 80

F of a standard solution of double-stranded DNA (10 μg/mL) = 100

What is the concentration of double-stranded DNA in solution A in μg/mL?

Solution: We begin by making several assumptions. The total fluorescence of solution A is due to double-stranded DNA and some contaminating RNA. Single-stranded DNA and RNA do not bind ethidium bromide. The concentration of double-stranded DNA is directly proportional to the fluorescence reading. The fluorescence intensity due to double-stranded DNA is 80. To calculate the concentration of double-stranded DNA in solution A, compare the fluorescence intensity (80) with that obtained for the standard DNA solution:

[DNA-A]/80 = [10 μg/mL]/100

[DNA-A] = 8 μg/mL

➤ **CAUTION**

Ethidium bromide must be used with great care as it is a potent mutagen. Gloves should be worn at all times while handling solutions of the dye. When finished with experiments, used and excess solutions of the dye must be disposed as directed by your instructor or laboratory director.

Agarose Gel Electrophoresis

Several techniques for the characterization of nucleic acids have been introduced in this chapter, but the standard method for separation and analysis of plasmids and other smaller DNA molecules is agarose gel electrophoresis (Chapter 6, Section B). The method has several advantages, including ease of operation, sensitive staining procedures, high resolution, and a wide range of molecular weights ($0.6–100 \times 10^6$) that can be analyzed. As discussed in Chapter 6, the mobility of nucleic acids in agarose gels is influenced by the agarose concentrations, the molecular size of the DNA, and the molecular shape of the DNA. In general, the lower the agarose concentration in the gels, the larger the DNA that can be analyzed. A practical lower limit of agarose concentration is reached at 0.3% agarose, below which gels become too fragile for ordinary use. This lower limit of agarose in the gels allows analysis of linear double-stranded DNA within the range of 5 and 60 kilobase pairs (up to 150×10^6 in molecular weight). Gels with an agarose concentration of 0.8% can separate DNA in the range of 0.5–10 kilobase pairs, and 2% agarose are used to separate smaller DNA fragments (0.1–3 kilobase pairs).

Nucleic acids migrate in an agarose medium at a rate that is inversely proportional to their size (kilobase pairs or molecular weight). In fact, a near-linear relationship exists between mobility and the logarithm of kilobase pairs (or molecular weight) of a DNA fragment. A standard curve may be prepared by including on the gel a sample containing DNA fragments of known molecular weights (see Figure 9.10). Several types of molecular-weight ladders are commercially available to mark and help estimate the sizes and concentrations of unknown nucleic acids.

Agarose gel electrophoresis is an ideal technique for analysis of DNA fragments. In addition to the positive characteristics discussed, the technique is simple, rapid, and relatively inexpensive. Fragments that differ in molecular weight by as little as 1% can be resolved on agarose gels, and as little as 1 ng of DNA can be detected on a gel. Nucleic acids are visualized after electrophoresis by treatment with ethidium bromide or one of the less toxic SYBR dyes mentioned in Chapter 6.

Sequencing DNA Molecules

When it was determined that genetic information in DNA was coded in the form of nucleotide base sequence, the direction of research turned to the design of experimental procedures for determining the base order. Early sequence studies used the inefficient, insensitive, and labor-intensive methods of acid-, base-, and nuclease-catalyzed hydrolysis of DNA and analysis of fragments by electrophoresis.

Two sequencing methods that are both simple and inexpensive are now widely used:

1. The **Maxam-Gilbert chemical cleavage method** (also called the chemical degradation method) uses chemical modification of bases which promotes cleavage at selected phosphodiester linkages. Cleavage products, which have one common end and vary in length, are analyzed by PAGE.

Figure 9.10

Standard curve obtained by electrophoresis of λ phage DNA fragments from EcoR1 cleavage.

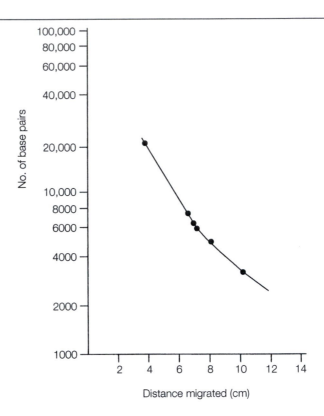

2. The **Sanger chain-terminating method** (also called the dideoxy method), uses the enzyme, DNA polymerase, coupled with the presence of all four nucleoside triphosphates and one of the 2′,3′-dideoxynucleoside triphosphates, to synthesize a DNA chain complementary to an added template. Replication stops when a dideoxynucleotide is in place on the new chain. Fragments of different length are produced and analyzed by PAGE.

In both sequencing methods, electrophoretic procedures are capable of separating nucleic acid fragments that differ in size by only one nucleotide. Detection of the fragments is done with chemiluminescent-labeling or ^{32}P-labeling of deoxynucleoside triphosphate (see Figure 6.16).

Sequencing procedures that are automated and computer controlled have now made possible the sequencing of entire genomes (Table 9.4). The first complete genome of a free-living organism, the prokaryote *Haemophilus influenzae*, was published in 1995. This genome contains 1,830,137 bases and 1740 genes. Complete sequencing of the human genome was announced in 2001. The human genome contains over 3.2 billion base pairs and about 20,000–25,000 genes.

Table 9.4		
Sequenced Genomes		
Organism	Genome Size (Kilobase Pairs)	Number of Chromosomes
Borrelia burgdorferi (carrier of Lyme disease)	1444	1
Haemophilus influenzae (human pathogenic bacterium)	1830	1
Mycobacterium tuberculosis (cause of tuberculosis)	4412	1
Escherichia coli (bacterium)	4639	1
Saccharomyces cerevisiae (yeast)	11,700	17 (in haploid chromosomes)
Drosophila melanogaster (fruit fly)	137,000	4 (in haploid chromosomes)
Oryza sativa (rice)	430,000	12
Homo sapiens (human)	3,200,000	23 (in haploid chromosomes)

Genomic sequences are now made available in repositories on the Internet. Three primary resources, combined in the International Nucleotide Sequence Database Collaboration, are listed in Table 9.5. Database resources for genome projects are also listed in Table 9.5. Readers may practice the use of the databases by working Study Problems 11 and 12.

Isolation and Characterization of RNA

The methods used for the isolation of RNA are similar to those described for DNA at the beginning of Chapter 9, Section B. However, some procedural changes must be made in the purification of RNA because of the presence of heterogeneous populations of molecules and the extra susceptibility to cleavage of the phosphodiester bonds in RNA. Here are some general comments on laboratory methods for the isolation and characterization of RNA:

- RNA molecules are naturally shorter than DNA. This makes RNA less susceptible to physical shearing so more vigorous conditions may be used for cell lysis.

- The phosphodiester bonds in DNA are stable in the pH range of 3–12; however, the presence of the 2'-OH groups in RNA makes it especially susceptible to base- and enzyme-catalyzed (RNase) cleavage. Special precautions must be taken to avoid chemical agents that degrade RNA and chemicals must be added to isolation buffers that destroy endogenous proteins that act as nucleases. The addition of chelating agents like EDTA and citrate ties up metal ions that are essential for RNase activity. Strong detergents may be added to

Table 9.5

International Databases for DNA Sequences

Organization	Website
The International Nucleotide Sequence Database Collaboration	
1. Genbank of the National Center for Biotechnology Information	http://www.ncbi.nlm.nih.gov/Genbank
2. EMBL (European Molecular Biology Laboratory)	http://www.ebi.ac.uk/
3. DDBJ (DNA Data Bank of Japan)	http://www.ddbj.nig.ac.jp/
Genome Project Databases	
1. Human Mapping Database Johns Hopkins University USA	http://gdbwww.gdb.org
2. dbEST (cDNA and partial sequences)	http://www.ncbi.nih.gov
3. Genethon Genetic maps based on repeat markers	http://www.genethon.fr
4. Whitehead Institute (YAC and physical maps)	http://www-genome.wi.mit.edu

denature proteins especially nucleases. The chemical reagent, guanidinium thiocyanate, is a useful addition as it acts as an RNase inhibitor and protein denaturant.

- It is essential for lab workers to wear plastic gloves to avoid transfer of nucleases from bare skin.

- Gradient centrifugation procedures as described in Chapter 4 may be used to isolate, purify, and characterize all types of RNA.

- Isolated RNA samples may be analyzed for purity by agarose gel electrophoresis procedures similar to those for DNA.

- Specific types of RNA may be isolated and purified using affinity chromatography. For example, eukaryotic mRNAs, which make up only about 2–5% of cellular RNA, have a poly (A) segment at their 3' ends. This RNA may be purified using an affinity matrix consisting of poly (T) or poly (U) on agarose or cellulose (see Chapter 5, Section G).

Study Problems

> 1. During the isolation of DNA, the solution becomes more viscous after treatment with lysozyme and SDS. Explain why.

> 2. How can RNA be removed from a DNA preparation?

> 3. Explain the action of SDS in disrupting the cell membrane in the extraction procedure of DNA.

> 4. A solution of purified DNA gave an absorbance of 0.55 at a wavelength of 260 nm. The absorbance was measured in a quartz cuvette with a path length of 1 cm. What is the concentration of DNA? (See Chapter 3, Section C).

5. The concentration of a purified DNA solution is 35 μg/mL. Predict the absorbance of the solution at 260 nm.

6. Why is bromophenol blue dye added to a gel-loading buffer before electrophoresis?

7. What assumptions must be made about the relative mobility of bromophenol blue dye and DNA fragments during electrophoresis?

8. What is the purpose of the ethanol precipitation step in the preparation of plasmids?

9. Predict and explain the action of each of the following conditions on the T_m of native DNA. Will the T_m be raised or lowered relative to the DNA in Tris-HCl buffer, pH 7.5?
   (a) Measure T_m in pH 12 buffer.
   (b) Measure T_m in distilled water.
   (c) Measure T_m in 50% methanol water.
   (d) Measure T_m in standard Tris-HCl, pH 7.5 buffer solution containing SDS.

10. The polyamines, 1,4-diaminobutane, 1,5-diaminopentane, spermine, and spermidine are metabolic products found in many cells. Although the specific function of these compounds has not been defined, they are known to bind to the nucleic acids. Describe how you would combine the ethidium bromide fluorescence assay described in this chapter with ligand binding plots defined in Chapter 8, Section A, to characterize the binding of the polyamines to DNA or RNA. Hint: The polyamines are found to bind to some of the same sites on the nucleic acids as does ethidium bromide.

11. Use one of the DNA sequence databases in Table 9.5 to obtain the sequence of cytosolic tRNA-phenylalanine from yeast (*Saccharomyces cerevisiae*).

12. Use one of the DNA sequence database repositories to find the DNA sequence for human insulin.

13. A solution of purified RNA gave an absorbance of 0.75 at 260 nm. The absorbance was measured in a quartz cuvette with path length of 1 cm. What is the concentration of RNA in the solution?

Further Reading

B. Alberts, et al., *Molecular Biology of the Cell,* 4th ed. (2002), Garland Press (New York).

J. Berg, J. Tymoczko, and L. Stryer, *Biochemistry,* 5th ed. (2002), Freeman (New York). Chapters 27–29 cover DNA and RNA.

H. Birnboim, *Methods Enzymol.* **100,** 243–255 (1983). "A Rapid Alkaline Extraction Method for the Isolation of Plasmid DNA."

R. Boyer, *Concepts in Biochemistry,* 2nd ed. (2002), John Wiley & Sons (New York). Chapters 10–13 cover the concepts in molecular biology.

F. Collins et al., *Nature,* **409,** 745–708 (2001). The human genome.

R. Garrett and C. Grisham, *Biochemistry,* 3rd ed. (2005), Brooks/Cole (Belmont, CA). Chapters 10 and 28–32 cover the structure and function of DNA and RNA.

J. Hammond and G. Spanswick, *Biochem. Educ.* **25,** 109–111 (1997). "A Demonstration of Genomic DNA Profiling by RAPD Analysis."

D. Holmes and M. Quigley, *Anal. Biochem.* **114,** 193–197 (1981). "A Rapid Boiling Method for the Preparation of Bacterial Plasmids."

J. Marmur, *J. Mol. Biol.* **3,** 208 (1961). "A Procedure for the Isolation of Deoxyribonucleic Acid from Microorganisms."

H. Miller, *Methods Enzymol.* **152,** 145–172 (1987). "Practical Aspects of Preparing Phage and Plasmid DNA: Growth, Maintenance, and Storage of Bacteria and Bacteriophage."

D. Nelson, and M. Cox, *Lehninger Principles of Biochemistry,* 4th ed. (2005), Freeman (New York). Chapters 8, 9, 24–28 cover DNA and RNA.

R. Rapley, Editor, *The Nucleic Acid Protocols Handbook* (2000), Humana Press (Totowa, NJ).

W. Ream and K. Field, *Molecular Biology Techniques: An Intensive Laboratory Course* (1998), Academic Press (San Diego).

J. Sambrook and D. Russell, *Molecular Cloning: A Laboratory Manual,* 3rd ed. (2001), Cold Spring Harbor Laboratory Press (Cold Spring Harbor, NY).

C. Tsai, *Computational Biochemistry* (2002), Wiley–Liss (New York). Chapter 9 covers nucleotide sequencing.

J. Venter, *Science,* **291,** 1304 (2001). The sequence of the human genome.

D. Voet and J. Voet, *Biochemistry,* 3rd ed. (2004), John Wiley & Sons (Hoboken, NJ). Chapters 5–7 and 29–35 cover DNA and RNA.

D. Voet, J. Voet, and C. Pratt, *Fundamentals of Biochemistry,* 2nd ed. (2006). John Wiley & Sons (Hoboken, NJ). Chapter 3; 23–27 cover DNA and RNA.

J. D. Watson et al., *Molecular Biology of the Gene,* 5th ed. (2004), Benjamin/Cummings (San Francisco). Excellent coverage of DNA and RNA. Review especially Chapters 2, 3, 6, 20 (methods).

http://www.cgcsci.com/cminfo.shtml

     Restriction map of pBR322 in circular and linear format. Designed by CGC Scientific, a provider of software tools for biologists.

http://www.fermentas.com

     Click on Product Profiles; then Restriction endonucleases for a listing of enzymes.

Click on individual enzymes for specific data including cleavage site and reaction conditions.

http://www.nobel.se/

Under the heading: "Educational", use "The Virtual Biochemistry Lab" to move around the lab, perform experiments on DNA and proteins, and listen to lectures. Also, read the report, "DNA–the double helix."

CHAPTER 10

MOLECULAR BIOLOGY II: RECOMBINANT DNA, MOLECULAR CLONING, AND ENZYMOLOGY

Investigation of the nucleic acids during the first three quarters of the 20th century led to the significant discovery that all of the cell's attributes and activities have their origin in DNA. For example, we now believe that the information needed to make all the proteins in a cell and organism resides in DNA. This discovery prompted biochemists and molecular biologists to develop laboratory procedures for the manipulation of DNA, because it was suspected that changes to DNA could lead to molecular and genetic changes in cells and organisms. A culmination of this research activity was the recent announcement of the sequence of the human genome (Human Genome Project, 2001). Medical scientists are now applying the newly developed lab methods to the area of gene therapy, where, in the future, dysfunctional genes that lead to disease will be replaced with functional genes that produce viable proteins.

In this chapter, we will discuss some experimental procedures that have moved us to the point where the nucleic acids, especially DNA, are now among the easiest biomolecules to work with. Some of the concepts we will encounter include **molecular cloning, recombinant DNA, restriction enzymes, nucleic acid blotting,** and the **polymerase chain reaction.**

A. RECOMBINANT DNA BIOTECHNOLOGY

Students often perceive biochemistry as a very research oriented discipline. It is easy to see how this idea can develop, because every day students listen to their instructors talk about research activities, attend research seminars, and watch their professors at work on basic research in the laboratory. Students may even participate in their own research projects. Biochemistry is very dependent on research activities that are basic and theoretical, but biochemistry has always had a very "practical" or "applied" side. Many of the early biochemical activities were encouraged and funded by companies involved in fermentation of beer and wine, baking, nutrition, and pharmaceuticals. We often refer to this practical side of biochemistry as **biotechnology,** which is defined as "the practical use of biological cells, biomolecules, and biological processes." We see many examples of the daily use of applied biochemistry:

1. Using enzymes to catalyze reactions in the industrial production of specialty chemicals.
2. The use of gene-replacement therapy to treat individuals with diseases caused by dysfunctional genes.
3. Identification of biological specimens at crime scenes.
4. Using bacterial cells to produce large quantities of proteins for medical or industrial use.
5. Using bacteria to clean up chemical waste sites.

Today, one of the very active areas of biotechnology, and also the most controversial, is the preparation of **recombinant DNA** for use in molecular cloning. This activity has bioethical considerations because it has the power to change the genetic characteristics of fundamental life forms, including humans.

Molecular Cloning

Our knowledge of DNA structure and function has increased at a gradual pace over the past 50 years, but some discoveries have had a special impact on the progress and direction of DNA research. Although DNA was first discovered in cell nuclei in 1869, it was not confirmed as the carrier of genetic information until 1944. This major discovery was closely followed by the announcement of the double-helix model of DNA in 1953. The more recent development of technology that allows manipulation by insertion of "foreign" DNA fragments into the natural, replicating DNA of an organism may well have a greater impact on the direction of DNA research than the earlier discoveries. In the short time since the first construction and replication of plasmid recombinant DNA, several scientific and medical applications of the new technology have been announced. This new era of genetic engineering has captivated the general public and scientists alike. Some of the predicted achievements in recombinant DNA research have been slow in coming; however, future workers in biochemistry and related fields will continue to see advances in the use of recombinant DNA and molecular cloning (genetic engineering).

Recombinant DNA or **molecular cloning** consists of the covalent insertion of DNA fragments from one type of cell or organism into the replicating DNA of another type of cell. Many copies of the hybrid DNA may be produced by the progeny of the recipient cells; hence, the DNA molecule is **cloned.** If the inserted fragment is a functional gene carrying the code for a specific protein, many copies of that gene and translated protein may be produced in the host cell. This process has become important for the large-scale production of proteins (insulin, somatostatin, bovine growth hormone, and other molecules) that are of value in medicine and basic science, but are difficult and expensive to obtain by other methods (see Chapter 11, Section B and Table 11.3).

Our current state of knowledge of recombinant DNA is the result of several recent biotechnological advances. The first major breakthrough was the isolation of mutant strains of *E. coli* that are not able to degrade or restrict foreign DNA. These strains are now used as host organisms for the replication of recombinant DNA. The second advance was the development of bacterial extrachromosomal DNA **(plasmids)** and bacteriophage DNA as cloning vehicles to carry the DNA to be cloned into host cells. The final, necessary advance was the development of methods for insertion of the foreign DNA into the natural vehicle. Manipulation of DNA was greatly aided by the discovery of **restriction endonucleases,** enzymes that catalyze the hydrolysis of phosphodiester bonds at selected sites in DNA (for details, see this chapter, Section B). The enzymes recognize specific base sequences (usually 4–6 bases) and catalyze hydrolytic cleavage of both DNA strands in or near the base sequence region (Reaction 10.1):

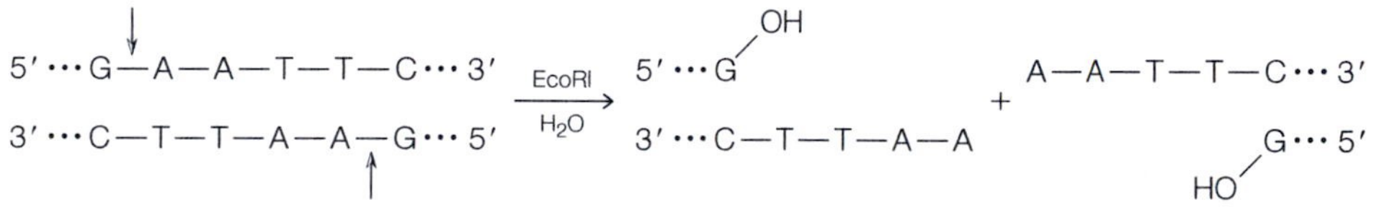

$$5'\cdots G \overset{\downarrow}{-} A - A - T - T - C \cdots 3' \quad \xrightarrow[H_2O]{EcoRI} \quad 5'\cdots G^{OH} \quad + \quad A - A - T - T - C \cdots 3'$$
$$3'\cdots C - T - T - A - A \underset{\uparrow}{-} G \cdots 5' \qquad 3'\cdots C - T - T - A - A \qquad \qquad G \cdots 5'$$

Reaction 10.1

The cleavage leads to one of two types of ends in the DNA (Figure 10.1):

1. **Cohesive (or sticky) ends,** where a few bases can remain weakly associated by hydrogen bonding, or

2. **Blunt ends** that do not overlap.

Figure 10.1

Results of the cleavage of DNA by a restriction endonuclease. Two types of ends after DNA cleavage: **A** Cohesive ends may remain weakly associated by hydrogen bonds between complementary bases. **B** Blunt ends do not overlap.

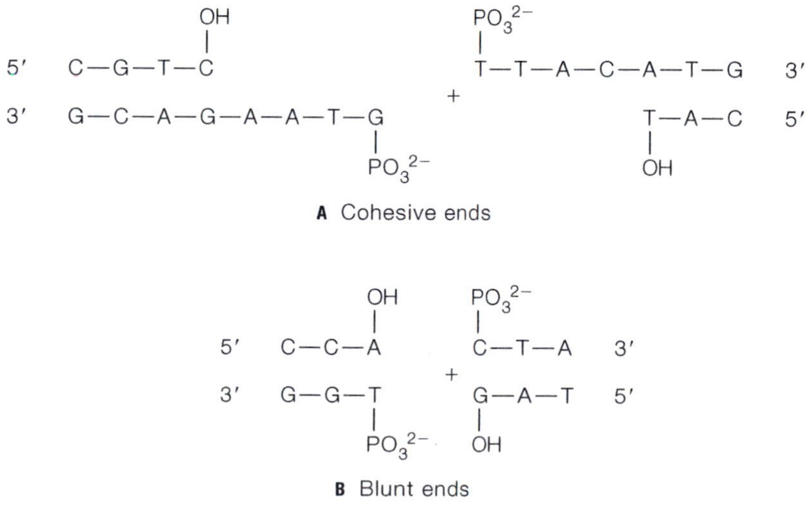

$$5'\quad C-G-T-\overset{OH}{C} \qquad \overset{PO_3^{2-}}{T}-T-A-C-A-T-G \quad 3'$$
$$3'\quad G-C-A-G-A-A-T-G \qquad + \qquad T-A-C \quad 5'$$
$$\underset{PO_3^{2-}}{} \qquad \qquad \underset{OH}{}$$

A Cohesive ends

$$5'\quad C-C-\overset{OH}{A} \qquad \overset{PO_3^{2-}}{C}-T-A \quad 3'$$
$$3'\quad G-G-T \qquad + \qquad G-A-T \quad 5'$$
$$\underset{PO_3^{2-}}{} \qquad \underset{OH}{}$$

B Blunt ends

Figure 10.2

Insertion of a DNA fragment into a linearized plasmid using DNA ligase. DNA ligase catalyzes the formation of phosphodiester bonds.

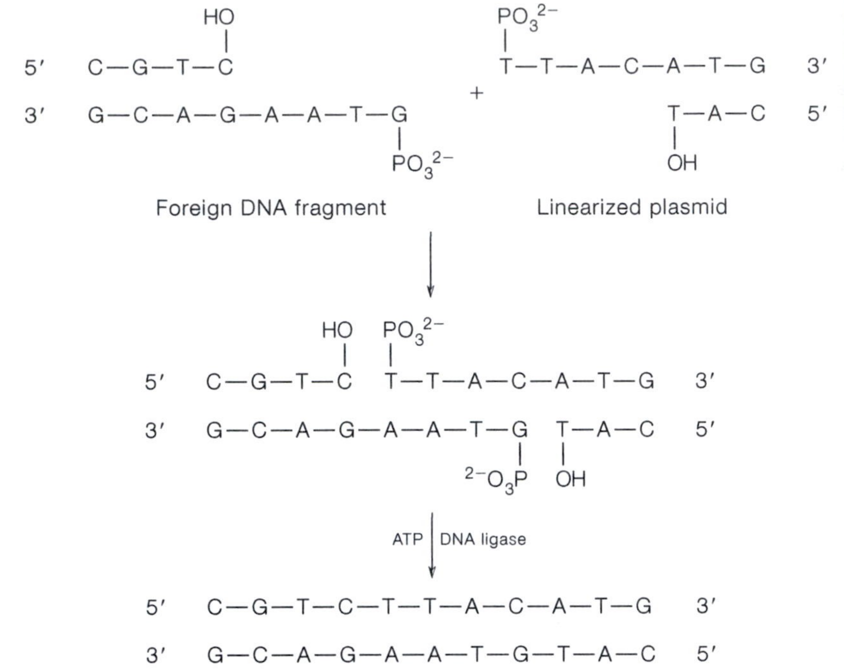

$$
\begin{array}{lll}
& \overset{\displaystyle HO}{\overset{|}{}} & \overset{\displaystyle PO_3^{2-}}{\overset{|}{}} \\
5' & C-G-T-C & T-T-A-C-A-T-G \quad 3' \\
& & + \\
3' & G-C-A-G-A-A-T-G & \qquad\qquad T-A-C \quad 5' \\
& \overset{|}{PO_3^{2-}} & \overset{|}{OH}
\end{array}
$$

Foreign DNA fragment   Linearized plasmid

$$
\begin{array}{lll}
& \overset{HO}{|}\ \ \overset{PO_3^{2-}}{|} \\
5' \quad C-G-T-C\ \ T-T-A-C-A-T-G \quad 3' \\
3' \quad G-C-A-G-A-A-T-G\ \ T-A-C \quad 5' \\
\qquad\qquad\qquad\quad \underset{^{2-}O_3P}{|}\ \ \underset{OH}{|}
\end{array}
$$

ATP | DNA ligase

$$
\begin{array}{ll}
5' \quad C-G-T-C-T-T-A-C-A-T-G \quad 3' \\
3' \quad G-C-A-G-A-A-T-G-T-A-C \quad 5'
\end{array}
$$

Restriction enzymes provide a gentle and specific method for opening (linearizing) circular vectors at predetermined sites, and for preparing fragments of DNA to be cloned.

Methods for covalent joining of the foreign DNA fragment to the vector ends and closure of the circular hybrid plasmid were then developed. The enzyme **DNA ligase,** which can catalyze the ATP-dependent formation of phosphodiester linkages at the insertion sites, is used for final closure (Figure 10.2).

Steps for Preparing Recombinant DNA

The basic steps involved in performing a recombinant DNA experiment are listed below and are outlined in Figure 10.3:

1. **Select and prepare a DNA fragment (X) that is to be incorporated into a host cell where it will be replicated and translated into a protein product**   Often the DNA fragment is a specific gene that carries the message for synthesis of a particular protein or proteins. As shown in Figure 10.3, the foreign DNA to be cloned may be prepared by chemical synthesis (usually by the polymerase chain reaction and sequence specific primers, see Section B of this chapter), action of restriction endonucleases, or transcription of mRNA catalyzed by reverse transcriptase.

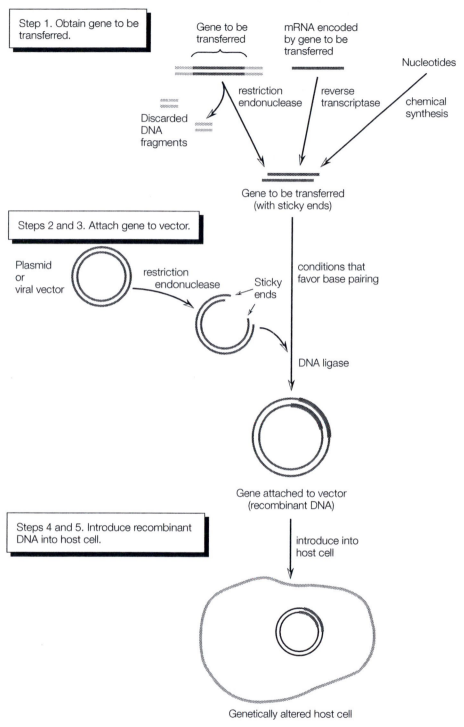

Step 1. Obtain gene to be transferred.

Gene to be transferred

mRNA encoded by gene to be transferred

Nucleotides

restriction endonuclease

reverse transcriptase

chemical synthesis

Discarded DNA fragments

Gene to be transferred (with sticky ends)

Steps 2 and 3. Attach gene to vector.

Plasmid or viral vector

restriction endonuclease

Sticky ends

conditions that favor base pairing

DNA ligase

Gene attached to vector (recombinant DNA)

Steps 4 and 5. Introduce recombinant DNA into host cell.

introduce into host cell

Genetically altered host cell

Figure 10.3

Outline of a typical experimental procedure for constructing recombinant DNA.

2. **Choose a vector or vehicle to carry X, the DNA fragment, into the host cell** The vector may be plasmid DNA, DNA from a phage, or yeast artificial chromosomes. If a circular vector is used (i.e., a plasmid), it must be linearized (broken open) in order to accept X. Ideally, the same restriction enzyme should be used to prepare X and the vehicle so there is the possibility of overlapping cohesive ends.

3. **Insert the DNA fragment, X, into the vector by overlapping cohesive ends or by modifying blunt ends using homopolymer tails** The final covalent bonds to hold X into the vector are formed by the action of **DNA ligase.** This enzyme catalyzes the ATP-dependent formation of phosphodiester bonds (Figure 10.2). The final product represents a recombinant or hybrid DNA.

4. **Introduce the hybrid DNA into a host organism (usually a bacterial cell), where it can be replicated** This process is called **transformation.** It has been demonstrated that incorporation of a hybrid plasmid or phage DNA into host *E. coli* cells is enhanced by the addition of calcium ions (Ca^{2+}).

5. **Develop a method for identifying and screening for host cells that have accepted and are replicating the hybrid DNA** This is usually accomplished by screening for antibiotic resistance (Figure 10.4).

Cloning Vectors

Four types of cloning vectors are in common use today: plasmids and bacteriophage DNA for prokaryotic cells, and yeast artificial chromosomes and baculovirus for eukaryotic cells.

Plasmids

Many bacterial cells contain self-replicating, extrachromosomal DNA molecules called **plasmids.** This form of DNA is closed circular, double-stranded, and much smaller than chromosomal DNA; their molecular weights range from 2×10^6 to 20×10^6, which corresponds to between 3000 and 30,000 base pairs. Bacterial plasmids normally contain genetic information for the translation of proteins that confer a specialized and sometimes protective characteristic (phenotype) on the organism. Examples of these characteristics are enzyme systems necessary for the production of antibiotics, enzymes that degrade antibiotics, and enzymes for production of toxins. Plasmids are replicated in the cell by one of two possible modes. **Stringent replicated** plasmids are present in only a few copies and **relaxed replicated** plasmids are present in many copies, sometimes up to 200. In addition, some relaxed plasmids continue to be produced even after the antibiotic chloramphenicol is used to inhibit chromosomal DNA synthesis in the host cell. Under these conditions, many copies of the plasmid DNA may be produced (up to 2000 or 3000) and may accumulate to 30 to 40% of the total cellular DNA.

Figure 10.4

Procedure for selecting
bacterial cells that contain hybrid
DNA.

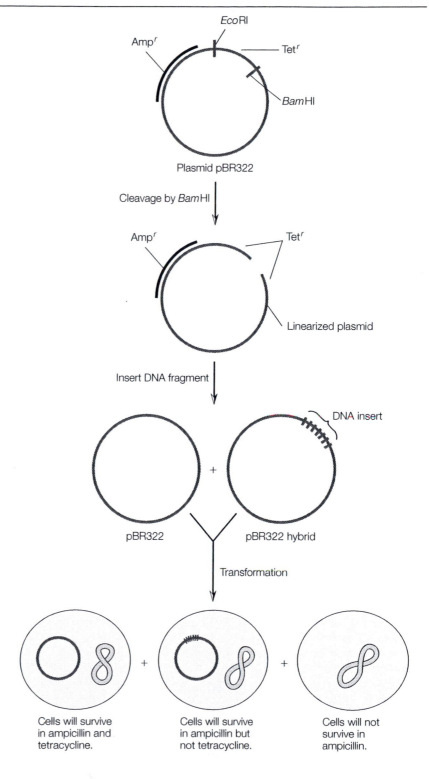

Plasmid pBR322

Cleavage by *Bam*HI

Linearized plasmid

Insert DNA fragment

pBR322 + pBR322 hybrid

Transformation

Cells will survive
in ampicillin and
tetracycline.

Cells will survive
in ampicillin but
not tetracycline.

Cells will not
survive in
ampicillin.

The ideal plasmid cloning vector has the following properties:

1. The plasmid should replicate in a relaxed fashion so that many copies are produced.

2. The plasmid should be small; then it is easier to separate from the larger chromosomal DNA, easier to handle without physical damage, and probably contains very few sites for attack by restriction endonucleases.

3. The plasmid should contain identifiable markers so that it is possible to screen progeny for the presence of the plasmid. At least two selective markers are desirable, a primary one to confirm the presence of the plasmid, and a secondary marker to confirm the insertion of foreign DNA. Resistance to antibiotics is a convenient type of marker.

4. The plasmid should have only one cleavage site for a specific restriction endonuclease. This provides only two "ends" to which the foreign DNA can be attached. Ideally, the single restriction site should be within a gene, so that insertion of the foreign DNA will inactivate the gene (called insertional marker inactivation).

E. coli Plasmids

Among the most widely used *E. coli* plasmids are derivatives of the replicon **plasmid ColE1.** This plasmid carries a resistance gene against the antibiotic colicin E. The plasmid is under relaxed control, and up to 3000 copies may be produced when the proper *E. coli* strain is grown in the presence of chloramphenicol. One especially useful derivative plasmid of ColE1 is pBR322. It has all the properties previously outlined; in addition, its nucleotide sequence of 4363 base pairs is known, and it contains several different restriction endonuclease cleavage sites where foreign DNA can be inserted. For example, pBR322 has a single restriction site for the restriction enzyme *Eco*RI. *In vitro* insertion of a foreign DNA fragment into the *Eco*RI cleavage site and incorporation into a host cell (transformation) lead to immunity of the host cell to colicin E1, but the cell is unable to produce colicin.

Useful modifications of the pBR322 plasmid are the **pUC plasmids** (developed at The University of California). The pUC plasmids have a resistance gene for tetracycline and sites for restriction endonuclease cleavage are concentrated in a region called the **multiple cloning site.**

Strains of *E. coli* are most often used as host cells because they are easy to grow and maintain. The rate of growth is exponential and can be monitored by measuring the absorbance of a culture sample at 600 nm. One absorbance unit corresponds to a cell density of approximately 8×10^8 cells/mL. Some *E. coli* strains that harbor the ColE1 plasmids are RR1, HB101, GM48, 294, SK1592, JC411Thy/ColE1, and CR34/ColE1. The typical procedure for growth and amplification of plasmids is, first, to establish the cells in normal medium for several hours. An aliquot of this culture is then used to inoculate medium containing the appropriate antibiotic. After overnight growth, a new portion of medium containing the antibiotic is inoculated with an aliquot of overnight culture. After the culture has been firmly established, a solution of chloramphenicol is added to inhibit chromosomal DNA synthesis. The ColE1 plasmids continue to replicate. The culture is then incubated for 12 to 18 hours and harvested by centrifugation.

Other Cloning Vectors

Another useful and widely used cloning vector is the DNA from **bacteriophage λ.** The DNA from λ phage is a double-stranded molecule with about 50,000 base pairs. It has many advantages as a cloning vector for prokaryotic proteins:

1. Many copies of recombinant phage DNA may be replicated in a host cell.

2. The recombinant phage DNA may be packaged as a viral particle for infecting the host bacteria.

3. Because λ phage is larger than plasmid DNA, it is possible to insert larger fragments of DNA including some smaller fragments of eukaryotic DNA. The inserted DNA may be as large as 25,000 base pairs.

The insertion of larger DNA, including eukaryotic DNA that may contain over a million base pairs, is more efficiently done with **yeast artificial chromosomes (YACs)** or the **baculovirus expression vector system (BEVS).** YACs containing DNA inserts of 100,000 to 1,000,000 base pairs may be prepared for cloning. The baculoviruses are double-stranded DNA viruses that have as their natural hosts different insect species. They are not known to infect vertebrate hosts. The BEVS may be used to express complex proteins that require special processes like proteolytic cleavage, intron splicing, chemical modifications (phosphorylation or glycosylation of hydroxyl group), formation of disulfide bonds, and proper protein folding.

B.   IMPORTANT ENZYMES IN MOLECULAR BIOLOGY AND BIOTECHNOLOGY

The Restriction Endonucleases

Bacterial cells produce many enzymes that act to degrade various forms of DNA. Of special interest are the **restriction endonucleases,** or restriction enzymes, that recognize specific base sequences in double-stranded DNA and catalyze hydrolytic cleavage of the two strands in or near that specific region. The biological function of these enzymes is to degrade or restrict foreign DNA molecules. Host DNA is protected from hydrolysis because some bases near the cleavage sites are methylated. The action of the restriction enzyme *Eco*RI is shown in Reaction 10.1.

Table 10.1

Specificity and Optimal Conditions for Several Restriction Endonucleases

Name	Recognition Sequence 5′........3′	T (°C)	pH	Tris (mM)	NaCl (mM)	MgCl$_2$ (mM)	DTT[2] (mM)
AatII	G—A—C—G—T↓C	37	7.9	20	50	10	10
AluI	A—G↓C—T	37	7.5	10	50	10	10
BalI	T—G—G↓C—C—A	37	7.9	6	–	6	6
BamHI	G↓G—A—T—C—C	37	8.0	20	100	0.7	1
BclI	T↓G—A—T—C—A	60	7.5	10	50	10	1
EcoRI	G ↓A—A—T—T—C	37	7.5	10	100	10	1
HaeII	Pu[1]—G—C—G—C↓Py[1]	37	7.5	10	50	10	10
HindIII	A↓A—G—C—T—T	37–55	7.5	10	60	10	1
HpaI	G—T—T↓A—A—C	37	7.5	10	50	10	1
MseI	T↓T—A—A	37	7.9	10	50	10	1
NotI	G—C↓G—G—C—C—G—C	37	7.9	10	150	10	–
SalI	G↓T—C—G—A—C	37	8.0	10	150	10	1
ScaI	A—G—T↓A—C—T	37	7.4	10	100	10	1
TaqI	T↓C—G—A	65	8.4	10	100	10	10

[1]Pu = *a purine base*; Py = *a pyrimidine base*.

[2]DTT = *dithiothreitol*.

The site of action of *Eco*RI is a specific hexanucleotide sequence. Two phosphodiester bonds are hydrolyzed (see arrows), resulting in fragmentation of both strands. Note the twofold rotational symmetry feature at the recognition site and the formation of cohesive ends. The weak base pairing between the cohesive ends is not sufficient to hold the two fragments together.

Several hundred restriction enzymes have been isolated and characterized. Nomenclature for the enzymes consists of a three-letter abbreviation representing the source (*Eco* = *E. coli*), a letter representing the strain (R), and a roman numeral designating the order of discovery. *Eco*RI is the first to be isolated from *E. coli* (strain R) and characterized. Table 10.1 lists several other restriction enzymes, their recognition sequence for cleavage, and optimum reaction conditions. A useful Web site for the restriction enzymes is REBASE: rebase.neb.com/rebase/rebase.htm.

Applications of Restriction Enzymes

Restriction enzymes are used extensively in nucleic acid chemistry. They may be used to cleave large DNA molecules into smaller fragments that are more amenable to analysis. For example, λ phage DNA, a linear, double-stranded molecule of 48,502 base pairs (molecular weight 31 × 10⁶), is cleaved into six fragments by *Eco*RI or into more than 50 fragments by *Hinf* I (*Haemophilus influenzae,* serotype f). The base sequence recognized by a restriction enzyme is likely to occur only a very few times in any particular DNA molecule; therefore, the smaller the DNA molecule, the fewer the number of specific sites. The λ phage DNA is cleaved into 0 to 50 or more fragments, depending on the restriction enzyme used, whereas larger bacterial or animal DNA will most likely have many recognition sites and be cleaved into hundreds of fragments. Smaller DNA molecules, therefore, have a much greater chance of producing a unique set of fragments with a particular restriction enzyme. It is unlikely that this set of fragments will be the same for any two different DNA molecules, so the fragmentation pattern is unique and can be considered a "fingerprint" of the DNA substrate. The fragments are readily separated and sized by agarose gel electrophoresis.

Restriction endonucleases are also valuable tools in the construction of hybrid DNA molecules. Several restriction enzymes act on bacterial plasmid vehicles that have only a single site of cleavage. This linearizes the circular plasmid and allows the insertion of a foreign DNA fragment. For example, the popular plasmid vehicle pBR322 has a single restriction site for *Bam*HI (*Bacillus amyloliquefaciens,* H) that is within the tetracycline resistance gene. The enzyme not only opens the plasmid for insertion of a DNA fragment, but also destroys a phenotype; this fact aids in the selection of transformed bacteria.

Restriction enzymes can be used to obtain physical maps of DNA molecules. Important information can be obtained from maps of DNA, whether the molecules are small and contain only a few genes, such as viral or plasmid DNA, or large and complex, such as bacterial or eukaryotic chromosomal DNA molecules. An understanding of the genetics, metabolism, and regulation of an organism requires knowledge of the precise arrangement of its genetic material. The construction of a **restriction enzyme map** for a DNA molecule provides some of this information. The map displays the sites of cleavage by restriction endonucleases and the number of fragments obtained after digestion with each enzyme.

A map is constructed by first digesting plasmid DNA with restriction endonucleases that yield only a few fragments. Each digest of DNA obtained with a single nuclease is analyzed by agarose gel electrophoresis, using standards for molecular weight determination. These are referred to as the primary digests. Second, each of the primary digests is treated with a series of additional restriction enzymes, and the digests are analyzed by agarose gel electrophoresis. The restriction map is then constructed by combining the various fragments by trial and error and logic, much as in sequencing a protein by proteolytic digestion by several enzymes and searching for overlap regions in the fragments. In the case of the restriction endonuclease digests, two characteristics of the DNA fragments are known: the approximate molecular weights (from electrophoresis) and the nature of the fragment ends (from the known selectivity of the individual

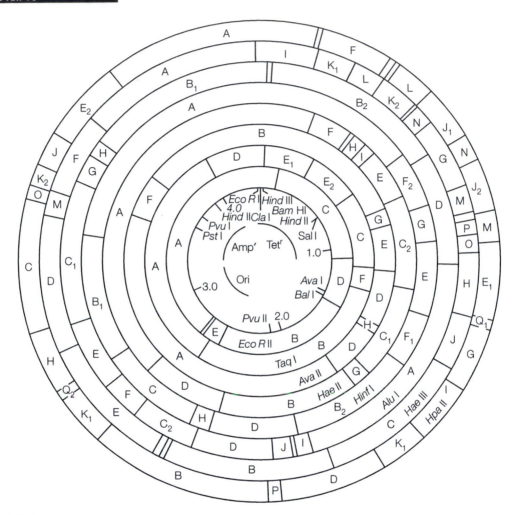

Figure 10.5

A restriction enzyme map for the plasmid pBR322. *From* Molecular Cloning: A Laboratory Manual, *by T. Maniatis, E. Fritsch, and J. Sambrook, Cold Spring Harbor Laboratory (Cold Spring Harbor, NY), 1982.*

restriction enzymes). The fragments may also be sequenced by the Sanger or Maxam-Gilbert method. A restriction enzyme map for the plasmid pBR322 is shown in Figure 10.5.

Practical Aspects of Restriction Enzyme Use

Restriction enzymes are heat-labile and expensive biochemical reagents. Their use requires considerable planning and care. Each restriction nuclease has been examined for optimal reaction conditions in regard to specific pH range, buffer composition, and incubation temperature. This information for each enzyme is readily available from the commercial supplier of the enzyme or from the literature. Table 10.1 gives important reaction information for several enzymes. The temperature range and pH optima for most restriction nucleases are similar (37°C, 7.5–8.0); however, optimal buffer composition is variable. Typical buffer components are Tris, NaCl, $MgCl_2$, and a sulfhydryl reagent (β-mercaptoethanol or dithiothreitol). Proper reaction conditions are crucial for optimal reaction rate; but, more important, changing reaction conditions have been shown to alter the specificity of some restriction enzymes.

Although restriction enzymes are very unstable reagents, they can be stored at -20°C in buffer containing 50% glycerol. They are usually prepared in an appropriate buffer and shipped in packages containing dry ice.

Disposable gloves should be worn when you are handling the enzyme container. Remove the enzyme from the freezer just before you need it. Store the enzyme in an ice bucket at all times when it is outside the freezer. The enzyme should never be stored at room temperature. Because of high cost, digestion by restriction enzymes is carried out on a microscale level. A typical reaction mixture will contain about 1 μg or less of DNA and 1 unit of enzyme in the appropriate incubation buffer. One unit is the amount of enzyme that will degrade 1 μg of λ phage DNA in 1 hour at the optimal temperature and pH. The total reaction volume is usually between 20 and 50 μL. Incubation is most often carried out at the recommended temperature for about 1 hour. The reaction is stopped by adding EDTA solution, which complexes divalent metal ions essential for nuclease activity.

Reaction mixtures from restriction enzyme digestion may be analyzed directly by agarose gel electrophoresis. This technique combines high resolving power and sensitive detection to allow the analysis of minute amounts of DNA fragments.

The Polymerase Chain Reaction

Producing multiple copies of a particular DNA fragment need not always require the tedious and time-consuming procedures of molecular cloning as described in Section A of this chapter. If at least part of the sequence of a DNA fragment is known, it is possible to make many copies using the **polymerase chain reaction (PCR).** The PCR method was conceived by Kary Mullis (Nobel Prize in chemistry in 1993) during a moonlit drive through the mountains of northern California.

Fundamentals of the PCR

The fundamentals of this relatively simple process are outlined here with the use of a fragment of double-stranded DNA. The DNA template is divided into five regions designated I to V in Figure 10.6. The complementary regions

Figure 10.6

The polymerase chain reaction has three steps: **A** Strand separation by heating at 95°C. **B** Hybridization of the primers. **C** Extension of the primers by DNA synthesis. Segments are labeled I, II, III, IV, V on the original DNA strand and I′, II′, III′, IV′, V′ on the complementary strand. Primer II has diagonal lines and primer IV′ has dots. Newly synthesized DNA is shown in black.

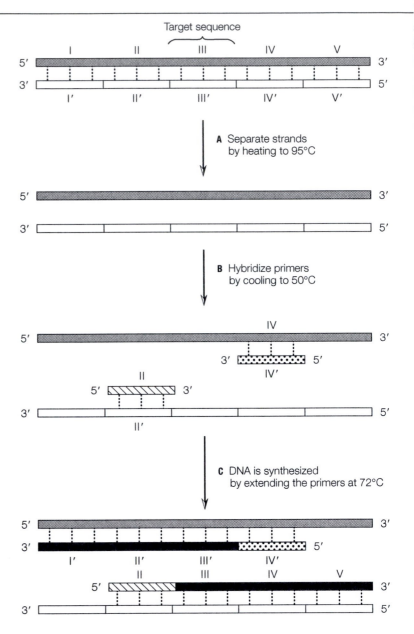

are I′ to V′. The region targeted for amplification is III (and III′), which can be reproduced by PCR if the nucleotide sequences of the flanking regions, II and IV, are known. Experimental requirements for the PCR include:

1. Two synthetic oligonucleotide primers of about 20 base pairs each, which are complementary to the flanking sequences II and IV, and also have similar binding ability.

2. A heat-stable DNA polymerase.

3. The four deoxyribonucleoside triphosphates, dATP, dGTP, dCTP, and dTTP.

The PCR is performed in cycles of three steps. Each cycle consists of:

1. Denaturation to achieve template DNA strand separation is done by heating a mixture of all components at 95°C for about 15 s.

2. Abrupt cooling of the mixture to the range of 37–55°C allows the primers to hybridize with the appropriate flanking regions. The primers are oriented on the template so their 3′ ends are directed toward each other. Synthesis of DNA extends across regions III and III′. Note the dual roles played by the complementary oligonucleotides: they locate the starting points for duplication of the desired DNA segment and they serve as 3′-hydroxy primers to initiate DNA synthesis. A large excess of primers to DNA is added to the reaction mixture to favor hybridization and prevent reannealing of the DNA template strands.

3. Synthesis of the targeted DNA is catalyzed by *Taq* DNA polymerase. The temperature is raised to 72°C to enhance the rate of the polymerization reaction. The enzyme extends both primers, producing two new strands of DNA, II-III-IV and II′-III′-IV′. The chosen polymerase is from a thermophilic bacterium, *Thermus aquaticus,* an organism originally discovered in a Yellowstone National Park hot spring. Other useful polymerases have been isolated from bacteria found in geothermal vents on the ocean floor. These enzymes are heat stable, so the reaction can be carried out at a high temperature, leading to a high rate of DNA synthesis. The DNA synthesis reaction is usually complete in about 30 s.

The usefulness of the PCR lies in the three steps–denaturation, hybridization, and DNA synthesis–that can be repeated many times simply by changing the temperature of the reaction mixture. Each newly synthesized strand of DNA can serve as a template, so the target DNA concentration increases at an exponential rate. In a process consisting of 20 cycles, the amplification for the DNA fragment is about a millionfold. Thirty cycles, which can be completed in 1–3 hours, provide a billionfold amplification. Theoretically, one could begin with a single molecule of target DNA and produce in 1 hour enough DNA for the Sanger dideoxy chain-terminating sequence procedure. Another benefit of the PCR is its simplicity. Instruments called **thermocyclers** are commercially available that allow the laboratory technician to mix the reagents and insert the reaction mixture in the cycler, which then automatically repeats the reaction steps by changing temperatures.

The PCR has become a routine tool in basic research carried out in universities, hospitals, and pharmaceutical companies. It is especially useful to amplify small amounts of DNA for sequencing. It was used widely in the

Human Genome Project, which in 2001 announced the sequence of the 3 billion base pairs in the human genome.

Applications of the PCR

Diagnostic Medicine

PCR may be used to detect the presence of infectious bacteria and viruses in an individual even before symptoms begin to appear. HIV infection is usually confirmed using tests that detect serum antibody proteins made against the viral proteins. However, there may be a period of at least 6 months between infection and when the anti-HIV antibodies reach a detectable concentration. PCR can be used to detect proviral DNA. An especially important application is the identification of HIV-infected infants. The antibody-based test cannot be done on infants because they have maternal antibodies for a period up to 15 months after birth. Other medical applications of PCR include early detection of tuberculosis and cancers, especially leukemias, and analysis of small samples of amniotic fluid to detect genetic abnormalities in fetuses.

Forensics

Recovery of fingerprints from a crime scene has been a long-standing and well-accepted method in forensics. A new procedure, **DNA fingerprinting,** is being developed and achieving widespread use. DNA fingerprinting is the biochemical analysis of DNA in biological samples (semen, blood, saliva) remaining at a crime scene. Since every individual possesses a unique hereditary composition, each individual has a characteristic phenotype that is reflected in his or her DNA sequence. DNA fingerprinting attempts to show these genetic variations that are found from person to person. These differences can be detected by analysis of an individual's DNA. Two experimental methods are currently used in DNA fingerprinting:

1. **Restriction Fragment Length Polymorphisms (RFLPs)** In this method, DNA samples are digested with restriction enzymes. The resulting fragments are separated by gel electrophoresis, blotted onto a membrane, hybridized to a DNA probe, and analyzed by autoradiography (see Section C on blotting, this chapter). Results vary from individual to individual (except identical twins, who have identical DNA) because of their differences in DNA sequence. Since each person's DNA has a unique sequence pattern, the restriction enzymes cut differently and lead to different-sized fragments.

2. **PCR-based Analysis** The second method of DNA fingerprinting involves the use of PCR-amplified DNA. Allele-specific oligonucleotide (ASO) primers are used for amplification of specific sequences. PCR-based DNA fingerprinting has many advantages over RFLP methods, including speed, simplicity, no requirements for radioactive probes, and greater sensitivity. DNA from a single hair or minute samples of blood, saliva, and semen can be analyzed.

C.   NUCLEIC ACID BLOTTING

Many of the procedures for isolating and characterizing DNA and RNA molecules discussed in Chapters 9 and 10 depend on the separation of nucleic acids by an electrophoresis step followed by some kind of analysis of the gel bands. Although electrophoresis does provide excellent separation of molecules, the method gives no indication as to the identity of the molecules at each band. As described in Chapter 6, Section C, **membrane blotting techniques** are now available for functional analysis of the molecules at each electrophoresis band.

The first blotting technique was reported by E. Southern in 1975. Using labeled complementary DNA probes, he searched for certain nucleotide sequences among DNA molecules blotted from the gel. This technique of detecting DNA-DNA hybridization is called **Southern blotting.** The general blotting technique has now been extended to the transfer and detection of specific RNA (for example mRNA) using labeled complementary DNA probes **(Northern blotting)** and the transfer and detection of proteins that react with specific antibodies (**Western blotting;** Chapter 6, Section C).

Blotting techniques have many applications, including mapping the genes responsible for inherited diseases by using restriction fragment length polymorphisms (RFLPs), screening collections of cloned DNA fragments (DNA libraries), and DNA fingerprinting for analysis of biological material remaining at the scene of a crime.

Study Problems

1. Explain the action of ampicillin as an inducer of plasmid replication.

2. How does chloramphenicol inhibit protein synthesis?

3. Why is a wavelength of 600 nm used to measure growth of bacteria? Could other wavelengths be used?

4. Why is polyacrylamide gel electrophoresis not suitable for analysis of most plasmid DNA?

5. How does the addition of an EDTA solution stop a restriction enzyme reaction?

6. Why is polyacrylamide electrophoresis not suitable for analysis of restriction enzyme fragments of most DNA?

7. Which of the following base sequences are probably not recognition sites for cleavage by restriction endonucleases? Why not?
   (a) 5′ GAATTC 3′
   (b) 5′ CATTAG 3′
   (c) 5′ CATATG 3′
   (d) 5′ CAATTG 3′

8. How many DNA fragments result from the action of restriction enzyme *Hae*II on the plasmid pBR322?

9. Assume the reaction digest from Problem 8 is analyzed by agarose gel electrophoresis. Draw an electrophoresis gel and show the approximate locations of each fragment.

10. Use the REBASE Web site (rebase.neb.com) to search for information about the restriction enzymes listed below. Find for each enzyme the recognition sequence, the cleavage site, and determine if the DNA cut ends are blunt or cohesive.

    a) *Apa*I   b) *Spm*I   c) *Alu*I

Further Reading

W. Backman, *Trends Biochem. Sci.* **26,** 268–270 (2001). "The Advent of Genetic Engineering."

J. Berg, J. Tymoczko, and L. Stryer, *Biochemistry,* 5th ed. (2002). Freeman (New York). Chapter 5 covers molecular cloning.

R. Boyer, *Concepts in Biochemistry,* 2nd ed. (2002), John Wiley & Sons (New York). Chapter 13 covers biotechnology.

J. Brooks, *Methods in Enzymol.* **152,** 113–129 (1987). "Properties and Uses of Restriction Endonucleases."

T. Brown, *Gene Cloning and DNA Analysis,* 4th ed. (2001), Blackwell Science (Malden, MA).

F. Collins et al., *Nature,* **409,** 708–745 (2001). The human genome.

C. Dieffenbach and G. Dveksier, Editors, *PCR Primer: A Laboratory Manual,* 2nd ed. (2003), Cold Spring Harbor Laboratory Press (Cold Spring Harbor, NY).

R. Garrett and C. Grisham, *Biochemistry,* 3rd ed. (2005), Brooks/Cole (Belmont, CA). Chapter 12 covers molecular cloning.

C. Hardin, et al., Editors, *Cloning, Gene Expression and Protein Purification: Experimental Procedures and Process Rationale* (2001), Oxford University Press (New York).

K. Mullis, *Sci. Am.* **263(4),** 56–65 (1990). "The Unusual Origin of the Polymerase Chain Reaction."

D. Nelson, and M. Cox, *Lehninger Principles of Biochemistry,* 4th ed. (2005), Freeman (New York). Molecular cloning is covered in Chapter 9.

D. Nicholl, *An Introduction to Genetic Engineering,* 2nd ed. (2003), Cambridge University Press (Cambridge).

L. Perez-Pons and E. Querol, *Biochem. Educ.* **24,** 54–56 (1996). A laboratory experiment illustrating basic principles of DNA cloning and molecular biology techniques.

W. Ream and K. Field, *Molecular Biology Techniques: An Intensive Laboratory Course* (1998), Academic Press (San Diego).

J. Sambrook and D. Russell, *Molecular Cloning: A Laboratory Manual,* 3rd ed. (2001), Cold Spring Harbor Laboratory Press (Cold Spring Harbor, NY).

C. Tsai, *Computational Biochemistry* (2002), Wiley-Liss (New York). Chapters 9 and 10 cover nucleotide sequencing and recombinant DNA.

J. Venter, *Science,* **291,** 1304 (2001). "The Sequence of the Human Genome."

D. Voet and J. Voet, *Biochemistry,* 3rd ed. (2004), John Wiley & Sons (Hoboken, NJ). Chapters 5, 6, and 29–34 cover DNA and RNA.

J. Watson et al., *Molecular Biology of the Gene,* 5th ed. (2004), Benjamin/Cummings (San Francisco). Chapter 3, 6, 20 (Methods).

L. Yount, Editor, *Cloning* (2000), Greenhaven Press (San Diego).

B. Zimm, *Trends Biochem. Sci.* **24,** 121–123, (1999). "One Chromosome: One DNA Molecule."

http://www.nobel.se/

Under the heading "Educational," use "The Virtual Biochemistry Lab" to perform experiments on DNA and proteins, and listen to lectures. Also read the report, "DNA–the Double Helix."

http://www.fermentas.com

Click on Product Profiles; then Restriction Endonucleases for a listing of enzymes. Click on individual enzymes for specific data, including cleavage site and reaction conditions.

http://rebase.neb.com/

A resource site for restriction enzymes.

PROTEIN PRODUCTION, PURIFICATION, AND CHARACTERIZATION

Proteins are often considered the workhorses among biomolecules, as they are responsible for the general maintenance and daily functioning of the cell and organism. The proteins are extremely versatile functional molecules, as they serve roles in biological catalysis (enzymes), immune response (antibodies), structural integrity, muscle contraction, cell regulation, and storage/transport processes. Because of their roles as essential biomolecules, there has always been intense interest in the extraction of proteins from their natural sources and in their characterization in terms of structure and biological function. In this chapter, we will focus on the purification procedures that are used to obtain native protein molecules that may be further characterized. The new discipline of **proteomics,** the study of the thousands of proteins expressed in cells and organisms and how the proteins interact with each other, has brought a renewed interest in the development of methods for their isolation and characterization. Because the normal concentration of proteins in cells is quite low, there is a strong interest in designing methods to enhance their production. Recombinant DNA procedures now play a major role in the production of relatively large quantities of scarce proteins for basic scientific study and for medical and industrial applications. In the medical technique of **gene therapy,** a hybrid DNA containing the genes for the desired protein is transferred into a host cell, which acts as a factory to produce the desired protein.

A. PROCEDURES FOR THE PURIFICATION OF PROTEINS

Protein purification is an activity that has occupied the time of biochemists and molecular biologists throughout the last two centuries. In fact, a large percentage of the biochemical literature is a description of

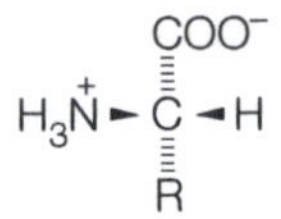

Figure 11.1

The general structure of the zwitterionic form of an L-amino acid. R represents the side chain.

how specific proteins have been separated from thousands of other proteins and other biomolecules in tissues, cells, and biological fluids. Biochemical investigations of all biological processes require, at some time, the isolation, purification, and characterization of a protein. In contrast to the repetitive procedures for isolation of DNA and RNA (Chapter 9, Section B), there is no single technique or sequence of techniques that can be followed to isolate and purify all proteins. It is sometimes still necessary to proceed by trial and error. Fortunately, the experiences and discoveries of thousands of scientists have been combined so that, today, a general, systematic approach is available for protein isolation and purification.

Composition of Proteins

All proteins found in nature are constructed by amide linkages between α-amino acids. The amino acids are selected from a group of 20 molecules that have common structural characteristics. They each have at least one carboxyl group and at least one amino group (Figure 11.1). The distinctive physical, chemical, and biological properties associated with an amino acid are the result of the side chain, R group, which is unique for each amino acid. A list of the 20 common amino acids and their three-letter and one-letter abbreviations are given in Table 11.1. If you are not yet familiar with

Table 11.1

Abbreviations of the 20 Common Amino Acids Found in Proteins

Name	Abbreviation	
	One-letter	Three-letter
Glycine	G	Gly
Alanine	A	Ala
Valine	V	Val
Leucine	L	Leu
Isoleucine	I	Ile
Methionine	M	Met
Phenylalanine	F	Phe
Proline	P	Pro
Serine	S	Ser
Threonine	T	Thr
Cysteine	C	Cys
Asparagine	N	Asn
Glutamine	Q	Gln
Tyrosine	Y	Tyr
Tryptophan	W	Trp
Aspartate	D	Asp
Glutamate	E	Glu
Histidine	H	His
Lysine	K	Lys
Arginine	R	Arg

the general structures and properties of amino acids and proteins, refer to your general biochemistry text.

➤   **Study Exercise 11.1**   Draw the structure of the peptide: Asp-Phe-Ala-Lys-Trp.

Amount of Protein versus Purity of Protein versus Expense

The procedure selected for protein purification is dependent primarily on how much of the protein is needed and how pure the protein must be; in other words, what is the purpose for isolating the protein? In terms of amount, one usually thinks on two scales: **preparative,** isolation of larger quantities (several milligrams or grams) to use for further characterization or structure and function studies; or **analytical,** isolation of microgram or milligram quantities in order to make a few precise measurements such as molecular weight, sequencing, or structure determination by NMR or X-ray diffraction.

In terms of purity, a protein sample is considered pure when it contains a single type of protein. It is not practically possible to obtain a protein in 100% purity, but most studies can be done on samples that are 90–95% pure. It often is not worth the extra time and reagent expense to purify a protein greater than 90% unless absolutely necessary. In general, the best purification scheme is one that yields the maximum (or appropriate) amount of the protein, of desired purity, with a minimum amount of time and expense. The following discussion outlines the basic steps available to develop a protein purification scheme (Table 11.2).

Table 11.2

Typical Sequence for the Purification of a Protein

1. Develop an assay for the desired protein.
2. Select the biological source of the protein including use of recombinant DNA techniques.
3. Release the protein from the source and solubilize it in an aqueous buffer system.
4. Fractionate the cell components by physical methods (centrifugation).
5. Separate the biomolecules present by differential solubility.
6. Apply chromatographic procedures.
    (a) Gel filtration chromatography
    (b) Ion exchange chromatography
    (c) Affinity chromatography
7. Isoelectric focusing, if desired.
8. Determine purity by electrophoresis or HPLC.

Basic Steps in Protein Purification

Development of Protein Assay

First and foremost in any protein purification scheme is the development of an assay for the protein. This procedure, which may have a physical, chemical, or biological basis, is necessary in order to determine quantitatively and/or qualitatively the presence of the specific protein. During the early stages of the purification, the particular protein desired must be distinguished from thousands of other proteins present in the crude cell homogenate. The desired protein may be less than 0.1% of the total protein. It should be obvious that the general protein assays (spectrophotometric, Bradford, Lowry, etc., Chapter 3, Section B) are not useful as assays for a specific protein, because they simply indicate the total protein present. If the desired protein is an enzyme, the obvious assay will be based on biological function, that is, a measurement of the enzymatic activity after each isolation–purification step. If the protein to be isolated is not an enzyme or if the biological activity is unknown, physical or chemical methods must be used. One of the most useful analytical methods is electrophoresis. If an antibody has been prepared against the desired protein, it may be analyzed by electrophoresis and monitored during purification using Western blotting (Chapter 6, Section C).

Source of the Protein

We next consider the selection of the source from which the protein will be isolated. If the objective is simply to obtain a certain quantity of a protein for further study, you would choose a source that contains large amounts of the protein. A possible choice is an organ from a large animal that can be obtained from a local slaughterhouse. Microorganisms are also good sources because they can be harvested in large quantities. If, however, you desire a specific protein from a specific type of organism, cell, tissue, cell organelle, or biological fluid, the source is limited. For example, if one desires to study the enzymes involved in glycolysis in alfalfa plant, one would begin with leaves of the plant.

If the desired protein is known to be located in an organelle or subcompartment of the cell, partial purification of the protein is achieved by isolating the organelle by fractional centrifugation (Chapter 4, Section C, Figure 4.11). If the protein is, instead, in the soluble cytoplasm of the cell, it will remain dissolved in the final supernatant obtained after centrifugation at $100,000 \times g$.

With the introduction of new techniques in recombinant DNA research, it is now possible to synthesize relatively large quantities of specific proteins in selected host cells. One begins by preparing the recombinant DNA–inserting the specific gene that contains the code for the desired protein, along with appropriate promoters and markers, into a plasmid or other vector. The recombinant DNA may then be transferred to a host cell (*E. coli,* or other bacteria, or yeast) and the host cell is allowed to grow and to synthesize the desired protein (molecular cloning, Chapter 10, Section A). This new procedure has several advantages compared to the traditional method of isolating and purifying proteins as described in the above paragraphs and in Table 11.2. Some of the benefits of using recombinant DNA include:

- Overexpression of the protein is possible, which produces relatively large amounts; this is especially helpful when dealing with scarce proteins, or with proteins for commercial or medical applications.

- Conditions can be controlled so there is less degradation of the desired protein by proteases and less contamination by other biomolecules. This makes it easier to isolate and purify the protein from the host cell. Some steps in Table 11.2 may be eliminated.

- With the use of genetic engineering techniques, including mutation, it is possible to produce modified proteins, i.e., proteins with one or more amino acid changes.

Because the techniques for purifying a protein using recombinant DNA procedures are somewhat different compared to traditional isolation and purification, a special section will be devoted to protein production by cloning (Section B of this chapter).

Preparation of the Crude Extract

Once the protein source has been selected, the next step is to release the desired protein from its natural cellular environment and solubilize it in aqueous solution. This calls for disruption of the cell membrane without damage to the cell contents. Proteins are relatively fragile molecules and only gentle procedures are allowed at this stage. The gentlest methods for cell breakage are osmotic lysis, gentle grinding in hand-operated glass homogenizers, and disruption by ultrasonic waves. These methods are useful for "soft tissue" as found in green plants and animals. When dealing with bacterial cells, where rigid cell walls are present, the most effective methods are grinding in a mortar with an inert abrasive such as sand or alumina, treatment with lysozyme, and/or a detergent. Lysozyme is an enzyme that catalyzes the hydrolysis of polysaccharide moieties present in cell walls. When very resistant cell walls are encountered (yeast, for example), the French press must be used. Here the cells are disrupted by passage, under high pressure, through a small hole.

Osmotic lysis consists of suspending cells in a solution of relatively high ionic strength. This causes water inside the cell to diffuse out through the membrane. The cells are then isolated by centrifugation and transferred to pure water. Water rapidly diffuses into the cell, bursting the membrane.

Another alternative, gentle **grinding,** is best accomplished with a glass or Teflon homogenizer. This consists of a glass tube with a close-fitting piston. Several varieties are shown in Figure 11.2A, B. The cells are forced against the glass walls under the pressure of the piston, and the cell components are released into an aqueous solution.

Ultrasonic waves, produced by a sonicator, are transmitted into a suspension of cells by a metal probe (Figure 11.2C). The vibration set up by the ultrasonic waves disrupts the cell membrane, releasing the cell components into the surrounding aqueous solution.

A less gentle, but widely used, solubilization device is the common electric blender. This method may be used for plant or animal tissue, but it is not effective for disruption of bacterial cell walls. A tool that is more scientifically designed is the **rotor stator homogenizer** (Figure 11.2D). The stator is a hollow tube and the rotor, attached to the stator, is a rapidly turning knife blade. Cells are torn apart by the turbulence and shear generated by the rotor.

The most recently developed homogenizer is the so-called **cell bomb,** which makes use of high pressure and decompression to disrupt cells. In this technique, a gas, usually nitrogen, helium, or air, is forced into cells under high pressure. When the pressure is released, expanding bubbles of the gas (nebulization) rupture the cell membrane. Cell bombs are available with pressure ranges of 250 to 25,000 psi. Cell bombs have an advantage over sonicators in that they cause no temperature increase that can denature proteins. A disadvantage of cell bombs is the potential for explosions because of the high-pressure conditions.

Since so many cell disruption methods are available, the experimenter must, by trial and error, find a convenient method that yields the maximum quantity of the protein with minimal damage to the molecules.

Stabilization of Proteins in a Crude Extract

Continued stabilization of the protein must always be considered in protein purification processes. While the protein is inside the cell, it is in a highly regulated environment. Cell components in these surroundings are protected against sudden changes in pH, temperature, or ionic strength and against oxidation and enzymatic degradation. Once the cell wall/membrane barrier is destroyed, the protective processes are no longer functional and degradation of the desired protein is likely to begin. An artificial environment that mimics the natural one must be maintained so that the protein retains its chemical integrity and biological function throughout the purification procedure. What factors are important in maintaining an environment in which proteins are stable? Although numerous factors must be considered, the following are the most critical:

1. **Ionic strength and polarity** The standard cellular environment is, of course, aqueous; because of the presence of inorganic salts, though, its

Figure 11.2

Tools for the preparation of a crude cell extract. **A** Hand-operated homogenizers, *courtesy of Ace Glass, Inc. Vineland, NJ.* **B** Homogenizer with electric motor, *courtesy of Sargent-Welch/VWR; www.sargentwelch.com/.* **C** Ultrasonic homogenizer, *courtesy of BioLogics, Inc., www.biologics-inc.com/.* **D** Rotor-stator, *courtesy of Omni International, Inc. www.omni-inc.com/.*

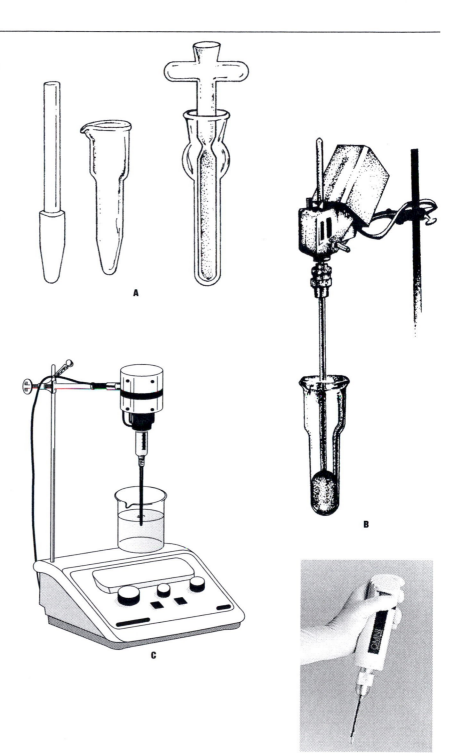

ionic strength is relatively high. Addition of KCl, NaCl, or $MgCl_2$ to the cell extract may be necessary to maintain this condition. Proteins that are normally found in the hydrophobic regions of cells (i.e., membranes) are generally more stable in aqueous environments in which the polarity has been reduced by the addition of 1 to 10% glycerol or sucrose.

2. **pH** The pH of a biological cell is controlled by the presence of natural buffers. Since protein structure is often irreversibly altered by extremes in pH, a buffer system must be maintained for protein stabilization. The importance of proper selection of a buffer system cannot be overemphasized. The criteria that must be considered in selecting a buffer have been discussed in Chapter 3. For most cell homogenates at physiological pH values, Tris and phosphate buffers are widely used.

3. **Metal ions** The presence of metal ions in mixtures of biomolecules can be both beneficial and harmful. Metal ions such as Na^+, K^+, Ca^{2+}, Mg^{2+}, and Fe^{3+} may actually increase the stability of dissolved proteins. Many enzymes require specific metal ions for activity. In contrast, heavy metal ions such as Ag^+, Cu^{2+}, Pb^{2+}, and Hg^{2+} are deleterious, particularly to proteins that depend on sulfhydryl groups for structural and functional integrity. The main sources of contaminating metals are buffer salts, water used to make buffer solutions, and metal containers and equipment. To avoid heavy-metal contamination, you should use high-purity buffers, glass-distilled water, and glassware specifically cleaned to remove extraneous metal ions (Chapter 1, Section C). If metal contamination still persists, a chelating agent such as ethylenediaminetetraacetic acid (EDTA, 1×10^{-4} M) may be added to the buffer.

4. **Oxidation** Many proteins are susceptible to oxidation. This is especially a problem with proteins having free sulfhydryl groups, which are easily oxidized and converted to disulfide bonds. A reducing environment can be maintained by adding mercaptoethanol, cysteine, or dithiothreitol (1×10^{-3} M) to the buffer system.

5. **Proteases** Many biological cells contain degradative enzymes (proteases) that catalyze the hydrolysis of peptide linkages. In the intact cell, functional proteins are protected from these destructive enzymes because the enzymes are stored in cell organelles (lysosomes, etc.) and released only when needed. The proteases are freed upon cell disruption and immediately begin to catalyze the degradation of protein material. This detrimental action can be slowed by the addition of specific protease inhibitors such as phenylmethylsulfonyl fluoride or certain bioactive peptides. These inhibitors are to be used with extreme caution because they are potentially toxic.

6. **Temperature**   Many of the above conditions that affect the stability of proteins in solution are dependent on chemical reactions. In particular, metal ions, oxidative processes, and proteases bring about chemical changes in proteins. It is a well-accepted tenet in chemistry that lower temperatures slow down chemical processes. We generally assume that proteins are more stable at low temperatures. Although there are a few exceptions to this, it is fairly common practice to carry out all procedures of protein isolation under reduced-temperature conditions (0 to 4°C).

After cell disruption, gross fractionation of the properly stabilized, crude cell homogenate may be achieved by physical methods, specifically centrifugation. Figure 4.11, outlines the stepwise procedure commonly used to separate subcellular organelles such as nuclei, mitochondria, lysosomes, and microsomes.

Separation of Proteins Based on Solubility Differences

Proteins are soluble in aqueous solutions primarily because their charged and polar amino acid residues are solvated by water. Any agent that disrupts these protein-water interactions decreases protein solubility because the protein-protein interactions become more important. Protein-protein aggregates are no longer sufficiently solvated, and they precipitate from solution. Because each specific type of protein has a unique amino acid composition and sequence, the degree and importance of water solvation vary from protein to protein. Therefore, different proteins precipitate at different concentrations of precipitating agent. The agents most often used for protein precipitation are (1) inorganic salts, (2) organic solvents, (3) polyethylene glycol (PEG), (4) pH, and (5) temperature.

The most commonly used inorganic salt, ammonium sulfate, is highly solvated in water and actually reduces the water available for interaction with protein. As ammonium sulfate is added to a protein solution, a concentration of salt is reached at which there is no longer sufficient water present to maintain a particular type of protein in solution. The protein precipitates or is "salted out" of solution. The concentration of ammonium sulfate at which the desired protein precipitates from solution cannot be calculated, but must be established by trial and error. In practice, ammonium sulfate precipitation is carried out in stepwise intervals. For example, a crude cell extract is treated by slow addition of dry, solid, high-purity ammonium sulfate in order to achieve a change in salt concentration from 0 to 25% in ammonium sulfate, is gently stirred for up to 60 minutes, and is subjected to centrifugation at $20,000 \times g$. The precipitate that separates upon centrifugation and the supernatant are analyzed for the desired protein. If the protein is still predominantly present in the supernatant, the salt concentration is increased from 25 to 35%. This process of ammonium sulfate addition and centrifugation is continued until the desired protein is salted out.

Organic solvents also decrease protein solubility, but they are not as widely used as ammonium sulfate because they sometimes denature proteins. They are thought to function as precipitating agents in two ways: (1) by dehydrating proteins, much as ammonium sulfate does, and (2) by decreasing the dielectric constant of the solution. The organic solvents used (which, of course, must be miscible with water) include methanol, ethanol, and acetone.

A relatively new method of selective protein precipitation involves the use of nonionic polymers. The most widely used agent in this category is polyethylene glycol. The polymer is available in a variety of molecular weights ranging from 400 to 7500. The biochemical literature reports successful use of different sizes, but lower-molecular-weight polymer has been shown to be the most specific. The principles behind the action of PEG as a protein precipitating agent are not completely understood. Possible modes of action include (1) complex formation between protein and polymer and (2) exclusion of the protein from part of the solvent (dehydration) followed by protein aggregation and precipitation. The most attractive advantage of PEG is its ability to fractionate proteins on the basis of size and shape as in gel filtration.

Finally, changes in pH and temperature have been used effectively to promote selective protein precipitation. A change in the pH of the solution alters the ionic state of a protein and may even bring some proteins to a state of charge neutrality. Charged protein molecules tend to repel each other and remain in solution; however, neutral protein molecules do not repel each other, so they tend to aggregate and precipitate from solution. A protein is least soluble in aqueous solution when it has no net charge, that is, when it is isoelectric. This characteristic can be used in protein purification, since different proteins usually have different isoelectric pH values (Chapter 6, Section B). An increase in temperature generally causes an increase in the solubility of solutes. This general rule is followed by most proteins up to about 40°C. Above this temperature, however, many proteins aggregate and precipitate from solution. If the protein of interest is heat stable and still water soluble above 40°C, a major step in protein purification can be achieved because most other proteins precipitate at these temperatures and can be removed by centrifugation.

Selective Techniques in Protein Purification

After gross fractionation of proteins, as discussed above, and listed in Table 11.2 Step 5, more refined methods with greater resolution can be attempted. These methods, in order of increasing resolution, are gel filtration, ion-exchange chromatography, affinity chromatography, and isoelectric focusing. Since the basis of protein separation is different for each of these techniques, it is often most effective and appropriate to use all of the techniques in the order given.

Chromatographic methods for protein purification have been discussed in Chapter 5. However, we must consider another important topic, preparation of protein solutions for chromatography. Fractionation of heterogeneous protein mixtures by inorganic salts, organic solvents, or PEG usually precedes column chromatography. The presence of the precipitating agents will interfere with the later chromatographic steps. In particular, the presence of ammonium sulfate increases the ionic strength of the protein solution and damps the ionic protein–ion-exchange resin interactions. Procedures that are in current use to remove undesirable small molecules from protein solutions include ultrafiltration (Chapter 3), dialysis (Chapter 3), and gel filtration (Chapter 5).

Chromatography is now and will continue to be the most effective method for selective protein purification. The more conventional methods (ion exchange and gel filtration) rely on rather nonspecific physicochemical interactions between a stationary support and protein molecule. These techniques, which separate proteins on the basis of net charge, size, and polarity, do not have a high degree of specificity. The highest level of selectivity in protein purification is offered by affinity chromatography—the separation of proteins on the basis of specific biological interactions (Chapter 5).

To some individuals, especially those who recall their experiences in the organic chemistry laboratory, the ultimate step in purification of a molecule is crystallization. The desire to obtain crystalline protein has long been strong, and many proteins have been crystallized. However, there is a common misconception that the ability to form crystals of a protein ensures that the protein is homogeneous. For many reasons (entrapment of contaminants within crystals, aggregation of protein molecules, etc.), the ability to crystallize a protein should not be used as a criterion of purity. The interest in protein crystals today has its origin in the demand for X-ray crystallographic analysis of protein structure (Chapter 7, Section E).

➤ | **Study Exercise 11.2**   List at least three types of chromatography useful for protein purification and identify the basis of separation for each type.

B.   PRODUCTION OF PROTEINS BY EXPRESSION OF FOREIGN GENES

It is now possible to produce large quantities of selected proteins by using recombinant DNA technology. In general, a hybrid DNA is prepared by inserting the gene to be expressed (carries the code for the desired protein) into a vector (plasmid, etc., see Chapter 10, Section A) and transferring it into a host cell where the desired protein is synthesized. These procedures

Table 11.3

Recombinant Proteins and their Use

Protein	Use
Human insulin	Treatment of diabetes
Human somatotropin (growth hormone)	Treatment of dwarfism
Bovine somatotropin (BST)	Enhances milk production in dairy cattle
Porcine somatotropin (PST)	Enhances growth in pigs
Pulmozyme (DNase)	Treatment of cystic fibrosis
Tissue plasminogin activator (TPA)	Treatment of heart attack, stroke victims; dissolves blood clots
Erythropoietin	Stimulates erythrocyte production in anemia
Interferons	Treatment of cancers and antiviral agent
Atrial natriuretic factor	Reduces high blood pressure
Leptin	Treatment of obesity
Hepatitis B vaccine	Treatment of hepatitis
Herceptin	Monoclonal antibody to treat metastatic breast cancer
Amylase and cellulase	Removal of carbohydrate precipitates from fruit juice

have been used in the production of hundreds of proteins for scientific, medical, agricultural, and industrial purposes. Table 11.3 lists some of the important protein products now made by recombinant DNA methods.

Gene Expression in Prokaryotic Organisms

Bacterial cells, especially *E. coli*, have been widely used as host cells for the large-scale production of proteins. Special requirements must be considered while preparing the recombinant DNA containing the "foreign" gene. The cloned structural gene that contains the message for the desired protein is incorporated into an expression vector that has all the necessary transcriptional and translational control sequences (Figure 11.3). If a relaxed plasmid is used as vector (Chapter 10, Section A), and the proper signals for gene expression present, it is possible to overexpress the desired protein so that its concentration may reach up to 40% of the host bacteria's total cellular protein. Although this provides a high yield of the expressed protein, the bacterial cells often concentrate the foreign protein into insoluble **inclusion bodies** in which the protein is present in a denatured, aggregated form. The protein may be renatured by isolating the inclusion bodies by centrifugation and incubating them with a denaturing agent (6 M guanidinium ion or urea) and then slowly removing the denaturing agent by ultrafiltration or dialysis (Chapter 3, Section D). At first thought, the presence of the desired protein in the inclusion bodies seems to be a disadvantage in the production of the protein, but it turns out sometimes to be a benefit, as it may actually save time in further purification steps. Because the inclusion bodies contain the desired protein in the absence of most other cellular proteins, this serves as

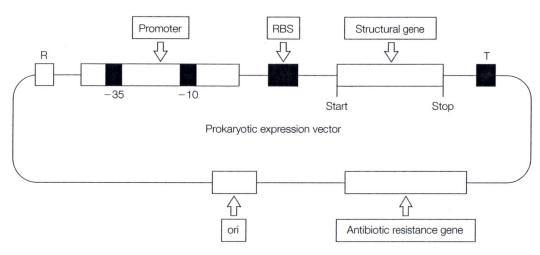

Figure 11.3

Components of a typical recombinant DNA prokaryotic expression vector. The structural gene contains the coding sequence for the protein. Essential components include the promoter (P), ribosome binding site (RBS), regulatory gene (R), and transcription terminator (T). ori refers to the origin of replication.

a major purification step. The renatured protein often is ready for further purification using the final steps of chromatography in Table 11.2. However, the renaturing process is very slow, and percent recovery of the protein may be as low as 50%.

Host Cell Secretion of Protein

The formation of inclusion bodies and other technical problems of gene expression can be avoided if the coding gene is genetically engineered to induce the host cell to secrete the protein into the periplasmic space (the area between the cell wall and the plasma membrane). The protein can be released from this space using the procedure of osmotic lysis (Chapter 11, Section A). These procedures greatly facilitate purification of the protein as they avoid the preparation of a crude cell homogenate where the protein is in a mixture of all other cellular molecules. The host bacterial cell can be coaxed into secreting the desired protein by incorporating a signal sequence for protein targeting into the expression vector.

The yield of biologically active protein can also be increased by reducing its intracellular degradation. Because the host bacterial cell recognizes the cloned protein as "foreign" (and, indeed, some proteins are toxic to the bacterium), it attempts to remove the protein by protease-catalyzed hydrolysis. Degradation by protease action is lessened if the protein is secreted. Protein degradation can also be reduced if the expression vector is modified to add or label the desired protein with a protease-resistant sequence of amino acids.

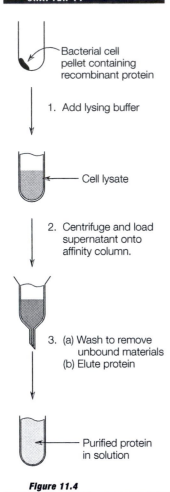

Bacterial cell
pellet containing
recombinant protein

1. Add lysing buffer

Cell lysate

2. Centrifuge and load
supernatant onto
affinity column.

3. (a) Wash to remove
unbound materials
(b) Elute protein

Purified protein
in solution

Figure 11.4

Purification of a His-tagged
protein using immobilized metal
affinity chromatography.
See text for details.
www.clontech.com/clontech.

Many of the above problems, including inclusion body formation, can be minimized or even eliminated by tagging the desired protein with a protein of bacterial origin. This is accomplished by inserting into the vector a DNA fragment carrying the code for a bacterial protein. The bacteria can be tricked into recognizing the conjugated protein (bacterial protein covalently linked to the desired protein) as "native." The "label" is often added to the N-terminus of the cloned protein. Such modified proteins are called **hybrid** or **fusion proteins.** The label may be removed from the desired protein, after extraction, by using a proteolytic enzyme.

Histidine-Tagged Proteins

A unique approach to eliminating some of the problems outlined above and also facilitating protein purification is to program the expression vector to add a string of six or more histidine residues to the N-terminus of the desired protein. The histidine residues (along with the protein) will bind to an affinity chromatography column containing nickel ions immobilized to a nitrilotriacetic acid solid support. Protein purification is achieved by passing the cell lysate (prepared by treatment of the host cells with lysozyme and/or a detergent) directly through the affinity column (Figure 11.4; see immobilized metal-ion affinity chromatography, Chapter 5, Section G). Equipment and reagents for these procedures are available in kit form from QIAGEN Inc. and Clontech Laboratories Inc. A modified version of this procedure, the MagneHis Protein Purification System is available from Promega. The MagneHis system involves the use of paramagnetic precharged nickel particles to isolate His-tagged protein directly from a crude cell lysate. In all procedures described above, the His-tagged protein is recovered from the solid support by eluting with imidazole, which binds to the nickel ions and displaces the protein.

Other peptide fusion labels that facilitate protein purification include:

• **Polyarginine**–a series of arginine residues is added to the C-terminus of the protein. This makes the protein highly basic and thus may be purified using a cation exchange resin.

• **Flag**–a short amino acid sequence (Asp-Tyr-Lys-Asp-Asp-Asp-Asp-Lys–) is attached to the N-terminus of the protein. The protein may then be purified by immunoaffinity chromatography with the use of a commercially available monoclonal antibody that recognizes the hydrophilic peptide sequence.

An extremely useful application of recombinant DNA technology is the ability to modify protein structures by changing the identity of amino acid residues. This provides a powerful tool for monitoring which amino acid residues are important in a protein's biological function. Amino acid changes may be made by a technique called **site-directed mutagenesis,** where an altered gene carrying the message for different amino acids is inserted into a recombinant DNA.

Gene Expression in Eukaryotic Cells

Expression of a eukaryotic gene in a bacterial host cell often leads to many technical problems. Among these problems are:

1. Bacterial cells are unable to carry out gene splicing processes that remove introns from the eukaryotic mRNA.

2. Bacterial cells do not recognize and respond to eukaryotic transcriptional and translational regulatory elements.

3. Bacterial cells do not have the enzyme systems to direct post-translational modifications of proteins as eukaryotic cells are capable of doing. These protein modifications include chemical changes such as addition of carbohydrates (glycosylation), formation of correct disulfide bonds, proteolytic cleavage of inactive precursors, phosphorylation of amino acid hydroxyl groups (tyrosine, serine, etc.), and other changes that assist the protein to fold into its native conformation and become biologically active.

The challenges of incorporating all of these messages into an expression vector for bacterial cells are great, if not insurmountable. The best approach is to use eukaryotic cells such as yeast as hosts.

The many challenges presented above, such as preparing complex recombinant DNA with all the proper signals and messages, and maintaining and using bacterial and eukaryotic host cell cultures, have encouraged scientists to search for more efficient systems for producing proteins.

cDNA

One procedure for making recombinant proteins that is growing in popularity is the use of **complementary DNA (cDNA).** All of the proteins in a cell are made from their specific mRNAs by translation, so it would seem logical to isolate the mRNA associated with a specific protein and then use it to make that protein. However, mRNA is too difficult to purify and too unstable to use directly for protein synthesis. Instead, the mRNA is converted into a DNA form using the RNA-viral enzyme system **reverse transcriptase (RT).** RT is able to read an RNA template and catalyze the synthesis of a complementary strand of DNA (cDNA). (RT is the enzyme that retroviruses use to copy their genes, originally in the form of RNA, into DNA, in order to "take over" a cell or organism.) Researchers now can isolate and purify the mRNA associated with the desired protein, transform it into cDNA with RT, and insert the cDNA into a suitable vector for cloning.

Cell-Free Systems

Another promising alternative for protein production is to use cell-free transcription and translation systems for expression of the recombinant DNA. Two *in vitro* systems currently used are wheat-germ extracts and rabbit reticulocyte lysates. The *in vitro* systems require the addition of the isolated mRNA fragment that contains the coded message for the protein, as well as amino acids, tRNAs, and ribosomes.

C.   PROTEIN CHARACTERIZATION

Now that the sought-after protein has been obtained in a highly purified state, it is usually suitable for further characterization and study. The characterization techniques applied to the protein depend on what is already known about the molecule and what kind of protein it is. One is, of course, interested in knowing the biological role the protein plays in the organism; however, those studies need to wait until more is known about its physical and chemical properties. Here is an outline of procedures one should apply to the protein for general characterization and for specific analysis. References are made to coverage of the characterization procedures in this book (See Appendix I for table of Web sites).

- Concentration of protein solutions–general assays like the Bradford, Lowry, spectrophotometric, etc. (Chapter 3, Section B).
- Purity of protein–PAGE and SDS-PAGE (Chapter 6, Section B), HPLC (Chapter 5, Section F), capillary electrophoresis (Chapter 6, Section B), isoelectric focusing (Chapter 6, Section B), and 2D electrophoresis (Chapter 6, Section B).
- Molecular weight, subunit structure–gel filtration chromatography (Chapter 5, Section E), SDS-PAGE (Chapter 6), and MS (Chapter 7, Section D).
- Identity and structure determination (including cofactors, coenzymes, metal ions, carbohydrates, etc.)–UV/VIS spectroscopy (Chapter 7, Section A), NMR spectroscopy (Chapter 7, Section C), MS (Chapter 7, Section D), X-ray diffraction (Chapter 7, Section E), and protein databases (Table 11.4).
- Sequence of amino acids in the protein and nucleotides in the coding gene–Web sites, databases (Table 11.4 and Chapters 2, 9, and 10). For analysis of the amino acid sequence (primary structure), see this chapter, Section D.

Table 11.4

Protein Database Resources

Name	Description	URL
Swiss-Prot	European protein sequence database	http://expasy.hcuge.ch/sprot/sprot-top.html
TREMBL	European protein sequence database	http://www.ebi.ac.uk/pub/databases/trembl
PIR	U.S. protein information resource	http://www.pir.georgetown.edu
PDB	Brookhaven protein structure database	http://www.rcsb.org
NRL-3D	Protein structure database	http://www.gdb.org/Dan/proteins/nrl3d.html

- Biological function:

  Enzyme–substrate specificity, kinetics, inhibition, regulation (Chapter 8, Section B).

  Transport protein–ligand binding (fluorescence spectroscopy, Chapter 7, Section B and Chapter 8, Section A).

  Other type of protein–procedures depend on function.

- Proteomics: identity of unknown proteins by amino acid composition and prediction of structure (Web sites and databases in Table 11.4 and Chapter 2), analysis by 2D electrophoresis (Chapter 6, Section B), interactions with other proteins–MALDI-MS (Chapter 7, Section D).

D.   DETERMINATION OF PRIMARY STRUCTURE

Structural elucidation of natural macromolecules is an important step in understanding the relationships between the chemical properties of a biomolecule and its biological function. The techniques used in organic structure determination (NMR, IR, UV, and MS; Chapter 7) are quite useful when applied to biomolecules, but the unique nature of natural molecules also requires the application of specialized chemical procedures. Proteins, polysaccharides, and nucleic acids are polymeric materials, each composed of hundreds or sometimes thousands of monomeric units (amino acids, monosaccharides, and nucleotides, respectively). But there is only a limited number of these types of units from which the biomolecules are synthesized. For example, only 20 different amino acids are found in proteins, but these different amino acids may appear several times in the same protein molecule. Therefore, the structure of a peptide or protein can be recognized only after the amino acid composition *and* sequence have been determined.

Amino Acid Composition

The amide bonds in peptides and proteins can be hydrolyzed in a strong acid or base. Treatment of a peptide or protein under either of these conditions yields a mixture of the constituent amino acids. Neither acid- nor base-catalyzed hydrolysis of a protein leads to ideal results because both tend to destroy some constituent amino acids. Acid-catalyzed hydrolysis destroys tryptophan and cysteine, causes some loss of serine and threonine, and converts asparagine and glutamine to aspartic acid and glutamic acid, respectively. Base-catalyzed hydrolysis leads to destruction of serine, threonine, cysteine, and cystine, and also results in racemization of the free amino acids. Because acid-catalyzed hydrolysis is less destructive, it is often the method of choice. The hydrolysis procedure consists of dissolving the protein sample in aqueous acid, usually 6 M HCl, and heating the solution in a sealed, evacuated vial at 100°C for 12 to 24 hours.

Now that the free amino acids present in a peptide or protein have been released by hydrolysis, they must be separated and identified. The most versatile, economical, and convenient techniques for separation are based on chromatographic methods (see Chapter 5 for a review). Earlier workers relied on paper chromatography (PC) and thin-layer chromatography (TLC); however, more sensitive techniques are now available. Automated ion-exchange chromatography (amino acid analyzers), capillary electrophoresis (CE), and high-performance liquid chromatography (HPLC) are now powerful tools for the qualitative and quantitative analysis of amino acids and derivatives. Because there is still some demand for rapid, qualitative, routine analysis of amino acids, thin-layer and paper chromatographic methods are still being developed and improved. Amino acids and derivatives may be analyzed directly by PC and TLC without further derivatization. Several support materials are available, but most analyses are carried out on silica gel or cellulose. The free amino acids can be detected on the developed chromatographic plates by reaction with ninhydrin. A pink-purple color is obtained for all amino acids except proline. A yellow color develops with proline.

High-performance liquid chromatographic techniques have been applied with success to the analysis of phenylthiohydantoin, 2,4-dinitrophenyl, and dansyl amino acid derivatives.

If only very small samples of amino acids are available for analysis, fluorescence is used for detection. One of the most sensitive methods of microanalysis is based on the reaction of amino acids with o-phthalaldehyde and β-mercaptoethanol (Equation 11.1). The isoindole derivative is fluorescent and amounts as small as 10^{-12} mole may be measured.

o-Phthalaldehyde   Amino acid β-Mercaptoethanol   *Equation 11.1*

Isoindole derivative
of amino acid

Another important derivatizing reagent for amino acids in 9-fluorenyl-methyl chloroformate (FMOC) (Equation 11.2). This reagent has the advantages that (1) the reaction with amino acids is very fast, occurring in less than 1 min; (2) the products are stable for long periods of time; (3) the derivatives can be separated by reversed-phase column chromatography and capillary electrophoresis procedures; and (4) the products are fluorescent for easy detection.

FMOC   Amino acid   N-FMOC
                                            derivative

Equation 11.2

Sequential Analysis

Sequential analysis of amino acids in purified peptides and proteins is best initiated by analysis of the terminal amino acids. A peptide has one amino acid with a free α-amino group (NH$_2$-terminus) and one amino acid with a free α-carboxyl group (COOH-terminus). Many chemical methods have been developed to selectively tag and identify these terminal amino acids.

N-Terminal Analysis

NH$_2$-terminal amino acid analysis is achieved by the use of (1) 2,4-dinitrofluorobenzene (Sanger reagent), (2) 1-dimethylaminonaphthalene-5-sulfonyl chloride (dansyl chloride), or (3) phenylisothiocyanate (Edman reagent). Figure 11.5 shows the structures of these reagents. Although the chemistry is different for each of these reagents, the same general concept is used. These compounds react with the NH$_2$-terminal amino acid of a peptide or protein to produce covalent derivatives that are stable to acid-catalyzed hydrolysis. The process is illustrated in Figure 11.6. The mixture of modified amino acid and free amino acids is separated, and the NH$_2$-terminal amino acid is identified by thin-layer, paper, high-performance liquid chromatography, or capillary electrophoresis. The first NH$_2$-terminal method to be widely used was based on the Sanger reagent, which produces yellow-colored 2,4-dinitrophenyl (DNP) derivatives of NH$_2$-terminal amino acids. The Sanger method has several disadvantages, including poor yield of the DNP-peptide derivative, low sensitivity of analysis, and instability of some DNP-amino acids during acid hydrolysis. The dansyl chloride method has largely replaced the Sanger method because very sensitive fluorescence techniques may be used for detection and analysis of the dansyl amino acid derivatives and the derivatives are more stable during acid hydrolysis.

Figure 11.5

Reagents useful in amino-terminal analysis of proteins.

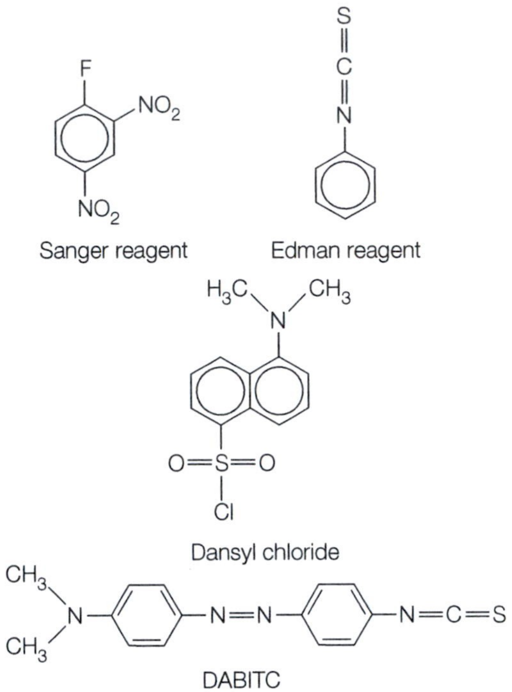

Sanger reagent   Edman reagent

Dansyl chloride

DABITC

Figure 11.6

Scheme to show analysis of amino-terminal residue of protein.

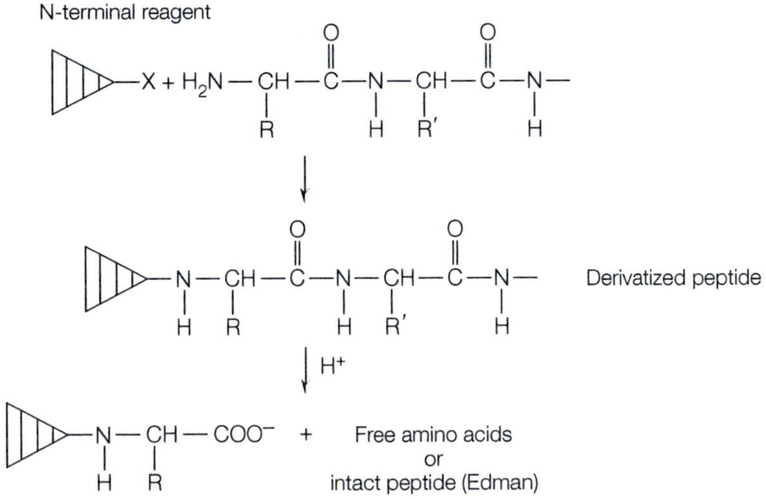

N-terminal reagent

Derivatized peptide

Free amino acids
or
intact peptide (Edman)

Edman Sequence Method

Even more versatile than the dansyl method is the Edman method (Figure 11.7). The NH$_2$-terminal amino acid is removed as its phenylthiohydantoin (PTH) derivative under anhydrous acid conditions, while all other amide bonds in

Figure 11.7

The Edman method for
analysis of proteins.

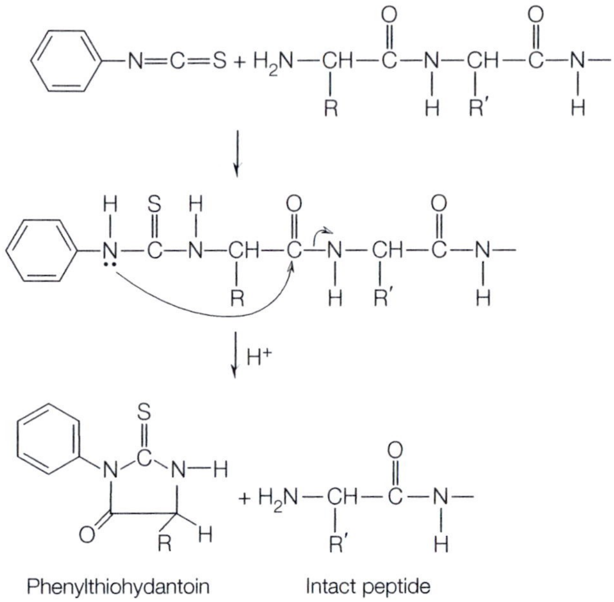

Phenylthiohydantoin   Intact peptide

the peptide remain intact. The derivatized amino acid is then extracted from
the reaction mixture and identified by paper, thin-layer, or high-performance
liquid chromatography. The intact peptide (minus the original NH_2-terminal
amino acid) may be isolated and recycled by reaction with phenylisothiocyanate. Since this method is nondestructive to the remaining peptide (aqueous
acid hydrolysis is not required) and results in good yield, it can be used for
stepwise sequential analysis of peptides.

Microsequencing

The development of new chemical reagents and instrumentation has now
made it possible to achieve end-group analysis and sequencing on extremely
small samples of protein (microsequencing). By using a chromophoric derivative of phenylisothiocyanate, 4-*N,N*-dimethylaminoazobenzene- 4'-isothio-
cyanate (DABITC; see Figure 11.5), and analyzing the derivatives by HPLC,
it is possible to sequence peptides and proteins at the nanomolar level.

C-Terminal Analysis

The COOH-terminal amino acid of a peptide or protein may be analyzed by
enzymatic methods. The method of choice is peptide hydrolysis catalyzed by
carboxypeptidases A and B. These two enzymes catalyze the hydrolysis of
amide bonds at the COOH-terminal end of a peptide (Equation 11.3), since
carboxypeptidase action requires the presence of a free α-carboxyl group in
the substrate.

Carboxypeptidase A catalyzes the hydrolysis of carboxyl-terminal acidic or neutral amino acids; however, the rate of hydrolysis depends on the structure of the side chain R'. Amino acids with nonpolar aryl or alkyl side chains are cleaved more rapidly. Carboxypeptidase B is specific for the hydrolysis of basic COOH-terminal amino acids (lysine and arginine). Neither peptidase functions if proline occupies the COOH-terminal position or is the next to last amino acid.

$$-\overset{\overset{\displaystyle O}{\|}}{C}-\underset{\underset{\displaystyle H}{|}}{N}-\underset{\underset{\displaystyle R}{|}}{CH}-\overset{\overset{\displaystyle O}{\|}}{C}-\underset{\underset{\displaystyle H}{|}}{N}-\underset{\underset{\displaystyle R'}{|}}{CH}-COO^- + H_2O \xrightarrow{\text{carboxypeptidase}}$$

Equation 11.3

$$-\overset{\overset{\displaystyle O}{\|}}{C}-\underset{\underset{\displaystyle H}{|}}{N}-\underset{\underset{\displaystyle R}{|}}{CH}-COO^- + H_2N-\underset{\underset{\displaystyle R'}{|}}{CH}-COO^-$$

Sequencing DNA Instead of the Protein

New techniques for protein sequence analysis have emerged from recombinant DNA research. It is now possible to clone long stretches of DNA. Genes carrying the message for selected proteins can be isolated in quantities sufficient for nucleotide sequencing. By using the genetic code, the structure of the protein product from a specific gene can be deduced from the sequence of nucleotides in the DNA. Techniques for DNA sequencing are faster and more reliable than for protein sequencing; however, determining protein sequences indirectly from DNA sequences will never completely replace direct analysis of the protein product. An amino acid sequence determined from DNA will be that of the initial protein product synthesized. That initial form (which is often not biologically active) is changed by post-translational modification processes. These changes may include removal of short stretches of amino acid residues, formation of disulfide bonds, and chemical changes on some amino acid residues.

Study Problems

1. A solution of a purified protein yielded an A_{280}/A_{260} of 1.3. Estimate the protein concentration of the solution.

2. Describe the action of the enzyme lysozyme in breaking bacterial cells for protein isolation and purification.

3. What amino acid residues in a protein absorb light of 280 nm?

4. Use a flowchart format to show how you would isolate a protein from an inclusion body.

▶ 5. A protein you wish to study is present in the mitochondrion of the cell. Show how you would use fractional centrifugation to isolate mitochondria to begin purification of the protein.

▶ 6. What type of electrophoresis would be better to study the subunit structure of a protein, PAGE or SDS-PAGE? Explain.

7. When would you use the technique of isoelectric focusing in the isolation and characterization of a protein?

8. During the expression of a protein in a bacterial host cell, what are some of the advantages of fusing a natural, bacterial protein to the "foreign" protein?

Further Reading

K. Backman, *Trends Biochem. Sci.* **26,** 268–270 (2001). "The Advent of Genetic Engineering."

J. Berg, J. Tymoczko, and L. Stryer, *Biochemistry,* 5th ed. (2002), Freeman (New York), pp. 77–116; 151–170. Introduction to protein purification and expression of foreign genes.

J. Bornhorst and J. Falke, *Met. Enzymol.* **326,** 245–254 (2000). "Purification of Proteins Using Polyhistidine Affinity Tags."

R. Boyer, *Concepts in Biochemistry,* 2nd ed. (2002), John Wiley & Sons (New York), pp. 88–89. Protein purification and analysis.

A. Campbell and L. Heyer, *Discovering Genomics, Proteomics, and Bioinformatics* (2003), Benjamin Cummings (San Francisco).

F. Carvalho and M. Gillespie, *Biochem. Mol. Biol. Educ.* **31,** 46–51 (2003). "Bringing Proteomics into the Lab."

E. Christodoulou and C. Vorgias, *Biochem. Mol. Biol. Educ.* **30,** 189–191 (2002). "Understanding Heterologous Protein Overproduction Under the T7 Promoter."

R. Garrett and C. Grisham, *Biochemistry,* 3rd ed. (2004), Brooks/Cole (Belmont, CA), pp. 112–114, 148–152. Introduction to protein purification and characterization.

D. Nelson and M. Cox, *Lehninger Principles of Biochemistry,* 4th ed. (2005), Freeman (New York), pp. 89–94. Introduction to protein purification.

B. Nielsen and S. Echols, *Biochem. Mol. Biol. Educ.* **30,** 408–413 (2002), "Use of Chloroplast rRNA Gene to Introduce Basic Molecular Biology Techniques."

L. Plesniak and E. Bell, *Biochem. Mol. Biol. Educ.* **31,** 127–130 (2003). "Teaching Proteomics."

S. Roe, Editor, *Protein Purification Techniques,* 2nd ed. (2001), Oxford University Press (Oxford).

S. Roe, Editor, *Protein Purification Applications,* 2nd ed. (2001), Oxford University Press (Oxford).

C. Sommer et al., *Biochem. Mol. Biol. Educ.* **32,** 7–10 (2004). "Teaching Molecular Biology to Undergraduate Biology Students."

C. Tsai, *An Introduction to Computational Biochemistry* (2002), John Wiley & Sons (New York).

M. Vijayalakshmi, Editor, *Biochromatography: Theory and Practice* (2002), Taylor & Francis (Hamden, CT).

D. Voet and J. Voet, *Biochemistry,* 3rd ed. (2004), John Wiley & Sons (Hoboken, NJ), pp. 116–160. Introduction to protein purification and molecular cloning.

D. Voet, J. Voet, and C. Pratt, *Fundamentals of Biochemistry,* 2nd ed. (2006). John Wiley & Sons (Hoboken, NJ), pp. 97–108 cover purification of proteins.

J. Walker, Editor, *The Protein Protocols Handbook,* 2nd ed. (2002), Humana Press (Totowa, NJ).

J. Watson et al., *Molecular Biology of the Gene,* 5th ed. (2004), Benjamin/ Cummings (San Francisco). Chapters 3, 5, 6, 20 (Methods).

http://www.piercenet.com
A discussion on protein purification.

http://www.promega.com
A discussion of the MagneHis Protein Purification System.

http://www.nobel.se/
Under the heading "Educational," use "The Virtual Biochemistry Lab" to perform experiments on protein chromatography, electrophoresis, NMR, sequencing, X-ray diffraction, and folding.

LIST OF SOFTWARE PROGRAMS AND WEB SITES USEFUL FOR EACH CHAPTER

This table lists Web sites that are focused on specific chapters. The list is not all-inclusive, but it does include popular and especially helpful URLs. Individual chapters also have detailed lists of associated Web sites.

Chapter No.	Name	Description	URL
1	Microsoft Excel	Statistical analysis	http://www.microsoft.com
1	BioResearch	List of journals	http://bioresearch.ac.uk Click on "Biochemistry"
1, 2	Medline	National Library of Medicine	http://www.nlm.nih.gov/
1, 3	ChemFinder	Chemical structures, properties	http://www.chemfinder.com
1, 3	Klotho	Alphabetical compound list	http://ibc.wustl.edu/klotho/
1, 3	SPSS	Statistical analysis	http://www.spssscience.com
1, 3	Systat	Statistical analysis	http://www.spssscience.com
1, 3	Merck Index	Biochemical structures, properties	http://www.merck.com/pubs/ mmanual/
2	BioResearch	Catalog of Internet resources in Biochemistry	http://bioresearch.ac.uk/ Click on "Biochemistry"
3	Lab3D (University of Virginia)	Virtual biochemistry lab-Amino acid titration	http:lab3d.chem.virginia.edu
5, 6, 7	Nobel Foundation— The virtual biochemistry lab	Protein chromatography, electrophoresis, NMR, X-ray, and folding	http://www.nobel.se/
6	Lab3D (University of Virginia)	Virtual biochemistry lab-electrophoresis	http://lab3d.chem.virginia.edu
6	BioResearch	Techniques in electrophoresis	http://bioresearch.ac.uk/ Click on "Biochemistry"
7	SDBS	Spectral database systems	http://www.aist.go.jp/RIODB/ SDBS/menu-e.html
7	BMRB	Biochemical NMR bank	http://www.bmrb.wisc.edu/
8	Lab3D (University of Virginia)	Virtual biochemistry lab-enzyme kinetics	http://lab3d.chem.virginia.edu

Chapter No.	Name	Description	URL
8	DynaFit	Ligand binding	http://www.biokin.com/
8	Leonora	Enzyme kinetics	http://ir2lcb.cnrs-mrs.fr/~athel/leonora0.htm
8	BRENDA	Enzyme information	http://www.brenda.uni-koeln.de/
8	ENZYME	Enzyme information	http://expasy.ch/enzyme/
8	IUBMB	EC nomenclature for enzymes	http://www.chem.qmw.ac.uk/iubmb/enzyme/
8	EMP	The Enzymology Database	http://wit.mcs.anl.gov/
8	Enzyme Structures Database	Enzyme structures in Protein Data Bank	http://www.biochem.ucl.ac.uk/bsm/enzymes/index.html
9, 10	GenBank of the NCBI	Nucleotide sequence database	http://www.ncbi.nlm.nih.gov/Genbank
9, 10	EMBL	Nucleotide sequence database	http://www.ebi.ac.uk/
9, 10	DDBJ	Nucleotide sequence database	http://www.ddbj.nig.ac.jp/
9, 10	Human mapping database	Genome project database	http://gdbwww.gdb.org
9, 10	dbEST	cDNA and partial sequences	http://www.ncbi.nih.gov
9, 10	Genethon	Genetic maps	http://www.genethon.fr
9, 10	Whitehead Institute	YAC maps	http://www-genome.wi.mit.edu
10	REBASE	The Restriction Enzyme Database	http://rebase.neb.com/
10	BioResearch	Nucleic acids	http://bioresearch.ac.uk/ Click on "Biochemistry"
11	Swiss-Prot	Protein sequences	http://expasy.hcuge.ch/sprot/sprot-top.html
11	TREMBL	Protein sequences	http://www.ebi.ac.uk/pub databases/trembl
11	PIR	Protein Information Resource	http://pir.georgetown.edu
11	NRL-3D	Protein structure database	http://www.bis.med.jhmi.edu
11	PDB	Protein structure database	http://www.rcsb.org
11	BioResearch	Proteins	http://bioresearch.ac.uk/ Click on "Biochemistry"
11	Nobel Foundation-The virtual biochemistry lab	Protein sequencing and folding	http://www.nobel.se/

PROPERTIES OF COMMON ACIDS AND BASES

Compound	Formula	Molecular Weight	Specific Gravity	% by Weight	Molarity (M)
Acetic acid, glacial	CH_3COOH	60.1	1.05	99.5	17.4
Ammonium hydroxide	NH_4OH	35.0	0.89	28	14.8
Formic acid	$HCOOH$	46.0	1.20	90	23.4
Hydrochloric acid	HCl	36.5	1.18	36	11.6
Nitric acid	HNO_3	63.0	1.42	71	16.0
Perchloric acid	$HClO_4$	100.5	1.67	70	11.6
Phosphoric acid	H_3PO_4	98.0	1.70	85	18.1
Sulfuric acid	H_2SO_4	98.1	1.84	96	18.0

APPENDIX III

PROPERTIES OF COMMON BUFFER COMPOUNDS

Compound	Abbreviation	Molecular Weight	pK_1	(20°C) pK_2	pK_3	pK_4
N-(2-Acetamido)-2-aminoethanesulfonic acid	ACES	182.2	6.9	—	—	—
N-(2-Acetamido)-2-iminodiacetic acid	ADA	212.2	6.60	—	—	—
Acetic acid		60.1	4.76	—	—	—
Arginine	Arg	174.2	2.17	9.04	12.48	—
Barbituric acid		128.1	3.79	—	—	—
N,N-bis (2-Hydroxyethyl)-2-aminoethane-sulfonic acid	BES	213.1	7.15	—	—	—
N,N-bis (2-Hydroxyethyl)glycine	Bicine	163.2	8.35	—	—	—
Boric acid		61.8	9.23	12.74	13.80	—
Citric acid		210.1	3.10	4.75	6.40	—
Ethylenediaminetetraacetic acid	EDTA	292.3	2.00	2.67	6.24	10.88
Formic acid		46.03	3.75	—	—	—
Fumaric acid		116.1	3.02	4.39	—	—
Glycine	Gly	75.1	2.45	9.60	—	—
Glycylglycine		132.1	3.15	8.13	—	—
N-2-Hydroxyethylpiperazine-N'-2-ethanesulfonic acid	HEPES	238.3	7.55	—	—	—
N-2-Hydroxyethylpiperazine-N'-3-propanesulfonic acid	HEPPS	252.3	8.0	—	—	—
Histidine	His	209.7	1.82	6.00	9.17	—
Imidazole		68.1	6.95	—	—	—
2-(N-Morphollno) ethanesulfonic acid	MES	195	6.15	—	—	—
3-(N-Morphollno) propanesulfonic acid	MOPS	209.3	7.20	—	—	—
Phosphoric acid		98.0	2.12	7.21	12.32	—
Succinic acid		118.1	4.18	5.60	—	—
3-Tris (hydroxymethyl)aminopropanesulfonic acid	TAPS	243.2	8.40	—	—	—
N-Tris (hydroxymethyl)methyl-2-aminoethanesulfonic acid	TES	229.2	7.50	—	—	—
N-Tris (hydroxymethyl)methylglycine	Tricine	179	8.15	—	—	—
Tris (hydroxymethyl)aminomethane	Tris	121.1	8.30	—	—	—

pK_a Values and pH_I Values of Amino Acids

Name	Abbreviations		pK_1 (α-carboxyl)	pK_2 (α-amino)	pK_R (side chain)	pH_I
Alanine	Ala	A	2.3	9.7	–	6.0
Arginine	Arg	R	2.2	9.0	12.5	10.8
Asparagine	Asn	N	2.0	8.8	–	5.4
Aspartate	Asp	D	2.1	9.8	3.9	3.0
Cysteine	Cys	C	1.7	10.8	8.3	5.0
Glutamate	Glu	E	2.2	9.7	4.3	3.2
Glutamine	Gln	Q	2.2	9.1	–	5.7
Glycine	Gly	G	2.3	9.6	–	6.0
Histidine	His	H	1.8	9.2	6.0	7.6
Isoleucine	Ile	I	2.4	9.7	–	6.1
Leucine	Leu	L	2.4	9.6	–	6.0
Lysine	Lys	K	2.2	9.0	10.5	9.8
Methionine	Met	M	2.3	9.2	–	5.8
Phenylalanine	Phe	F	1.8	9.1	–	5.5
Proline	Pro	P	2.0	10.6	–	6.3
Serine	Ser	S	2.2	9.2	–	5.7
Threonine	Thr	T	2.6	10.4	–	6.5
Tryptophan	Trp	W	2.4	9.4	–	5.9
Tyrosine	Tyr	Y	2.2	9.1	10.1	5.7
Valine	Val	V	2.3	9.6	–	6.0

MOLECULAR WEIGHT OF SOME COMMON PROTEINS

Name (Source)	Molecular Weight
Albumin (bovine serum)	65,400
Albumin (egg white)	45,000
Carboxypeptidase A (bovine pancreas)	35,268
Carboxypeptidase B (porcine)	34,300
Catalase (bovine liver)	250,000
Chymotrypsinogen (bovine pancreas)	23,200
Cytochrome c	13,000
Hemoglobin (bovine)	64,500
Insulin (bovine)	5,700
α-Lactalbumin (bovine milk)	14,200
Lysozyme (egg white)	14,600
Myoglobin (horse heart)	16,900
Pepsin (porcine)	35,000
Peroxidase (horseradish)	40,000
Ribonuclease I (bovine pancreas)	12,600
Trypsinogen (bovine pancreas)	24,000
Trypsin (bovine pancreas)	23,800
Tyrosinase (mushroom)	128,000
Uricase (pig liver)	125,000

COMMON ABBREVIATIONS USED IN THIS TEXT

A	adenine or absorbance
A	absorbance
AMP, ADP, ATP	adenosine mono-, di-, or triphosphate
AP	alkaline phosphatase
BCA	bicinchoninic acid
bp	base pairs
Bq	becquerel
BSA	bovine serum albumin
C	cytosine
cDNA	complementary DNA
CE	capillary electrophoresis
Ci	Curie
CM cellulose	carboxymethyl cellulose
COSY	correlation spectroscopy
CPM	counts per minute
D	dalton
DANSYL	1,1-dimethylaminonaphthalene-5-sulfonyl chloride
2-DE	two-dimensional electrophoresis
DEAE cellulose	diethylaminoethyl cellulose
DNA	deoxyribonucleic acid
DNase	deoxyribonuclease
DNP	dinitrophenyl
DOPA	dihydroxyphenylalanine
E	absorption coefficient
EDTA	ethylenediaminetetraacetic acid
ELISA	enzyme linked immunosorbent assay
ESI	electrospray ionization
EtBr	ethidium bromide
FAB	fast atom bombardment
$FAD(H_2)$	flavin adenine dinucleotide (reduced form)
FAME	fatty acid methyl ester
$FMN(H_2)$	flavin mononucleotide (reduced form)
FOCSY	foldover-corrected spectroscopy
FPLC	fast protein liquid chromatography
FT	Fourier transform
g	gravitational force
G	guanine
GC	gas chromatography
GMP, GDP, GTP	guanosine mono-, di-, or triphosphate
Hb	hemoglobin
HMIS	Hazardous Materials Identification System
HPLC	high-performance liquid chromatography
HRP	horseradish peroxidase
HTML	hypertext markup language

IE	immunoelectrophoresis
IEF	isoelectric focusing
IHP	inositol hexaphosphate
IMAC	immobilized metal-ion affinity chromatography
IR	infrared
ISP	Internet service provider
IVS	intervening sequence
kb	kilobase pairs
L	ligand
M	macromolecule
M	molarity
MALDI	matrix-assisted laser desorption ionization
MS	mass spectrometry
MSDS	material safety data sheet
NAD(H)	nicotinamide adenine dinucleotide (reduced form)
NADP(H)	nicotinamide adenine dinucleotide phosphate (reduced form)
NMR	nuclear magnetic resonance
NOESY	nuclear Overhauser effect spectroscopy
OSHA	Occupational Safety and Health Administration
PAGE	polyacrylamide gel electrophoresis
PC	paper chromatography
PCR	polymerase chain reaction
PEG	polyethylene glycol
PFGE	pulsed field gel electrophoresis
PMT	photomultiplier tube
POPOP	1,4-bis[5-phenyl-2-oxazolyl]benzene
PPO	2,5-diphenyloxazole
PVDF	polyvinyldifluoride
RCF	relative centrifugal force
R_f	relative mobility in chromatography
RFLP	restriction fragment length polymorphism
RIA	radioimmunoassay
RNA	ribonucleic acid
RNase	ribonuclease
RT	reverse transcriptase
s	sedimentation coefficient
S	Svedberg (10^{-13} s)
SDS	sodium dodecyl sulfate
SECSY	spin-echo correlation spectroscopy
SMP	submitochondrial particles
T	thymine
TAG	triacylglycerol
TEMED	N,N,N', N'-tetramethylethylenediamine
TLC	thin-layer chromatography
TOCSY	2-D total correlation spectroscopy
TOF	time of flight
Tris	tris(hydroxymethyl)aminomethane
TROSY	transverse relaxation-optimized spectroscopy
U	uracil
URL	uniform resource locator
UV	ultraviolet
VIS	visible
WWW	World Wide Web

UNITS OF MEASUREMENT

The International System of Units (SI)

Quantity	Unit	Abbreviation
length	Meter	m
mass	Kilogram	kg
time	Second	s
temperature	Kelvin	K
electric current	Ampere	A
amount of substance	Mole	mol
radioactivity	Becquerel	Bq
volume	Liter	L

Metric Prefixes

Name	Abbreviation	Multiplication Factor (relative to "1")
atto	a	10^{-18}
femto	f	10^{-15}
pico	p ($\mu\mu$)	10^{-12}
nano	n (mμ)	10^{-9}
micro	μ	10^{-6}
milli	m	10^{-3}
centi	c	10^{-2}
deci	d	10^{-1}
deca	da	10
kilo	k	10^{3}
mega	M	10^{6}
giga	G	10^{9}

Units of Length

Name	Abbreviation	Multiplication Factor (relative to meter)
kilometer	km	10^3
meter	m	1
centimeter	cm	10^{-2}
millimeter	mm	10^{-3}
micrometer	μm	10^{-6}
nanometer	nm	10^{-9}
Angstrom	Å	10^{-10}

Units of Mass

Name	Abbreviation	Multiplication Factor (relative to gram)
kilogram	kg	10^3
gram	g	1
milligram	mg	10^{-3}
microgram	μg	10^{-6}
nanogram	ng	10^{-9}

Units of Volume

Name	Abbreviation	Multiplication Factor (relative to liter)
liter	L	1
deciliter	dL	10^{-1}
milliliter	mL	10^{-3}
microliter	μL	10^{-6}

TABLE OF THE ELEMENTS*

	Symbol	Atomic No.	Atomic Mass		Symbol	Atomic No.	Atomic Mass
Actinium	Ac	89	227.0278	Hassium	Hs	108	[270]
Aluminum	Al	13	26.98154	Helium	He	2	4.00260
Americium	Am	95	[243]†	Holmium	Ho	67	164.9304
Antimony	Sb	51	121.75	Hydrogen	H	1	1.0079
Argon	Ar	18	39.948	Indium	In	49	114.82
Arsenic	As	33	74.9216	Iodine	I	53	126.9045
Astatine	At	85	[210]	Iridium	Ir	77	192.22
Barium	Ba	56	137.33	Iron	Fe	26	55.847
Berkelium	Bk	97	[247]	Krypton	Kr	36	83.80
Beryllium	Be	4	9.01218	Lanthanum	La	57	138.9055
Bismuth	Bi	83	208.9804	Lawrencium	Lr	103	[262]
Bohrium	Bh	107	[264]	Lead	Pb	82	207.2
Boron	B	5	10.81	Lithium	Li	3	6.941
Bromine	Br	35	79.904	Lutetium	Lu	71	174.967
Cadmium	Cd	48	112.41	Magnesium	Mg	12	24.305
Calcium	Ca	20	40.08	Manganese	Mn	25	54.9380
Californium	Cf	98	[251]	Meitnerium	Mt	109	[268]
Carbon	C	6	12.011	Mendelevium	Md	101	[258]
Cerium	Ce	58	140.12	Mercury	Hg	80	200.59
Cesium	Cs	55	132.9054	Molybdenum	Mo	42	95.94
Chlorine	Cl	17	35.453	Neodymium	Nd	60	144.24
Chromium	Cr	24	51.996	Neon	Ne	10	20.179
Cobalt	Co	27	58.9332	Neptunium	Np	93	237.0482
Copper	Cu	29	63.546	Nickel	Ni	28	58.70
Curium	Cm	96	[247]	Niobium	Nb	41	92.9064
Darmstadtium	Ds	110	[281]	Nitrogen	N	7	14.0067
Dubnium	Db	105	[262]	Nobelium	No	102	[259]
Dysprosium	Dy	66	162.50	Osmium	Os	76	190.2
Einsteinium	Es	99	[252]	Oxygen	O	8	15.9994
Erbium	Er	68	167.26	Palladium	Pd	46	106.4
Europium	Eu	63	151.96	Phosphorus	P	15	30.97376
Fermium	Fm	100	[257]	Platinum	Pt	78	195.09
Fluorine	F	9	18.998403	Plutonium	Pu	94	[244]
Francium	Fr	87	[223]	Polonium	Po	84	[209]
Gadolinium	Gd	64	157.25	Potassium	K	19	39.0983
Gallium	Ga	31	69.72	Praseodymium	Pr	59	140.9077
Germanium	Ge	32	72.59	Promethium	Pm	61	[145]
Gold	Au	79	196.9665	Protactinium	Pa	91	231.0359
Hafnium	Hf	72	178.49	Radium	Ra	88	226.0254

*Atomic masses are based on carbon-12.

†A value given in brackets denotes the mass number of the longest-lived or best-known isotope.

	Symbol	Atomic No.	Atomic Mass		Symbol	Atomic No.	Atomic Mass
Radon	Rn	86	[222]	Technetium	Tc	43	[98]
Rhenium	Re	75	186.207	Tellurium	Te	52	127.60
Rhodium	Rh	45	102.9055	Terbium	Tb	65	158.9254
Rubidium	Rb	37	85.4678	Thallium	Tl	81	204.37
Ruthenium	Ru	44	101.07	Thorium	Th	90	232.0381
Rutherfordium	Rf	104	[261]	Thulium	Tm	69	168.9342
Samarium	Sm	62	150.4	Tin	Sn	50	118.69
Scandium	Sc	21	44.9559	Titanium	Ti	22	47.90
Seaborgium	Sg	106	[266]	Tungsten	W	74	183.85
Selenium	Se	34	78.96	Uranium	U	92	238.029
Silicon	Si	14	28.0855	Vanadium	V	23	50.9415
Silver	Ag	47	107.868	Xenon	Xe	54	131.30
Sodium	Na	11	22.98977	Ytterbium	Yb	70	173.04
Strontium	Sr	38	87.62	Yttrium	Y	39	88.9059
Sulfur	S	16	32.06	Zinc	Zn	30	65.38
Tantalum	Ta	73	180.9479	Zirconium	Zr	40	91.22

ANSWERS TO SELECTED
STUDY PROBLEMS

Chapter 1

2. Personal Protection Index: splash goggles, gloves, synthetic apron, vapor respirator.

4. (a) Add 75.1 g of zwitterionic glycine to 1-liter volumetric flask. Add purified water to mark.

   (b) Add 90 g of glucose to a 1-liter volumetric flask and add water to mark.

   (c) Add 0.46 g of ethanol to a 1-liter volumetric flask and add water to mark.

   (d) Add 6.5 mg of hemoglobin to a 1-liter volumetric flask and add water to mark.

5. (a) Use 0.75 g of glycine.

   (b) Use 0.9 g of glucose.

   (c) Use 0.0046 g of ethanol.

   (d) Use 0.065 mg of hemoglobin.

6. 100 mM.

7. (a) 0.56 mM; 560 μM.

   (b) 220 mM; 220,000 μM.

8. 58.5 mM; 20 mg/mL; 2%.

9. (a) 5.17 mM;

   (b) 4.17 mM;

   (c) 3.33 mM.

10. (a) Sample mean = +3.21°.

   (b) Standard deviation = ±0.043°.

   (c) 95% confidence limits = +3.21° ± 0.03° at a probability of 0.05.

Chapter 3

1. NaH_2PO_4: 0.31 mole; 37.2 g;
   Na_2HPO_4: 0.19 mole; 26.9 g

2. Weigh out 0.5 mole of NaH_2PO_4 and dissolve in about 900–950 mL of purified water. Monitor the pH of the solution and add dropwise, with stirring, a concentrated solution of NaOH until the pH is 7.0. Add water to the 1-liter mark. Check final pH and adjust to 7.0, if necessary.

3. Use 0.03 mole of glycine and 0.07 mole of sodium glycinate for 1 liter of solution.

4. Dissolve 24.2 g of Tris base in about 900–950 mL of purified water. Add concentrated HCl dropwise until the pH is 8.0. Add water to 1 liter and check pH.

5.

Tube No.	1	2	3	4	5	6
μg BSA	0	10	20	40	80	100
A_{595}	0.0	0.08	0.16	0.32	0.64	0.80

6. 65 μg/mL.

7. 17.5 μg/mL.

9. Primarily Tyr and Trp; no.

10. Most proteins that contain Phe, Try, and Trp.

11. (a) Citrate.

    (b) Imidazole.

    (c) Glycine.

12.

Stock (2M)	Final (M)	Stock (mL)	Water (mL)
	1.66	16.6	3.4
	1.33	13.3	6.7
	1.0	10.0	10.0
	0.66	6.6	13.4
	0.33	3.3	16.7
	0.00	0.0	20.0

13. Solution B: 0.2 M.

    Solution C: 0.02 M.

    Solution D: 0.002 M.

16. 100 minutes.

17. 0.49/day.

18. 77 days.

19. Must be β emitters.

Chapter 4

1. (a), (b), (d).

2. 48,000 rpm.

3. Hemoglobin.

4. When the substrate is bound, the enzyme molecule folds into a more compact or spherical shape.

5. There are four subunits of molecular weight 10,000 and two with molecular weight 30,000.

6. Cell nuclei would be present in the sediment after centrifugation at $600 \times g$ (see Figure 4.11).

7. Magnesium ions bind to DNA in place of smaller protons, causing it to spread out and thus become less dense.

8. Yes, the two forms have different densities. Supercoiled DNA is usually more compact.

9. $125,000 \times g$.

10. Most mitochondria sediment at $20,000 \times g$.

Chapter 5

1. (a) Asp, Gly, His.

   (b) Glu, Ala, Arg.

   (c) Glu, Phe, His.

2. Ser, Lys, Ala, Val, Leu.

3. Cyt *c*, myoglobin = hemoglobin, serum albumin, egg albumin, pepsin.

4. Myosin, catalase, serum albumin, chymotrypsinogen, myoglobin, cyt *c*.

5. Differential refractometer–essentially all molecules are detected, but this method is not especially sensitive. Photometric detector–molecules that absorb in the ultraviolet or visible light region. Fluorescence detector–molecules that fluoresce.

6. Malate dehydrogenase, alcohol dehydrogenase, glucokinase.

8. A dilute solution of NAD^+ should elute the enzyme from the affinity gel.

10. Hydrogen bonding, ionic bonds, hydrophobic interactions, van der Waals forces.

Chapter 6

1. Charge, size.

2. From top to bottom: serum albumin, egg white albumin, chymotrypsin, lysozyme.

5. The gel matrix in slab gels is more uniform than column gels, which are made individually.

6. Polyacrylamide gels may be used for nucleic acids up to 2000 base pairs.

11. Western blotting: used to identify a specific protein or group of proteins by immunoblotting (detection by antibodies). Southern blotting: used to identify a specific base sequence in DNA. Northern blotting: used to identify specific base sequences in RNA.

12. Before the protein detection process can begin, it is necessary to block protein binding sites on the membrane that are not occupied by blotted proteins. This is essential because antibodies used to detect blotted

proteins are also proteins and will bind to the membrane and interfere with detection procedures. Milk, a relatively inexpensive chemical reagent, contains casein proteins that act as blocking agents.

13. Nylon membranes, because they are cationic and strongly bind acidic proteins.

14. Infection by the AIDS virus will cause the production of antibodies in the patient's serum. Serum proteins are separated by PAGE and a detection system is then developed to recognize those antibodies.

15. If SDS-PAGE is used for separation, the proteins to be blotted are denatured. The antibodies used for the detection process must be able to recognize and bind to denatured proteins.

16. It must be assumed that the blue dye marker moves faster in electrophoresis than any of the proteins analyzed in the gel.

Chapter 7

1. (a) X-ray.
   (b) Ultraviolet.
   (c) Visible.
2. About $0.001\ M$
3. (1) a.
   (2) a.
   (3) c.
   (4) d.
   (5) b.
4. (a), (b), and (d).
6. Glass does not allow the transmission of UV light.
7. All the molecules contain alternating double bonds (are highly conjugated).
8. Fluorescing light is measured at right angles to the light irradiating the sample.

Chapter 8

1. First plot: $n = 2$; $K_f = 1.1$ mM. Second plot: $n = 2.5$; $K_f = 0.6$ mM.
2. The binding of the molecular probe must be reversible; the bound probe must have a different absorbance spectrum from the unbound.
3. The tryptophan residue must be close to a site on the human albumin where sugars bind. Bound sugars interact with the tryptophan.
6. From Michaelis-Menten plot: $V_{max} = 140\ \mu M$/min; $K_M = 40–45\ \mu M$.
7. Turnover number $= 280$/minute.

9.   Type of inhibition: competitive.

11.   The rate is linear for only about 2 minutes.

Chapter 9

1.   DNA, which forms viscous solutions, is released from the cells into the medium.

2.   Add ribonuclease to the preparation.

3.   SDS is a detergent that dissociates protein-lipid complexes in cell membranes.

4.   28 μg/mL.

5.   The absorbance at 260 nm $=$ 0.70.

8.   Ethanol is added to precipitate plasmid DNA from solution. Most RNA remains in solution; hence, ethanol precipitation provides an effective technique for removing contaminating RNA from DNA extracts.

9.   (a) Lowered.

     (b) Lowered.

     (c) Lowered.

     (d) Probably little or no effect.

13.   30 μg/mL.

Chapter 10

1.   Many plasmids contain genes that carry messages for the synthesis of proteins that protect microorganisms against antibiotics. The presence of certain antibiotics is a signal to the microorganism that more of these proteins are necessary. The microorganism responds by increasing the rate of production of plasmids.

2.   See the section on protein synthesis in your biochemistry book.

4.   Natural DNA molecules and restriction fragments are too large to penetrate polyacrylamide gels. Even gels with low percentage cross linking are not useful.

5.   Restriction endonucleases require the presence of magnesium ions for activity. The quench buffer contains EDTA which complexes transition metal ions in the solution. The metal ions are no longer available for binding by the nuclease molecules and enzyme activity is inhibited.

7.   (b).

8.   Ten fragments.

Chapter 11

1.   The protein concentration is about 1.2 mg/mL.

5.   See Figure 4.11.

6.   SDS-PAGE.

Index

Note: An *f* indicates a figure and a *t* indicates a table.